ENERGY: CRISIS OR OPPORTUNITY?

An Introduction to Energy Studies

DIANA SCHUMACHER

with contributions from

Israel Berkovitch
Ross Hesketh
Judith Stammers

MACMILLAN

First published 1985 by
Higher and Further Education Division
MACMILLAN PUBLISHERS LTD
Houndmills, Basingstoke, Hampshire RG21 2XS and London
Companies and representatives
throughout the world

Typeset in Great Britain by
RDL Artset Ltd, Sutton, Surrey
Printed in Hong Kong

British Library Cataloguing in Publication Data
Schumacher, Diana
     Energy: crisis or opportunity?
     1. Power resources
     I. Title
     333.79   TJ163.2

ISBN 0-333-27191-2
ISBN 0-333-38425-3 Pbk

ENERGY: CRISIS OR OPPORTUNITY?

Other Macmillan titles of related interest

Coal Combustion and Conversion Technology D. Merrick
Computing Methods in Solar Heating Design J. R. Simonson
Fuel Cells Angus McDougall
Hydrogen and Energy Charles A. McAuliffe
Introduction to Nuclear Reactions G. R. Satchler

To Christiana, Katharine and all young people whose energy inheritance is our concern, responsibility and opportunity

Contents

Foreword

The evolution of Life and Intelligence on planet earth has reached a crisis point. *Homo sapiens*, perhaps arrogantly, thinks he is the culmination to date of an age-long evolutionary process. Certainly that process has resulted in Man having greatly increased his intelligence in a short period as measured on the geological timescale. But he has achieved a highly specialised intelligence in which reason predominates, reason being thought controlled by the discipline of accurate observation, measurements, and the collation of what are taken to be facts. At the same time he has developed an onlooker's consciousness, seeing himself as master and the physical and animal world as something to be shaped and controlled to suit his convenience.

Whether he wants it or not, he has now the power and the responsibility to guide, to some extent, the future path of evolution. Has he the wisdom for this task? He seems to have over-developed the reasoning side of his brain at the expense of the intuitive and imaginative side. As the Scriptures say: 'Where there is no vision, the people perish.'

This lack of vision, of the faculty of choosing the kind of future that would foster the deeper values of life for generations to come, has brought us dangerously close to destroying our earthly home. This predicament now stares Man in the face in many different guises, such as resource depletion, population explosion, irresponsible exploitation of discoveries, poisoning the environment, and perhaps a nuclear war.

Diana Schumacher's book is addressed to what is currently regarded as the 'energy crisis'. It deals with this issue with a wider sweep than any other book I know of among the spate of books on energy now pouring out of the press. She awakens the reader — and most of us need awakening — to the fact that the 'energy crisis' is but one facet of a global crisis of many interlocking parts. Our complex predicament must be recognised for what it is, metaphorically a seamless garment.

Diana Schumacher has put much painstaking work into marshalling a multitude of facts, figures, and trends. The main lines of the picture are clearly portrayed, soundly based, and of enduring value. Change is now so rapid that the details will soon become out of date; the situation is changing while the attempt is made to capture it in words, figures and graphs.

The abiding impression I get from the book is that we must no longer let this change occur by inadvertence, apathy, or culpable ignorance. But that, alas, is what is now happening. It is up to all of us to help fashion a decision-making process that will not only be consistent with growing knowledge but — much more to the point — will be guided by Vision and Wisdom.

Diana Schumacher has made a gallant effort to help us rise to the occasion so that we may help create a world to pass on to our children without shame. That is now the great hope. Courage, vision and emerging wisdom must transform that hope into determination and achievement.

Wootans, Branscombe, Devon

Sir Kelvin Spencer
(Formerly Chief Scientist,
Department of Energy, U.K.)

Acknowledgements

Owing to the diverse nature of the subject this book would never have been completed without the help, encouragement and ideas of a very wide spectrum of people, from several different countries. Alas it is impossible to name them all, but I am personally indebted to each. I should like to thank most warmly those who have given valuable time to read sections of the manuscript falling within their own particular expertise and all who have provided valuable advice and comments. Among them I am extremely grateful to Dr Clive Baker (Binnie and Partners), Dr Geoff Barnard (Earthscan and Imperial College, London), Lord Beaumont of Whitley, Professor Paul Blau (Vienna University), Dr Bill Carlyle (Binnie and Partners), Ms Alison Clarke (Incpen), Mr Robin Clarke (Energy Consultant), Dr Norman Connor (University of Salford), Dr Bill Coombes (formerly of Tate and Lyle Ltd), Mr Wouter van Dieren (Foundation for Applied Ecology, Netherlands), Mr Alex Eadie, M.P., Mr Jeremy Evans (BNOC), Professor Ian Fells (University of Newcastle-upon-Tyne), Dr John Garnish (Energy Technology Support Unit), Mr Robin Grove-White (CPRE), Professor David Hall (King's College, University of London), Mr Robert Hutchinson (Policy Studies Institute), Dr Ted Lawrence (Crown Agents), Mr Gerald Leach (IIED), Dr Peter Musgrove (Reading University), Mr Malcolm Peebles (Shell International Gas Ltd), Mr David Rodda, Mr Stephen Salter (University of Edinburgh), Dr Phil Wolfe (formerly of Lucas BP Solar Systems Ltd), to the Department of Energy and the OECD for all their assistance, and to the Macmillan Press. I am very appreciative of Vernon Hayes for producing most of the artwork under difficult circumstances!

Warm thanks are also due to Tim Sparrow (formerly of the London School of Economics) and Schumacher Projects; to Andrew Cox (University of Newcastle-upon-Tyne) who helped with updating, editing and refining; and to Frances Harrison who uncomplainingly typed through successive drafts of each chapter. I am deeply appreciative of the fact that Israel Berkovitch, Ross Hesketh and Judith Stammers provided their chapters at very short notice despite many previous commitments.

I am also incredibly grateful to Dr Leen Weeda for his constant encouragement and work in converting all the inconsistent sets of units into a standard form; to Martin Polden without whose involvement and practical support the book might never have seen the light of day, and above all to Christian and my family (it is not easy to see a wife and mother's energy so diversified!). Thank you!

D.S.

Introduction

'Watertight compartments of knowledge are dangerous. This is not yet taken into account sufficiently in practice by scientists engaged in energy work. Economics, sociology, political science, are fields of knowledge with which those who consider energy options must be conversant. Woe to him who devises a process of coal gasification without accounting for the climatic influences of its use two or three decades away'

J. O. Bockris (1980)

The title of this book was suggested by the Chinese symbol for crisis which includes the symbol depicting opportunity. *Energy: Crisis or Opportunity?* is intended as background reading for school sixth forms and first-year students of engineering, economics, geography, environmental studies, sociology and other energy-related courses as well as for the interested lay person. The aims of the book are five-fold:

(1) To offer a broad overview of traditional, current and emerging energy forms, their history, development, present and future potential, together with the practical advantages and disadvantages of each.
(2) To define patterns of energy use and development in the U.K. and, where appropriate, compare them with those in other countries including Europe, the U.S.A., other industrial nations and the Third World.
(3) To relate historical growth in energy-consumption patterns (summed up in the first chapter) to estimated future demand and the availability of supply including the life expectation of fossil fuels and the potential of new energy sources (see chapter 12).
(4) To draw together the various strands implicit in any coherent energy development policy. In future these will not only include resource availability, technology, and economics; but factors such as politics, education, conservation and employment will play an increasing part together with the geographic, sociological and environmental aspects.
(5) To offer the reader a summary of relevant criteria against which different energy options may be evaluated, and suggested further reading and addresses where more detailed information on particular topics may be obtained.

The book seeks to look at the total context of the energy problem from a multi-disciplinary standpoint. Unfortunately, owing to a necessary but drastic reduction in the length of the book many of the details, figures, references and examples have had to be omitted. Tables and statistics are therefore merely given as illustrations, since these change annually. In posing questions and choices it is hoped that the reader will gain an appreciation of many of the issues involved in formulating energy policy. To quote Amory Lovins: 'The energy problem is intimately related to all the great issues of our day and the disadvantage of the semi-quantitative approach to energy futures is that energy studies are done in isolation and do not emphasise the intimate relationship between energy policy and every other kind of policy. The most important, difficult and neglected questions of energy strategy are not mainly technical or economic but rather social and ethical.'

Decisions about energy options for the future will no longer inevitably depend only on what is available, accessible or most economic now. The choices, when all known factors are taken into account, may sometimes be simple. In some instances, however, they may be very finely balanced and may depend on choosing the lesser of two evils, for example short-term gain or long-term investment. After the twentieth century's propensity to live off energy capital rather than energy income, the twenty-first century will need to achieve an energy balance of permanence in order to convert crisis into opportunity.

Godstone, 1984   D.S.

1

Energy and Man—the Development of Demand

'One of Nature's laws, publicised by Parkinson, is that demand always rises to meet supply. Unfortunately he failed to state what happens when supply fails to meet demand'

HRH The Duke of Edinburgh
World Energy Conservation Month, Oct. 1979

Development of Energy Demand before the Industrial Revolution

From the dawn of civilisation energy use has been the most important single material factor influencing man's activities, ranging from physical comfort and food supplies, to every type of economic achievement. Without the use of energy the quality of life can hardly rise above subsistence level, or the necessities for bare survival. This fact is today readily apparent in many communities of the Third World.

Primitive Man

'Primitive man' consumed only the energy contained in the food that he gathered, amounting to just over 8 megajoules (MJ) or 2000 kilocalories (kcal) daily.[1] The first rise in man's standard of living was also the first step up the ladder of increasing energy consumption. During the mid-Palaeolithic period consumption increased to about 20 MJ or 5000 kcal daily with 'hunting man' who had learned to make some tools from stone and bones and burned wood for cooking and heating. Archaeological discoveries in the Near and Middle East show that by 10,000 B.C. man's energy was engaged in agriculture, corn grinding and hunting animals with bow and arrow. He was producing food as opposed to merely gathering it; he was using energy in order to enhance his food energy. He built shelters and began to form settlements. Around 5000 B.C. primitive 'agricultural man' had begun to use

1

animal power for crop cultivation with an estimated daily average energy consumption of about 50 MJ or 12,000 kcal. As technical progress continued, he began to
mine metals and, with the help of heat, to form them into tools and weapons. First
came copper, then bronze (a mixture of copper and tin), and then the Iron Age.

Bronze Age Man

There is evidence that the Bronze Age began in the Middle East, Greece, India and
Mesopotamia as early as 3000 B.C., although it only arrived in Britain about
1900 B.C. Bronze Age man showed a significant advance in energy use. First he
was beginning to combine and transform the raw materials that he mined from
below the ground. Second, such crafts as mining and the reduction of metals from
ores were more specialised and required training and skill as well as increased
amounts of heat energy in the form of wood combustion. Originally, Neolithic
society had been self-sufficient with each family or local group producing enough
food and materials to sustain its own requirements. Now families had to produce a
surplus to use in the barter system; the community as a whole had to make an
exportable surplus to exchange for raw materials from more distant mines. Widespread use of the ox-drawn plough vastly increased the amount of land one family
could cultivate while the invention of the wheel revolutionised transport and trade.

Iron Age Man

The Iron Age is believed to have started in the Near and Middle East with the
Sumerians and Mesopotamians. The royal tombs at Mari in Mesopotamia demonstrate that man had mastered the reduction of iron ore in a charcoal fire by 2500
B.C. By 1000–900 B.C. the full techniques of the iron age — carburising, quenching,
and tempering — were being employed by the Hittites, although the first evidence of
iron use in Europe from the Tyrol and Velem Szentvid in Hungary goes back only
to 700 B.C. The Forest of Dean is thought to be the first area in England to have
developed a local iron industry using charcoal. Yet by Roman times the use of iron
in everyday life, from weapons and armour to cooking and agriculture, was so
essential that Noricum, a centre of iron smelting has been termed 'the Sheffield of
antiquity'.
    To achieve these basic industries all early civilisations, including those of Greece
and Rome, depended primarily for their energy on wood, animal and human power
(often in the form of slaves). Even until very recently the term 'horsepower' was
used as a mechanical power measurement. However, wind and water power were
also being developed. As well as the widespread use of wind to power ships in all
seafaring countries, in Persia archaeologists have found evidence of the use of
wind-driven water pumps for irrigation dating from the fifth century B.C.
    Windmills are thought to have existed in China and Japan around 2000 B.C. and
their use later spread to neighbouring countries.[2] In the first century A.D. Hero of
Alexandria in his *Pneumatica* even describes an 'anemurion' or wind-driven musical
organ. From the ninth century writers described the remarkable windmills providing
irrigation in the arid region of Seistan, bordering on Persia and Afghanistan. It is

thought that knowledge of windmills was first brought to Western Europe by the returning Crusaders, the first references coming from England and France towards the end of the twelfth century. By 1400 their use was widespread, mainly for grinding corn, hoisting materials from mines and pumping water. By the end of the seventeenth century the largest windmills were estimated to have achieved outputs of up to the equivalent of 12 kW, 5 kW being about the average.

It is impossible to give an exact date for the earliest examples of *water wheels* but several types were in use during the first hundred years A.D. The Nora wheel had a horizontal shaft and flat paddles. Cups of wood, bamboo or pottery were fastened to the rim and as the wheel was turned by the river flow, these cups dipped in and emptied their contents into a trough at the highest point of wheel revolution. The water flowed from the trough to irrigation channels. This type of wheel was used only for lifting water and did not drive machinery. On the River Orantes in Syria (figure 1.1), there still exist very large iron water wheels thought to have been functioning for over 2000 years.

Figure 1.1   *Water wheels on the River Orantes, Hama, Syria*

In Egypt the Nile began to dry up after about 5000 B.C. forcing the local inhabitants to withdraw from the banks and to provide irrigation of arable land by building dykes, canals and water pumps. The Roman water mill, first described by Vitruvius in his *De Architectura* around 25 B.C. was mainly used for corn grinding. The water wheel had an horizontal shaft and flat buckets similar to the Nora. The shaft was carried in two bearings and drove the millstones through primitive right-angle gearing. These mills were not extensively used for the next 300 years, probably on account of the availability of animal and slave labour. By the end of the fourth century there was an increasing number of references to the use of water power by the Romans. The world's largest water-power installation until the Machine de Marly in the seventeenth century is believed to have been built by the Romans at Barbegal in the South of France. It consisted of 16 overshot wheels located in pairs down a steeply sloping hillside and made use of a head of more than 60 ft. It is known that the Romans also used water mills in Britain and the *Domesday Book* (*circa* 1086) records over 5000 working water mills serving about one million people.

Advanced Agricultural Man – The Middle Ages

By now, European man was already making use of all the known renewable energy resources on a wide scale – wind, water and, in some cases, direct sun and geo-thermal energy. Animals were still a significant mechanical-energy source, and wood the most important. Wood had long been used extensively for heating and cooking and, following the practice that had begun in the Middle East centuries earlier, charcoal was used for glass smelting and the production of metals. Despite these advances, by A.D. 1400 European man still consumed very little energy. According to Cook,[1] he used just over 100 MJ or 24,000 kcal per day per head (figure 1.2). This was only twice as much as his forebears in 5000 B.C., and over the 6400-year period represented an annual *per capita* increase of one-hundredth of 1 per cent. Overall energy use increased comparatively slowly because the population was relatively static and also because technology showed very little advance over that of the preceding centuries. Water pumps and corn mills described in books in the later Middle Ages show only marginal improvements on those known to have existed 500 years earlier. It is thought that few major innovations were made until the Renaissance although the mediaeval historian Jean Gimpel suggests that some of the machine designs attributed to Leonardo da Vinci were in use in earlier times. On the whole, however, new methods evolved very slowly in the world of crafts-manship with techniques usually passed down from father to son. Most machines had reached a point where comparatively little improvement could be made within the existing state of technology and energy use. Even by the beginning of the eighteenth century the 9.4 million population of Great Britain and Ireland was consuming an estimated 3 million tonnes of coal equivalent (Mtce) or 0.31 tonnes per person per year, a figure that is only marginally different from the consumption of many Third World countries today.

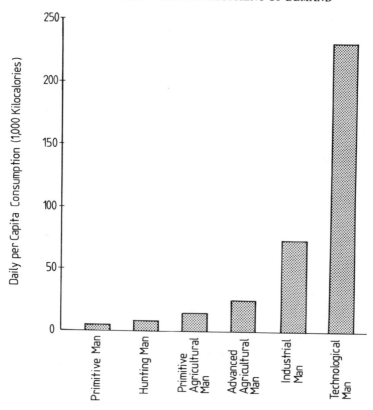

Figure 1.2 *Energy consumption in the development of man* (data derived from *Scientific American*, Vol. 225, No. 3, Sept. 1971)

Industrial Man

With the new scientific and technological developments, from about the eighteenth century men began to experiment with new energy forms and energy-harnessing equipment. By the early nineteenth century there was a massive increase in energy consumption (figure 1.3). The spectacular growth in demand was met mainly by an increased use of water power.

In England in the late-eighteenth century, John Smeaton demonstrated that the overshot water wheel, developed in the sixteenth century and now usually built of iron rather than wood, was twice as efficient as the undershot wheel described by Vitruvius. By early in the nineteenth century, however, water resources in key industrial areas of the U.K. were becoming inadequate to meet the increasing demand. This prompted the search for alternatives, the traditional European 'wood and water' industrial energy sources gradually being replaced by coal and steam. These changes in energy use subsequently spread and in turn helped to create an industrial revolution in other countries.

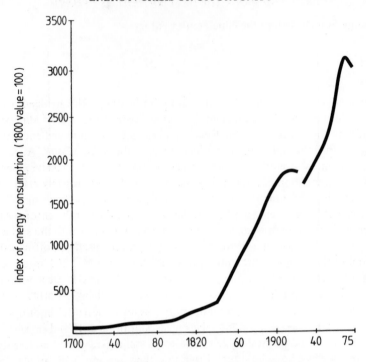

Figure 1.3 *Index of energy consumption in the U.K. (1700–1975)* (source: W.S. Humphrey and J. Stanislaw, 'Economic Growth and Energy Consumption in the U.K. 1700–1975', *Energy Policy*, March 1979, p. 31)

The interplay of causes leading to the dramatic growth in energy consumption has been amply covered by economic historians and historically minded economists. It is interesting to note, however, that to achieve such rapid development or change in the technology of energy use, whether during the Industrial Revolution or today, four conditions are generally required. The first is necessity, caused by scarcity or some other form of crisis acting as a catalyst. The second is the will or initiative to act. The third is an infrastructure supportive of change. The fourth is a natural resource to exploit or develop. In eighteenth century Britain all these conditions existed. A firewood crisis and charcoal shortage, together with an increase in the population and the demand for good-quality iron for improved tools and agricultural purposes helped to provide the need. The changing attitudes and values brought about by the Reformation, the discoveries of science, expanding colonialism, and later the war with France, provided the stimulus. There was already in existence a basically healthy economic infrastructure owing to the agrarian revolution, a century of peace and the expansion of overseas trade which stimulated demand and the desire for improvement. Finally, there existed massive indigenous coal and iron deposits to exploit. This last condition provided the main impetus to the increasing energy consumption, coal being the fossil fuel that underpinned the first century of economic growth in the Industrial Revolution.

Main Energy Sources before the Nineteenth Century

Coal

In Britain coal was used by the Romans mainly for heating baths. Evidence exists
that in the third century it also fuelled the sacred fire in the temple of Minerva at
Bath, but with the departure of the Romans it ceased to be employed and was not
rediscovered until the end of the twelfth century. In the *Bolden Book* (A.D. 1183)
there are references to smiths who were asked to 'find the coal' to make plough-
shares. The first mention of mining coal appears in 1243 and probably refers to
open cast mining. This practice then developed rapidly and it is likely that all the
English coal fields were being worked by the end of the thirteenth century. At that
time narrow shafts were sunk down to the coal, which was raised to the surface in
large baskets. Water collecting at the bottom of the shaft was drawn off by subterranean drains. By the mid-fourteenth century sea-trading of coal had begun between
English ports and during the reign of Elizabeth I the overseas coal trade was under
way. Records from 1591 indicate that 140,000 tons were being exported annually.

Coal was not widely used for domestic purposes until the Tudor invention of the
chimney, and even then it was unpopular on account of its dirty and smoky
qualities, the soot and its intolerable smell. It was still regarded as considerably
inferior to wood as a household fuel. From that time onwards, without planned
reforestation, increasing shortages of wood gradually forced householders to look
to coal for their heating and cooking fuels. Between 1450 and 1650 the price of
firewood rose in some parts of the country by over twice that of other general
commodities, the situation of many Third World countries today. By about 1700
approximately six times as much coal as wood was being used for domestic purposes.

The industrial use of coal also became common at the end of the eighteenth
century. It was gradually introduced into various manufacturing processes such as
the production of alum, copper, saltpetre, gunpowder, salt, soap, sugar, starch and
candles, the preparation of food and brewing.[3] Unfortunately, a method of using
coal for the smelting of metals had not been discovered, which explains why timber
shortages were especially acute in iron-making areas since the furnaces relied on
charcoal as their fuel and water power to maintain continuous blasts of air. By the
seventeenth century, the shortage and consequent price increase of wood in south-
east England, together with the permanent drying up of many streams (itself the
result of excessive deforestation), had caused a collapse of the iron industries of
Kent, Surrey and Sussex. The industry was revived only when coke was used in
Darby's historic blast furnace at the outset of the Industrial Revolution. Its location was from now on governed by the location of the coal sources, especially
where found in conjunction with water power such as those in the north of
England, the Midlands and Wales. Energy use and technological advance went hand
in hand, for to mine coal demanded the development of a new mining technology.
The main problem, avoiding flooding caused by natural water tables, was overcome
only with the invention of the steam-driven water pump.

Steam Power

Although the use of steam as a source of energy had been known to the early Greeks, the 'fire engine' of Thomas Savery which in 1699 produced steam from a coke-fired boiler, was the first practical steam pump to be invented. It was used for pumping water, partly by direct steam pressure and partly by vacuum, without the use of a piston. A few years later, in 1712, Thomas Newcomen erected the first steam engine with a piston and cylinder. This was followed by John Smeaton's improved engine in 1730 which increased the efficiency of energy conversion of Newcomen's engine from the extremely small 0.56 per cent to just over 1 per cent. Later still, in 1769, James Watt patented his condenser, which raised the efficiency of the steam engine to around 3.5 per cent. As a consequence, coal consumption increased further. In turn, plentiful coal gave the iron industry its much needed impetus to provide the materials for the numerous machines and devices that could form the basis of other industries such as textiles, chemicals and shipbuilding. Richard Trevithick's high-pressure steam engines, patented at the beginning of the nineteenth century, paved the way for subsequent rail and road transport. The steam engine itself, although developing in terms of conversion efficiency, was not widely used for purposes other than water pumping until after the 1820s. Consequently in the early decades of the Industrial Revolution, as demand for energy increased in the newly emerging industries, attempts were made to meet this demand by upgrading the performance of the traditional energy sources.

Wind and Water Power Developments

Already by the middle of the eighteenth century there were 10,000 or so water wheels and 2000 windmills in England.[4] These were mainly used to drain land and mines and to drive mills and grind corn. Other applications included fulling cloth, working forge hammers and furnace bellows, sawing and crushing. As industrialisation proceeded numerous larger watermills and windmills were built to provide power for the emerging industries, especially the new textile factories.

As energy requirements intensified, the renewable energies at their contemporary state of development were to prove incapable of providing the concentrated and constant high-grade energy which mechanisation now required. As Jevons commented:[5]

> 'No possible concentration of windmills. . .would supply the power required in large factories or ironworks. An ordinary windmill has the power of about 34 men, or at most 7 horses. Many ordinary factories would therefore require 10 windmills to drive them; and the great Dowlais ironworks employing a total engine power of 7308 horses, would require no less than 1000 large windmills.'

Moreover, wind and, to a lesser extent, water power, was an uncertain, limited, and irregular source compared to the constant and plentiful supply provided by the new mechanised steam engines.

Coal Gas

Closely associated with the growth in demand for coal was the increasing consumption of another fuel – coal gas. Coal gas is produced when coal is heated in the absence of air to give coke. Gradually this volatile matter came to be harnessed as another energy source. It was also found that the coke residue could be gasified in steam to make water gas. The earliest commercial use of coal gas in England was for street lighting at the beginning of the nineteenth century. By the 1850s domestic gas cooking had been established and by the 1880s gas fires were in use. Further details are given in chapter 4.

Growth of Energy Demand in the Nineteenth Century

If coal gas and steam ushered in the age of cheap and plentiful energy supply at the outset of the Industrial Revolution, on the demand side the consumption of non-renewable energy sources was stimulated by three main inter-related factors. The first was the growth of population.

Until the mid-seventeenth century, the population of England and Wales had never exceeded 5.6 million, a slight rise from the 2-4 million estimated at the beginning of the fourteenth century. Energy consumption was also relatively stable during this period since it generally moved in parallel with increases in population. However, as already mentioned, the existence of a healthy economic infrastructure in the second half of the seventeenth century was providing an increased national subsistence which would support a greater population. Consequently, from the eighteenth century onwards, the population started to rise rapidly, and by the census of 1801 it had totalled 8.9 million in England and Wales. Energy consumption followed suit, since much of the increased demand was from the domestic sector. Increasing quantities of fuel were also being consumed by industry. In 1800, mining and building already accounted for some 23 per cent of national output, although with the exception of iron-making, the type of industrialisation taking place was not yet particularly coal-intensive.

Another factor stimulating demand for fossil fuels, especially from 1830 onwards, was the rapidly increasing level of industrial mechanisation as steam power progressively replaced the muscle, wind and water power of pre-industrial workforms. The process has continued to the present day. It is strikingly illustrated by the example of the mechanisation of harvesting quoted by S. J. Wright.[6] He calculated the number of men required to harvest a six-quarter crop of wheat at the rate of 10 acres per day (the normal average in 1800) at different periods and levels of mechanisation. The results were as follows:

Harvesting Manning Requirements 1840–1940

Date		Men
1840	Entirely by hand	126
1870	Reaper (tying by hand) and steam thresher	32
1900	Reaper and binder and steam thresher	21
1920	Tractor-drawn reaper and binder with steam thresher	18
1940	Combine harvester	3

This example, which could be replicated in nearly every industry, is significant, not only because it illustrates the tremendous manpower savings that the new technologies could achieve, but also because the renewable energy provided by labour during the hand-operation period was later almost entirely replaced by mechanical energy obtained from non-renewable sources. In fact, since the Industrial Revolution the maximum possible power output of basic energy-converting machines has increased over ten-thousandfold from the hundred kilowatts of the water wheel in 1750 to the million kilowatts of today's steam turbine.

The third factor influencing energy demand from the beginning of the Industrial Revolution onwards to the present day was the growth of industrial output itself in the form of goods. Particularly from 1830, when steam began to be used on a large scale in the cotton, textile and textile-related industries, energy consumption started to rise at exponential rates with the increased industrial output. These rates would have appeared even more spectacular without the very significant advances in energy-conversion efficiency rates attained by successive generations of machines. For example, the coal consumption of Newcomen's engine was about 20 lb per unit of indicated horsepower (i.h.p.) and fuel wastage was over 99 per cent, leading critics to scoff that it took an iron mine to build it and a coal mine to keep it going. Watt's condenser saved three-quarters of the fuel that the Newcomen engine had required, reducing consumption to 5 lb of coal per i.h.p. Later engines improved upon these figures. Triple expansion engines required 2 lb of coal per i.h.p., large steam turbines 1.5 lb and suction gas engines 1 lb.

The efficiency of energy converters rose most steeply from about 1800 onwards, in line with the increases in power outputs. A simple unweighted average of efficiencies in the domestic-heating, process-heating, electricity and industrial-engine categories of energy use shows that energy efficiency rose from a level of about 1–3 per cent before 1850 to about 8 per cent in 1900, and 30 per cent in 1950.

In all, as the industrial infrastructure and spread of large-scale industry became established, energy consumption in the U.K. rose from 1800 to 1880 by the staggering factor of 10. This increase is accounted for by a doubling of the population together with a fivefold increase in *per capita* energy consumption. The latter is explained by a threefold increase in the level of output per person and an increase of 80 per cent in the energy content of each unit of output, largely owing to the increased production of heavy industry. Table 1.1 illustrates this situation.

Table 1.1   *Output and energy consumption in the UK: 1800-1975*

	Output		Energy consumption (mineral fuels and hydropower)				
Year	Index (1800 ≡ 100)	Index of output/ head (1800 ≡ 100)	Million tce	Index (1800 ≡ 100)	Consumption per head (tce/head)	Index of consumption per head (1800 ≡ 100)	Index of consumption per unit output (1800 ≡ 100)
1800	100	100	10.78	100	0.68	100	100
1810	122	108	13.73	127	0.76	112	104
1820	158	120	17.16	159	0.82	121	101
1830	226	149	21.86	203	0.91	153	90
1840	285	170	31.94	296	1.20	176	104
1850	339	197	45.78	425	1.67	246	125
1860	436	240	72.34	671	2.49	366	154
1870	534	268	97.64	906	3.09	454	170
1880	642	293	125.18	1161	3.59	528	181
1890	801	336	145.82	1353	3.86	568	169
1900	985	377	165.56	1536	3.99	587	156
1910	1104	387	188.56	1749	4.16	612	158
1920	1093	400	(193.81)	(1798)	(4.12)	(606)	(158)
	(1137)	(384)					
1925	1191	430	182.60	1694	4.14	609	142
1930	1281	454	184.70	1713	4.12	606	134
1938	1531	512	196.58	1824	4.14	609	119
1950	1855	583	225.70	2094	4.46	656	113
1960	2415	731	264.90	2457	5.06	744	102
1965	2854	834	296.90	2754	5.46	803	96
1970	3196	916	328.00	3043	5.91	869	95
1975	3423	964	319.70	2965	5.67	834	87

Note: All figures before 1920 include Eire, all those after 1920 exclude Eire. For 1920 the figures not in parentheses exclude Eire while those in parentheses include Eire.
Source: W. S. Humphrey and J. Stanislaw, 'Economic Growth and Energy Consumption in the UK 1700-1975', *Energy Policy*, March 1979, p. 30.

Surprisingly perhaps, U.K. energy consumption *per unit output* reached peak levels as early as 1880, after which it has been declining. The reason for the rising energy consumption per unit output between 1830 and 1880 lies in the pattern of growth at the time. This was the period in which most of the U.K. railway infrastructure was laid down, entailing an exceptional twofold demand for coal, both for the locomotives themselves and for the massive iron-smelting programme needed to produce the railway stock. Also the period witnessed a similarly massive expansion of steam shipping, with like effects on energy and iron consumption. Much architecture of the period – houses, bridges and public buildings – was conspicuous for the amount of iron used both structurally and for ornamentation. Abraham Darby's Iron Bridge at Coalbrookdale is an early example. Throughout the century, coal was the primary fuel in all these uses. By 1887, 19 per cent of all coal in the U.K. was being consumed by the iron industry, 8 per cent by mines, 15 per cent by steam navigation, 7 per cent by the gas and electricity industries, 31 per cent by other manufacturing processes and 20 per cent by domestic households.

There were two main reasons why the index of energy consumption per unit output began to show a decline again after 1880. The first was the increased thermal efficiency of heat engines mentioned earlier. The second was the changing economic structure which was becoming less dominated by iron and steel production as lighter industries took hold. For example the iron industry's share in the National Product of Great Britain fell from 11.6 per cent in 1871 to 6.4 per cent in 1907. Nevertheless, as industrial output went on expanding exponentially in absolute terms, energy consumption continued to rise dramatically. The consumption of energy in the ten years between 1880 and 1890 equalled the entire consumption in the 100 years up to 1790! By 1900, the annual U.K. energy consumption had reached over 15 times the annual rate of that a century earlier.

Technological Man – Growth of Energy Demand in the Twentieth Century

Coal

Coal continued to be by far the largest primary fuel used in the U.K. for the first half of the twentieth century, accounting for some 90 per cent of total inland primary energy consumption. Annual U.K. coal consumption in 1950 was 202.9 million tons compared with an average 155.7 million tons in the decade 1893–1902.

Numerous estimates of the coal resources of Great Britain have been made, one of the earliest being in 1861 when Hull calculated that there were 59,109 million tons in seams of not less than 2 ft in thickness situated within 4000 ft of the surface. A Royal Commission in 1866 upgraded this figure to 146,480 million tons in seams not less than 1 ft thick, and in 1901 a further Commission put proved coal fields at about 100,000 million tons with another 40,000 in unproved fields. A less optimistic survey by the Ministry of Fuel and Power in 1945 put total proved reserves in seams not less than 1 ft 6 in thick at 40,387 million tons. Before 1938 there were no statistics showing the annual breakdown of coal consumption by industry in the U.K. Since that year figures have become available and in 1945 out of a total inland consumption of 177.2 million tons, domestic consumers absorbed 31.2 million tons; authorised electricity undertakings, railway and transport authorities 23.5 million tons; gas undertakings 21.1 million tons; coke ovens 20.1 million tons; railways 14.9 million tons; collieries 10.6 million tons; the iron and steel industry 9.5 million tons; the engineering industries 4.0 million tons; and other industries 26.4 million tons.[7]

By 1945, 24 per cent of this coal was used for conversion into the secondary fuels, coal gas and electricity. Electricity, especially, transformed living standards through the growth of the domestic appliance industry. It transformed social behaviour, through the growth of communications equipment. Increasingly it transformed industry by facilitating the introduction of faster, cleaner and more efficient machines, giving rise to a whole range of light industries and, latterly, through the electronics revolution, bringing in the computer.

Electricity

(1) *Supply.* Following Michael Faraday's pioneering work in 1831, the first electricity generator was developed in 1834. By the 1850s electric arc lighting had begun to be used and in 1879 Thomas Edison invented the light bulb. Electric motors first appeared in 1873, electric railways in 1879 and power stations in 1882. By 1900 there were already about 90 electrical supply undertakings in the U.K. operating almost 200 generating stations. These ran independently and served local markets. There was no standardisation on such matters as the voltage of supply, whether direct or alternating current, and in the case of the latter on the frequency used. After 1900 there was a progressive attempt to standardise and integrate electricity supply and to diversify and stimulate electricity demand. On the supply side, the first step in the transition from a large number of local stations to a single supply organisation covering the whole country was the Electricity Lights (Amendment) Act of 1909. This empowered the Board of Trade to regulate regional electricity supplies.

After the First World War, the Electricity Supply Act of 1919 established Electricity Commissioners to promote, regulate and supervise the supply of electricity. In early 1925 a committee under Lord Weir recommended the setting up of the Central Electricity Board to construct a 'gridiron' system of high-voltage transmission lines across the country, later to become known as the National Grid. In 1926, standardisation of frequency had begun when the Central Electricity Board adopted the three-phase 50 Hz a.c. supply which was then already standard throughout Europe. By the end of 1935 the whole of England was linked except for the north-east. By 1946 the entire country was served by a single interconnected transmission system covering some 3700 miles of primary transmission and 1500 miles of secondary transmission. In 1945, a standard voltage – 240 V a.c. – was adopted throughout the country for the first time, obviating the need for electrical devices to possess their own internal transformers. This, in turn, cut costs and encouraged a much more widespread use. Country-wide frequency standardisation was completed in 1947. In 1948 a further step was taken to rationalise the electricity supply when it became a nationalised industry. The 1957 Electricity Act laid down the future structure of the U.K. electricity industry. This replaced the British Electricity Authority with the Central Electricity Generating Board which was responsible for electricity supply and the National Grid, and the Electricity Council which was responsible for co-ordinating CEGB policy and advising the Minister of Power on all salient matters affecting the industry.

(2) *Electricity demand.* Apart from a few electric trains, mains electricity was first used in the late-nineteenth century almost exclusively for lighting, taking over from gas lighting in towns and paraffin lamps or wind-generated electricity in some country districts. Later, domestic customers began to use it also for heating and cooking. By 1920, industry and commerce had become the most important customers – a position which they have held until the present day, as shown in table 1.2.

The changing patterns of energy use over the past 50 years have taken place in

Table 1.2 *Proportion of electricity sales to different consumers in U.K.*
1920-81

| | Percentage of total | | | | | | Total sales | |
	Industrial	Commercial	Farming	Domestic	Traction	Street lights	(GWh)	(10⁶ GJ)
1920	69	11	–	8	11	1.3	3,240	11.7
1925	66	12	–	11	10	1.6	5,036	18.1
1930	58	14	0.1	17	9	1.8	8,165	29.4
1935	53	16	0.2	22	7	1.9	13,191	47.5
1940	57	12	0.3	26	5	0.1	21,822	78.6
1945	56	11	0.5	28	4	0.5	28,126	101.3
1950–1	49	14	1.0	32	3	0.9	43,025	154.9
1960–1	48.6	14.1	1.5	33	1.6	0.8	96,053	345.8
1970–1	41.6	16.6	1.7	38	1.3	0.8	174,254	627.3
1980–1	39.0	19.8	1.6	37.4	1.2	1.0	196,429	707.1

Source: Handbook of Electricity Supply Statistics, CEGB, London, 1981, p. 66.

the context of a rapidly rising general demand for electricity, as the last column in table 1.2 shows. Until the slowing down of economic growth in the 1970s the average annual growth of electricity was between 6 and 7 per cent compared to an average growth of 2 per cent for energy as a whole. Today the electricity industry has a central place in the U.K. energy economy, although total electricity consumption is no longer growing — total consumption in 1980/81 was exactly the same as in 1977/78.

U.K. Energy Consumption 1950-73

From 1950 onwards, the sources of primary energy in the U.K. began to diversify from coal to oil and gas. By 1973 the proportion of total inland energy provided by coal had fallen from 90 per cent to 50 per cent. Oil, since the beginning of the century had been increasing its hold on the energy market and by the end of the Second World War enough successful exploration and capital investment had taken place to make its rapid replacement of coal and other energy forms secure.

Oil

Because of its unique versatility oil has perhaps had an even greater impact on the pattern of life in the U.K. than electricity. Mineral oil, or petroleum, has been known to man for thousands of years. Noah's ark was caulked with pitch, and crude oil from natural seepages from oil-impregnated ground was used to fuel primitive lamps many years before the Industrial Revolution. It was in the 1850s, however, that oil first became a commercially attractive proposition with the discovery that it could be distilled in the same way as alcohol to make a clean and highly combustible fuel. Soon efforts were being made to obtain oil from the ground by drilling wells. The first well is said to have been drilled by the self-styled 'Colonel' Drake on 27 August 1859 near Titusville in Pennsylvania, U.S.A. The distilled oil (called kerosene or paraffin) was mainly used for oil lamps and stoves and the residual liquid was thrown away. Before long it was discovered that this waste oil could be refined for use as a lubricant and in paints and varnishes. From the time of Daimler's experiments on high-speed gas engines in 1885 and Benz's contemporaneous work on the first three-wheeled vehicles, oil has been gradually developing its uses as a transport fuel. Towards the end of the nineteenth century the first oil-powered motor cars began to appear on the roads. Already between 1908 and 1927 Henry Ford had sold over 15 million Model T Ford cars. On the other hand, at the time oil was generally regarded as of minor importance. Dugald Clerk, in the 1915 Thomas Hawksley Lecture for the Institution of Mechanical Engineers, London, said: 'Oil does not materially affect our problems . . .even if the whole of the crude petroleum were employed as fuel, in steam raising, it would not replace, allowing for its high thermal value, much more than 5 per cent of the world's output of coal; while if used in internal combustion engines, it would be equivalent as a source of power to about 15 per cent of coal. . .'[8]

The error of these forecasts is now well known. The First World War saw the rapid development of aeroplanes; a revolutionary form of transport. Throughout the present century and especially since 1950 oil consumption has escalated and the number of uses for oil products other than for transport fuels has grown at unprecedented rates. On a worldwide scale the growth of oil-based transport has affected the patterns of life, the development of cities and the location of industries. By 1973 some 22 per cent of total world oil was being consumed by motor vehicles. Another 22 per cent was being converted into diesel fuel for use in diesel engines, as a burner fuel in heating installations such as furnaces, and for enriching water gas to increase its luminosity. A further larger amount (37 per cent of total) of the residual oils was left for use in shipping and industrial large-scale heating installations. In all by 1973 oil accounted for 46 per cent of world primary energy consumption and 47 per cent of consumption in the U.K. The subsequent development of the oil industry is described in more detail in chapters 2 and 3.

Natural Gas

Closely allied to oil is natural gas which is usually found in the same places. By 1975 it was in commercial production in many regions throughout the world including North Africa, the Middle East, New Zealand, Pakistan, the U.S.S.R., the Netherlands, Canada, France, Italy, Norway and the U.K. Natural gas first became a major fuel in the U.K. after 1970 as a result of the North Sea oil and gas discoveries. It was the policy of the British Gas Corporation to replace coal gas with the much cheaper natural gas as quickly as possible, and massive industrial and domestic conversion schemes were undertaken. By 1980, natural gas accounted for 20.3 per cent of U.K. primary energy consumption. On a world level, natural gas was responsible for some 18.6 per cent of world energy consumption in 1977, a significant contribution when compared with the 29.4 per cent supplied by coal.

Nuclear Energy

During the 1960s and 1970s commercial developments in nuclear-energy production showed that this new power source could also make a serious contribution to electricity supply.

The early history of atomic energy from Marie Curie's isolation of radium from uranium ore in France in 1902 is well known. In the U.K. interest in this new form of energy began with John Dalton's paper on atomic theory in 1803. This was developed by Ernest Rutherford and his successors, who postulated in the early decades of the twentieth century that the atom was like a miniature solar system in which negative particles (electrons) moved around a central nucleus which in turn was made up of positively charged particles (protons) and neutral particles (neutrons). It was also discovered that the atoms of certain heavy elements spontaneously emitted energy in the form of radiation. Later in 1934 in France, Irene Curie and Frederic Joliot discovered artificial radioactivity showing that lighter particles (in this case aluminium) could be made radioactive (that is, release energy) by being bombarded with nuclei (helium) from a naturally radioactive source. Enrico Fermi showed that it was much easier to bombard particles with the electrically neutral neutrons than with the positively charged protons, since the latter were repelled by the positive charge in the receiving nucleus. Energy release upon bombardment could come either because two light nuclei combined or because a heavy nucleus split into two of medium weight. The source of the energy derives from the fact, verified by Albert Einstein, that the new stable nuclei have a measured mass smaller than the sum of the original nuclei, the difference being accounted for by released energy.

In practice however, formidable difficulties lay in the way of controlling nuclear reactions to produce new nuclei. A breakthrough was the invention of particle accelerators such as the cyclotron (by Lawrence and Livingston in 1934) which made it possible to bombard materials with high-speed protons. Another method was to slow down the speed of neutrons by means of a material such as heavy water or graphite. The energy required to effect the necessary collisions was often greater than the energy released by the new nuclei and also there seemed no way of keeping the process going long enough for a sufficient number of new nuclei to be created, in order to build up a significant store of released energy.

At the end of 1938, O. Hahn and F. Strassman discovered that if uranium (the heaviest element) was bombarded with neutrons, it split into two nearly equal fragments. The energy released in this process was shown to be some 10 to 100 times as much as in any other nuclear disintegrations. Moreover, it was found that additional neutrons were also released in the process which meant that these were available spontaneously to bombard a further portion of the uranium thus setting off a chain reaction. This process was known as uranium fission, and it opened up the way for the use of nuclear energy as a fuel.

The first controlled production of atomic energy through uranium fission was achieved by Enrico Fermi at the University of Chicago on 2 December 1942, as part of the Manhattan project to develop a nuclear bomb in the Second World War.

After this the development of nuclear fission for power generation was mainly a matter of engineering, technical development and economics. By the 1950s reactors had been built for experimental work in the U.S.A., Canada, the U.K., Norway, France and the Netherlands.

Several countries, including the U.K., were also attempting to develop a 'Fast Breeder Reactor' to extract some 60 times more of the energy in natural uranium than these early reactor designs. A fuller account of nuclear developments is given in chapter 5.

Energy Demand and Gross National Product

At the end of 1972, at the height of a consumer boom, all the energy sources mentioned so far — notably oil, coal, gas and electricity (some of which was supplied by nuclear power) — were competing vigorously with one another for what appeared to most observers to be a market promising almost unlimited growth. Oil, together with natural gas, was supplying 63 per cent of the U.K. market, leaving a much-reduced coal industry to supply a further 33 per cent. Nuclear energy still accounted for only 3 per cent of U.K. energy supply in 1972. By this time oil had obtained a position of dominance, not only in the U.K. but in the world as a whole.

Only five years earlier world oil consumption accounted for 42 per cent and U.K. oil consumption for 45 per cent of the total, illustrating the unprecedented speed of the oil takeover. Moreover, world energy consumption as a whole was still rising at the rapid average rate of 5 per cent per year as economic growth continued unabated in industrial and developing countries alike. Each time an additional product was made, a car produced or a bridge built, more energy was consumed. Rapid economic growth and unrestrained energy consumption went hand in hand.

The statistical connection between *per capita* energy consumption and *per capita* Gross National Product is evident from figure 1.4, which shows the relationship for a number of industrialised and developing countries in 1972, the year before the oil crisis. Figure 1.4 also reveals the very wide differences in energy consumption between the industrialised and the developing countries. *Per capita* energy consumption in the U.S.A. (on the extreme top right of the figure), for example, was more than double that of every other country in the world except for Canada, Belgium. Luxembourg and the U.K.

These differences in energy consumption are the result of a complex interaction of factors including the energy intensiveness of the dominant industries in a country, the distribution of wealth, the geographical dispersion of the population and its social habits. The energy intensity of a given country's economy will also vary over time, as has already been pointed out in relation to the U.K. economy. In general, as has been shown by the history of energy development in the U.K., countries at earlier stages in their economic development may be expected to show higher energy/output ratios than countries at later stages. While the former are laying down their infrastructure with a preponderance of heavy, energy-intensive

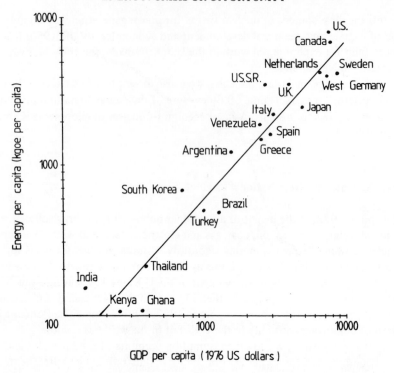

Figure 1.4 *Energy consumption versus GDP per capita for selected countries (1976)*
(source: World Bank, 1978, 1979. Quoted in R. Eden *et al.*, *Energy Economics*,
Cambridge University Press, 1982)

industries associated, the latter are diversifying into high technology and service
areas which are usually less energy-intensive. The relationship between energy consumption *per capita* and output *per capita* for the U.K. between 1700 and 1975 is
illustrated in figure 1.5. It exemplifies the initial acceleration of energy consumption
at early stages of economic development and its subsequent decline. Three main
phases may be discerned. The first, from 1700 to 1910, may be termed the 'heavy
energy' phase when rates of energy growth substantially exceeded those of economic
growth. The next, from 1910 to 1970, was a 'medium energy' phase when energy
and growth moved more or less in equilibrium; and the third phase from 1970 onwards may be called the 'light energy' phase when incremental energy use dropped
significantly in relation to growth in general.

This latter phase in the U.K. was primarily due to the fact that since 1970, 80
per cent of all economic growth has taken place in the service sector and in the non-
energy-intensive light industrial and telecommunications sectors. Another reason
was that, for the first time, industry was becoming aware that energy costs were a
significant variable in their profitability equations, and efforts towards better
energy utilisation were being made. Finally, some nagging troubles were beginning

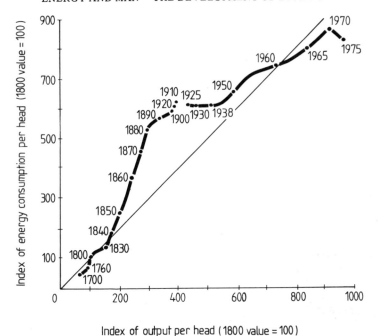

Figure 1.5 *Energy consumption and output in the U.K. (1700–1975)* (derived from W. S. Humphrey and J. Stanislaw, 'Economic Growth and Energy Consumption in the U.K. 1700–1975', *Energy Policy*, March 1979, p. 32)

to afflict the energy industries themselves. In certain quarters, questions were being asked about the long-term availability and security of oil and natural gas supplies and nuclear-energy development was hindered by higher costs and such problems as public disquiet about environmental issues, safety, proliferation, storage of long-life radioactive materials, and decommissioning procedures. At the same time the coal industry was experiencing growing labour problems, shortage of finance for new investments and the aftermath of a decade of decline. Other traditional energy forms – wood, peat, water and wind – had been allowed to decline in the era of abundant fossil fuels as they were limited both in magnitude and reliability. For the same reason the potential of solar, tidal, biomass and geothermal sources had not been developed. It was alleged that these renewables could not contribute significantly to the country's energy needs before the next century. Even the possibility that there might one day be an energy shortage was admitted – a strange thought after two and a half centuries of unmitigated expansion.

References

1. Cook, E., 'The Flow of Energy in an Industrial Society', *Scientific American*, Vol. 225, No. 3, September 1971

2. Minchington, W., *History Today*, Vol. 30, March 1980
3. Salzman, L. F., *English Industries of the Middle Ages*, Pordes, London, 1969, pp. 1–20
4. Laxton, P., 'The Geodetic and Topographical Evaluation of English County Maps, 1740–1840', *Cartographic Journal*, June 1976
5. Jevons, W. A., *The Coal Question*, Macmillan, New York, 1906, pp. 164–5
6. Rolt, L. T. C., *The Mechanicals*, Heinemann, London, 1967
7. *The British Fuel and Power Industries*, Political and Economic Planning, London, 1947, p. 27
8. Clerk, D., Thomas Hawksley Lecture, 29 October 1915

Further Reading

Gimpel, J., *The Medieval Machine: The Industrial Revolution of the Middle Ages*, Holt, Rinehart and Winston, New York, 1976
Matthias, P., *The First Industrial Nation*, Methuen, London, 1969
Adams, R. N., *Paradoxical Harvest – Energy and Explanation in British History, 1870–1914*, Cambridge University Press, Cambridge, 1982
Hubbert, M. K., *Resources and Man*, Freeman, San Francisco, 1969
Butti, K. and Perlieu, J., *A Golden Thread – 2500 Years of Solar Architecture and Technology*, Marion Boyars, London, 1981
Cook, E., *Man, Energy, Society*, Freeman, San Francisco, 1976

2

The 1973 Oil Crisis and its Aftermath

'It continues to affect the standard of living of the populations of oil importing countries, where economic growth is still hampered by the fourfold increase in oil prices that followed the Arab–Israeli conflict of 1973, and foreign policy, especially of the United States, is heavily influenced by the fear that supplies will be withheld'

Geoffrey Kirk

The 1973 Oil Crisis

By 1973, 36 per cent of the world's oil was supplied by the Middle East countries since oil from that quarter was more abundant and cheaper to produce than other known sources. Historically, however, the region has always been riddled with political and cultural tensions. As the result of Egypt crossing the Suez Canal and attacking Israel on 6 October 1973, which led to the Yom Kippur War, many countries in the industrial world had their oil supplies curtailed and by the following year oil prices had quadrupled. By mid-1980 the price of crude oil had risen nearly twelvefold. To what extent did the 1973 'oil crisis' prove a turning point or watershed in the history of energy use, and how did the outbreak of the Yom Kippur War affect energy policy not only in the U.K. but throughout the world?

Some believe that 1973 was decisive. The *Oil and Gas Journal* of 31 December 1973 described it as a 'turning point in history for the lay man as well as for the oil man, for neither will experience the same energy lifestyle again.' Against this, others argued that the 1973 'oil crisis' was merely a temporary interruption in supplies. In their view, people would continue to use energy in much the same ways as before, despite the shock induced by this interruption and *ad hoc* controls and other government measures to meet the sudden crisis.

Although in most OECD countries, 1974/5 saw the first decline in energy con-

sumption since the mid-1950s, it was assumed that this was partly due to the
'temporary' recession caused by higher oil prices. Also 1976/7 and 1977/8 saw two
exceptionally mild winters in Western Europe which similarly modified the demand
statistics. Figure 2.1 shows that oil consumption again continued to climb after
1975 both in OECD countries and in the rest of the world. Car and gasoline sales
continued to rise on a worldwide basis despite mounting costs, and there was a
growth in oil use for industrial and power-generation purposes. The weakness of
the dollar combined with the effects of inflation even caused the real price of some
petroleum products to fall. In 1979 the French newspaper *Le Figaro* pointed out
that in 1918 the price of a barrel of oil had been equivalent to the price of an ounce
of gold. This happened a second time in 1979. In the short term it would seem that
the industrial countries after 1973 were run in much the same way as before. The
U.K. in particular, with perhaps the best fuel mix and energy reserves in Europe,
could afford to ignore the need for appropriate investments in conservation and
renewable energy sources.

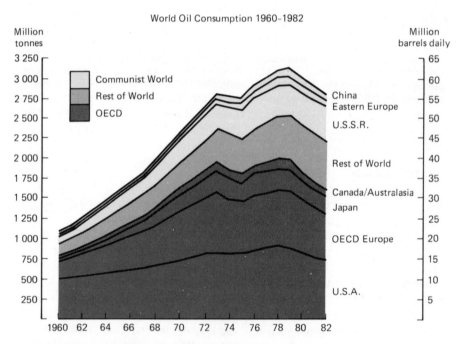

Figure 2.1   *World oil consumption 1960–82* (source: *BP Statistical Review of World Energy, 1982*)

Yet life was not the same. Cheap fuel had been replaced almost overnight by
expensive fuel. The 25 years prior to 1973 had seen an approximate 5 per cent
annual increase in world primary-energy consumption; in the 6 years after 1973
the rate averaged 3 per cent. Before 1973 oil prices to the consuming nations were
affected only by the costs of prospecting, production, refining, transport and the

profits of the oil companies. Indeed, improvements in exploitation techniques in the previous 25 years had driven fuel prices downwards. For example, these were 37 per cent lower in real terms in the U.S.A. than a quarter of a century earlier.[1] After 1973 oil price levels began to be influenced by the monopolistic power of the producers. Before 1973 alternative energies like hydroelectric, nuclear and geothermal energy contributed relatively small amounts to world supply. The post-1973 era saw a gradual change in these patterns of energy production and an intensifying search to develop and invest in other fuel sources. There was a growing awareness that the energy sources underpinning the entire lifestyle of the industrial nations, as well as the industrial growth of the Third World, were not only politically vulnerable but finite. The scarcer they became the more precarious was the way of life of any country without indigenous energy supplies. The tacit assumption of an exponential growth in energy consumption had come to an end.

Events leading up to 1973

A decade after the fuel shortages following the Second World War the OECD set up an energy commission which issued a cautionary document entitled *Europe's Energy Requirements – can they be met?* This pointed to the need for greater fuel efficiency, an expansion of the coal industry and the development of nuclear power. On the other hand, new oil reserves continued to be discovered as earlier wells became depleted. In 1938, world proven oil reserves stood at 31 billion barrels (15 years supply at the time); by 1950 the reserves had increased to 95 billion barrels (25 years supply); by the late-1960s the development of Middle East oil increased estimates to 529 billion barrels.[2] These continually revised estimates not only enabled production to be increased and encouraged consumers to switch from other power sources to oil, but also created a sense of false optimism and energy security. In the mind of the general public, industrialists and politicians alike, there was a belief that whatever warnings might be uttered, more oil would always be discovered somewhere in the world and that it would be available and cheap.

The Seven Sisters and the Formation of OPEC

An early warning was the Suez crisis in 1956 with the disruption of Middle East oil supplies on which the U.S.A. had come to rely. As a result, President Eisenhower introduced oil-import controls in the U.S.A. to the effect that 88 per cent of America's consumption would be met by indigenous resources and only 12 per cent imported. Although the U.S.A. was the largest oil producer at that time it was also importing oil from the Arab countries in order to sustain its vast consumption. One effect of these import controls was that when normal oil supplies were resumed, the major oil companies found themselves with millions of barrels of surplus oil with which they plied Europe, Japan and any other available market at prices even lower than those that had existed previously. The oil-exporting countries themselves, realising that their reserves were being depleted and that they were being exploited, set up the Organisation of Petroleum Exporting Countries (OPEC) in 1960 to safeguard their remaining supplies and to increase their negotiating power. In December of the same year the Organisation for Economic Co-operation and Development was established to achieve higher economic growth and promote world trade.

The OPEC group originally consisted of Iran, Iraq, Kuwait and Saudi Arabia (countries in the Middle East basin), and Venezuela. These were later joined by the United Arab Emirates and Qatar, Libya and Algeria in North Africa, and Nigeria, Indonesia, Ecuador and Abu Dhabi. OPEC was thus formed in direct response to the market power and extraction policies of the large oil companies. The most powerful of those were the seven multi-nationals consisting of Exxon, Royal Dutch Shell, British Petroleum, Texaco, Mobil, Socal and Gulf Oil (collectively known as 'The Seven Sisters').

The next significant developments took place with the Six Days War in 1967 between Israel and Egypt which resulted in a world oil shortage. This enabled some OPEC countries to raise the price of their oil exports and so counteract the combined strength of the Seven Sisters cartel. Following their success, OPEC states began to negotiate with the oil companies for 25 per cent participation agreements. In June 1972 Iraq began the nationalisation of its northern oil fields; in the following months the Gulf States and other OPEC members started to sign the participation agreements and in March 1973, in response to OPEC's success, the Shah of Iran announced that the National Iranian Oil Company would take over all consortium activities in his country. The movement for OPEC countries to take control of their oil fields instead of being satisfied with 25 per cent participation agreements continued to grow.

Colonel Gaddafi, who had overthrown the Libyan monarchy in 1969, exerted pressure on foreign oil companies operating the country's oil fields to limit their production rates, fully realising that oil was the country's main economic asset and that reserves were limited. When the oil companies failed to comply, early in 1973, he campaigned for Libya to gain control of its oil industry. First he served notice that the Libyan government wanted 50 per cent control of oil production operations, and 100 per cent by June of that year. Unfortunately Libya did not have the necessary technical expertise to operate its vast oil fields so on 1 September 1973, Gaddafi announced a unilateral takeover of 51 per cent of all the major oil companies' operations, with the exception of Esso's gas liquefaction plant at Brega.[3] Although some companies held out, gradually the other OPEC members also took over oil-producing facilities in their territories. Even before the outbreak of the Yom Kippur War between Israel and her Arab neighbours, the oil tide had turned. The oil companies, whose policy had hitherto been to create and saturate new markets, were to face increasing pressures from the bargaining power of the oil-producing countries. The initiative of posting market prices for crude oil passed to the OPEC group and the established oil companies had no choice but to accept the new levels.

After the formation of OPEC, the power of the Seven Sisters' cartel was further weakened by a number of new companies who had taken up concessions in the oil-producing countries, built refineries, and established market networks around the world. Examples included the Compagnie Française des Pétroles of France, the Ente Nazionale Idrocarburi (ENI) of Italy, Phillips Petroleum and Continental Oil.

Also, by 1972 some OPEC countries were beginning to invest in the shares of the oil companies themselves, a trend that increased after the Yom Kippur War. In another development, during the post-1973 era many of the producing countries themselves followed the early example of the Libyans, Iraqis and Iranians and set up their own national oil companies at the expense of the Western companies. They included Sonatrach (Algeria), Yacimientos (Argentina), Petrobras (Brazil), Indian Oil (India), Pertamina (Indonesia), KNPC (Kuwait), Pomex (Mexico), Petromin (Saudi Arabia), Turkiye Petrolleri (Turkey) and Petroven (Venezuela).

All these trends increased the economic pressures on the oil-importing countries. Moreover, when the tenuous Arab–Israeli peace, which had lasted since 1967, again gave way to conflict, a new political dimension was introduced: the oil weapon.

Effects of the 1973 Oil Crisis

For the first time, as a result of the 1973 Yom Kippur War, Arab strategy under the Saudi Arabian Oil Minister, Sheikh Ahmed Zaki Yamani, used oil as a political weapon when he attempted to pressurise the world powers to force Israel to withdraw from the occupied territories. Among other measures such as restricted deliveries, a complete oil embargo was imposed on certain 'unfriendly' nations who supported Israel, namely the Netherlands and the U.S.A. Also, in 1973 the OPEC oil ministers' meeting in Kuwait decided to raise official oil prices by 66–70 per cent, affirming that future prices would not be subject to negotiation but would be set by the governments of producing countries. By the time the oil embargo was withdrawn in 1974 the price of crude oil had quadrupled. To start with, the industrial countries were apprehensive as to the future. Some thought the 'oil weapon' would gravely threaten national security and economic prosperity and fragment the Western alliance. F. J. Gardener writing in the *Oil and Gas Journal* in December 1973 went as far as to say that the crisis had already 'split the E.E.C. down the middle, splintered the OECD, isolated Japan, crippled world shipping, sent shivers of apprehension through NATO and changed the everyday habits of nearly every oil-consumer in the world. And all without firing a shot. In a week's time the oil weapon achieved more than all the Arab armies could ever hope to have won.'[3]

The immediate effect of the oil embargo was to lead several industrial countries hastily to introduce emergency measures such as speed restrictions on roads and motorways, heating restrictions, bans on Sunday driving and limited petrol station opening times. Nevertheless for 3–5 years after 1973, broader national strategies to counter the immense dependency on oil were restrained. Apart from the establishment of the International Energy Agency in 1974 to co-ordinate the energy interests of OECD countries, few steps were taken to safeguard future supplies. Energy forecasts continued to be optimistic and the energy consumers continued to live in much the same way. Energy even continued to be squandered in homes, offices, industries and public institutions. Moreover, since the combined effects of the emergency measures and the recession mentioned earlier had lessened the rate of increase of oil demand, by the end of 1975 the emergency restrictions were able to be relaxed. Once more demand increased. By 1979, total world oil consumption (excluding the U.S.S.R., China and Eastern Europe) was 7 per cent higher than that in 1973 and 43 per cent of the total was still being supplied by the Middle East compared with 45 per cent in 1973.

In retrospect, it is evident that it took a second oil crisis in 1979, when the new revolutionary regime in Iran under Ayatollah Khomeni decided to cut their exports by 50 per cent to persuade most governments of the oil-consuming countries to take the energy problem seriously. While some European countries and countries with inadequate fossil-fuel resources (such as Japan and Israel) had gradually begun to respond after 1973, others, notably the U.S.A., failed to do so. Although the U.S.A. was the world's largest energy consumer with a *per capita* consumption of over twice that of the E.E.C. countries, and despite President Carter's constant warnings and appeals to the nation to reduce its dependency on oil, oil imports continued to increase at a rate faster than that of any other country.

Since 1979 efforts in many countries to reduce their oil dependency have intensified considerably. In most countries energy ministers and agencies responsible for energy supply were appointed. Various countries began to formulate energy policies for the first time, with the aim of reducing their dependence on oil and to stimulate public awareness of their energy problems. Some examples are given later in this chapter. Most previous energy-supply decisions had been taken on an *ad hoc* basis with the assumption that increased future demand would always be met by some form of supply, but often without specific data. Now governments, industries and academic institutions began systematically collecting energy data, both locally and on a worldwide scale. For the first time some long-range analyses were based on supply factors as opposed to demand and growth factors. It became dangerous for any government even within the OECD to pursue a policy of 'growth' on the assumption that energy would be readily available and reasonably priced. Although in the short-term 1973 only affected oil supplies, governments gradually became aware that reserves of other fossil fuels were also limited and began to impose pricing restraints on these also. Energy had now to be viewed with the eyes of the politician and economist as well as those of the engineer and marketing salesman.

The oil 'problem', however, had not been solved. In addition to the new power acquired by OPEC, it was becoming evident that oil reserves were not everlasting and sooner or later would run out. Already throughout the latter part of the 1970s Sheikh Yamani had consistently pointed out that Western countries should modify their life-styles to conserve the diminishing oil reserves as future supplies could not meet their increasing demands. The Secretary General of OPEC also directly approached the oil-importing countries urging them to show restraint. At current rates of oil-consumption growth, Middle East oil would last only a further 20–50 years and, as this was the main livelihood of the 250 million people of the Arab nations, it was in their interests to ensure that their wells did not run dry. The oil producers maintained that conservation must come from the oil users themselves. Enforced restrictions such as large price increases imposed by OPEC to protect their supplies would succeed only in sending the Western countries into a deep economic recession.

Figure 2.2 *Sheikh Yamani and delegation: Arab Oil Summit, June 1979*
(source: Central Press Photos Ltd)

Another long-term result of both the 1973 and 1979 oil crises was that conservation, a casualty of the affluent society, became a major feature of energy policy.
Although conservation measures were not introduced sufficiently quickly by most
governments, they were regarded as a major potential energy option. There was
also a re-evaluation of the potential energy contribution of previously neglected or
discontinued energy sources such as the renewable energies described in chapters
6–9 and a rapid expansion of coal and nuclear programmes in many industrialised
countries. Energy research and development was funded on a large scale and the
rate of exploration for alternative oil sources also intensified.

Effects of the 1973 Oil Crisis in the U.K. and Elsewhere

(1) In common with many other countries, the *U.K.* did not react with undue
vigour after the 1973 oil crisis. In 1974 the Department of Energy was created to
replace the Ministry of Fuel and Power and the first Minister responsible for energy
was appointed. In 1977 the Energy Commission was established and in 1978 the
government published its Green Paper *Energy Policy, a Consultative Document*.
The government, rather than the energy industries alone, now became directly
involved in funding and controlling energy development in a way that could not
have been envisaged before 1973. In 1978 the Commission on Energy and the
Environment was set up. Interest in energy research and development of alternative
energy supply sources also expanded but with less alacrity than in other industrialised nations such as Japan, the U.S.A., France and West Germany where fuel
shortages were felt more acutely. Examples of new bodies engaged in energy

research in the U.K. were the Watt Committee and the energy research units set up by the universities, the University of Sussex and the Open University being two of the first. There was also a growth in the flow of energy data and information from all the leading energy industries themselves, and the government-funded Energy Technology Support Unit (ETSU), which was established at Harwell to assist in research, particularly in the renewable energies. Energy courses found their way into school, polytechnic and university curricula. In 1979, the Open University offered the first course in energy studies.

A major plank of U.K. policy in the mid-1970s was to engage in an all-out exploitation of North Sea oil, despite the fact that this was more expensive to extract than OPEC oil and required vast capital investment. In fact U.K. oil production increased eightfold between 1977 and 1983. This was in sharp contrast to many OPEC countries who had to impose sharp cutbacks on production to help maintain oil prices. Further details are given in chapter 3.

The coal industry which had been declining since the 1960s was also a major target of revival and expansion. The National Coal Board's 1973 *Plan for Coal* aimed at an annual output of 135 million tonnes by 1985, with the creation of 42 million tonnes of new colliery capacity and a 15 million tonne increase in open cast production.[4] By 1982/3 production had reached 124.3 million tonnes.

The U.K. also committed itself to a rapid expansion of the 1965 nuclear programme. Although technical and other difficulties hindered progress, in 1978 two new Advanced Gas Cooled Reactors were ordered and in December 1979 the government announced further plans to build ten new nuclear reactors before the end of the century, at a cost of some £20,000 million, a figure later significantly revised upwards.

Because the immediate effects of the fuel shortages were perhaps less apparent in the U.K. than elsewhere in Europe on account of large reserves of fossil fuels, priority was given to the development of these rather than the renewable energies. Even by 1980 the government was allocating only just over £9 million for the development of renewable energy sources including solar, wind, wave, tidal, geothermal and biomass out of a total energy research and development budget that exceeded £365 million. This was comparatively little when set against investments in coal, North Sea oil and nuclear energy, and when compared with the funding of renewable energies in most other industrial countries. Although the U.K. budget allocation for renewable energy R & D was subsequently increased to just over £14 million per year, by April 1982 the government-sponsored Advisory Council for Research and Development for Fuel and Power (ACORD) was recommending that the budget ceiling be reduced once more to £11 million a year for the following 2–3 years and the funding of R & D in certain renewable resources be dropped completely even though some of the pilot projects were still in progress. In view of inflation the recommended cuts represented an even more significant sum in real terms and, in line with the U.S.A. in the early 1980s, a significant statement of central government policy.

(2) *The Federal Republic of Germany* imports virtually all its oil, which accounted for 48 per cent of Total Primary Energy (TPE) in 1980. The *Third Revision of the*

Energy Programme, adopted in November 1981, increased conservation measures, further reduced the share of oil in power plants and advocated a rapid expansion of the nuclear programme. The intention is to reduce the proportion of oil to 37 per cent of TPE by 1990. Other measures include increasing natural gas imports (especially from Eastern Bloc countries) and expanding both imports and the domestic production of coal. There is only limited investment in solar, wind and conservation technologies. In 1980 the proposed *Programme of Investment for the Future* provided federal financial assistance for the expansion of district-heating schemes, and a coal conversion and liquefaction programme was also presented.

(3) *Japan's* net energy imports accounted for 85 per cent of TPE in 1980, of which 69 per cent was oil. Since Japan does not have adequate fossil-fuel reserves to meet her competitive industrial expansion, energy conservation has long been a central goal in reducing this dependency. Japan's conservation measures date from 1951 when a Heat Control Law was passed leading to the appointment of 25,000 administrators responsible for controlling the heat in factories. The measures have been gradually reinforced and extended, more recently by a very comprehensive 1979 Energy Conservation Act. On the supply side the oil crises of 1973 and 1979 gave rise to an ambitious programme aimed at switching from oil to natural gas (mainly LNG) and expanding the nuclear programme, which is expected to supply 30 per cent of Japan's electricity by 1990. An important development was the introduction of 'Project Sunshine' following the 1973 crisis, which aimed at developing solar energy and researching all promising renewable energy technologies. This was followed by the Alternative Energy Development Act in 1980, and the establishment of the New Energy Development Organisation (NEDO). NEDO's remit was initially to concentrate on developing coal liquefaction, solar energy and geothermal power generation, to take over the administration of 'Project Sunshine' and to develop overseas coal mines.

(4) In the *Netherlands* during 1980, 70.2 Mtoe of indigenous natural gas were produced from the Groningen field, as against a total TPE of 65.5 Mtoe. 42.1 Mtoe were exported, leaving the Netherlands dependent on imported energy, notably oil (45 per cent of TPE). The country suffered from a total oil embargo by OPEC after 1973 and, after this crisis, energy policies were designed both to switch away from imported oil and to use gas for essential purposes only. During the 1970s measures were introduced to restrict demand through energy pricing, conservation and the diversification of supply sources. A premium was added to the price of gas to discourage unnecessary use, especially for electricity generation and boiler fuel which can be provided by coal. Gas exports, from being greater than oil imports, were also to be phased out, on the termination of existing contracts, by 1998. A comprehensive set of conservation measures has also been implemented, notably the 1974 National Insulation Plan for houses and several schemes combining heat and power and district heating. Grants for conservation in industry, fiscal measures and legislation to discourage energy use in transport have also been introduced. To make up the future energy shortfall, coal imports will be increased and new coal-combustion technologies promoted. No significant nuclear capacity has been installed in the Netherlands, with only two stations in operation by 1982.

(5) Imported oil in *Sweden* accounted for 55 per cent of TPE in 1980, the main indigenous energy resources being hydroelectric power (29 per cent), nuclear power (13 per cent) and solid fuels (7 per cent). Following the results of a referendum in 1980 on the controversial nuclear-power programme, it was decided that the current construction programme of 12 reactors should be completed after which nuclear power should be phased out at a rate consistent with the electricity requirement necessary to maintain employment and welfare, that is, by about the year 2010. In 1981 an Energy Bill was presented to Parliament outlining a strategy to encourage substitution of other energy forms for oil imports. Improved conservation was a key factor in improving Sweden's energy position, especially through combined heat and power and district-heating schemes. Grants and mandatory measures, publicity, an energy advice service, and training in conservation and appropriate energy use formed other strands of the policy. An Oil Substitution Fund, created in 1980, and financed through an increase in the retail duty on oil products, was designed to encourage the use of alternatives such as peat and wood, biomass, coal and solar energy. Sweden was also among the first countries in Europe to develop and extend the use of district-heating schemes, heat pumps and other energy-conserving devices for public and domestic buildings. Because Sweden has no known oil or gas fields and only limited coal, official measures concentrate on securing supplies from abroad. By 1990 about 10 per cent of the 1979 oil consumption is to be replaced by renewable and synthetic energy forms, notably biomass from the country's abundant forests.

(6) In 1982 the *U.S.A.* consumed a quarter of all the world's primary energy. By 1985 imported oil is expected to account for 18 per cent of TPE, and domestic oil and gas (which are likely to run out before the end of the century) for another 23 per cent and 22 per cent, respectively. Following the 1973 OPEC oil embargo on the U.S.A., the Nixon and Carter administrations were forcibly alerted to the energy vulnerability of the U.S.A. President Nixon's 'Project Independence' and President Carter's National Energy Plan of April 1977 set the pattern for future policy towards increased self-sufficiency. The National Energy Plan set the following goals for 1985:

1. A reduction in the annual growth of total energy demand to below 2 per cent.
2. A reduction in gasoline consumption of 10 per cent below current levels.
3. A reduction in oil imports from projected 1985 levels of well over 12 mbd to 7 mbd.
4. An increase in coal production from approximately 700 million tons per year to over one billion tons per year.
5. A national programme designed to bring 90 per cent of all residences and many other buildings to minimum energy efficiency standards.
6. A major programme for the use of solar energy in new and existing residences.

The first steps to implement the Plan were incorporated in the National Energy Act of October 1978 which comprised a variety of measures. The National Energy Conservation Policy Act provided weatherisation grants for low-income housing, loans for conservation and solar energy investments in schools, hospitals and public buildings. It also established efficiency standards for major energy-consuming appliances. Utilities were required to advise on conservation and where appropriate solar applications. The Power Plant and Industrial Fuel Use Act required most new industrial and power plants (with a provision for earning exemption) to stop using gas and possibly oil after 1990. The Public Utility Regulatory Policies Act required a review of rate structures to encourage conservation and conferred powers to ensure load-sharing and co-operation between power suppliers. The Natural Gas Policy extended federal regulation and provided for the progressive removal of price controls up to 1985. Associated with decontrol, a Windfall Profits Tax was introduced providing revenues to be used for business energy-investment credits, residential energy tax credits, income tax credits for heavy oil and non-conventional oil and gas production, and low-income energy assistance. The Energy Tax Act provided incentive tax credits for energy conservation and use of alternative energy sources, and made car manufacturers liable to increasing tax rates for high-consumption 'gas-guzzlers', starting with 1980 models.

In the mid-twentieth century U.S. energy demand per unit of output was more than twice that of almost every other country and, despite this initial battery of legislation, consumption continued at an extravagantly high level and energy prices remained relatively low. However, the consequences of the 1979 Iranian revolution once more focused American public attention on energy security. The 1980 Energy Security Act established a Synthetic Fuels Corporation to help fund demonstration plants and commercial synfuels projects. An Energy Conservation and Solar Bank was set up to subsidise conservation improvements and solar installations, and provision was made for the funding of development of biomass, alcohol fuels and geothermal energy.

After January 1981, however, when the new Administration under President Reagan took office, major changes in energy policy occurred reflecting the philosophy that market forces should determine energy supply and demand. Domestic crude oil and petroleum products were decontrolled, strong efforts were made to revitalise the domestic nuclear programme, and funding for conservation, solar and other renewable-energy projects was severely curtailed. Government grants in these latter areas as well as in public information and education pro-grammes had been steadily increasing since 1973.

(7) *The E.E.C.* also undertook to fund energy research and demonstration projects in member countries and supply analyses and evaluation of individual programmes, comparing levels of investment and measures undertaken in different areas.[5]

From the few sketched examples of energy policies in selected I.E.A. countries, several important features emerge. It is evident that, apart from differences in emphasis and vast differences in available funding, there are wide variations between countries' own natural resources and also their dependence on imported oil and fossil-fuel supplies. Australia, Canada, New Zealand, Norway and the U.K. are all extremely well endowed with a variety of energy resources, in contrast to the remainder which are all heavily dependent on imported oil. For some, such as West Germany, France, the U.S.A. and Japan, there is heavy reliance on nuclear energy together with investment in renewable resources. For others, a marginal or non-nuclear strategy has been adopted, for example in Denmark, Norway, Sweden and Austria. Some countries are relying mainly on the expansion of indigenous or imported conventional fuels such as coal and gas, while others (for example, Sweden) are putting increasing efforts into developing the renewables and hydropower.

Table 2.1 illustrates the changing patterns of oil consumption for 1973 and 1982 for different countries and regions of the world. These are compared with movements in total energy consumption between the two years. Almost without exception, the industrialised countries of the West have significantly reduced their total consumption of oil since 1973, while the developing countries and those of the Eastern Bloc have increased their consumption. The richest countries of the world have also decreased their total primary energy consumption in real terms, notably the Benelux countries, France, Sweden, the U.K., West Germany and the U.S.A. By contrast in the same decade the developing countries and the Eastern Bloc have shown overall increases in energy consumption of up to 85 per cent. In 1982, in the world as a whole, oil and natural gas together contributed 60 per cent of total consumption and coal 30 per cent, leaving only 10 per cent for water power and nuclear energy combined. It is interesting to note that of the renewable energies, only water power is regarded as significant enough for inclusion in a world total. Non-commercial energy sources such as firewood are not included because of the difficulty in assessing their contribution.

Oil Prices after 1973

For a decade after 1973, the initiative for setting market prices for crude oil was held by the OPEC group, the major Western companies having lost control of the Middle East oil wells. Yet within OPEC the debate during the post-1973 period was between the so-called 'moderate' states such as Saudi Arabia, who favoured small price increases, and the 'hard-liners', including Libya and Iraq, who pressed for ever larger price increases. In times of major political upheaval within an OPEC country, the hard-liners usually triumphed. A striking example was the 1979 political crisis in Iran when, after the February revolution, Iran 'officially' cut back its oil-production rate from 6 to 4 million barrels per day (mbd), two-thirds of the pre-revolution rate. This reduction sent shock waves throughout the oil-importing world since it represented a 4 per cent shortfall in non-communist bloc oil supplies and a notional 'gap' of 1.5 mbd between world oil supply and demand. Subsequently, Iraq and the United Arab Emirates indicated that they too would cut production. They agreed that they were unwilling to meet the 'insatiable' demands of the industrialised countries if the consequence was that they would damage their own economies by saturating them with too much oil-derived finance. It was also evident that, in a demand-inelastic suppliers' market, as much revenue could be earned by producing less oil at higher prices as by producing more oil at lower prices.

The restrictions by the hard-liners had an immediate effect on the oil market

Table 2.1 Total oil consumption and total primary energy consumption, major countries, 1973 and 1982

Country/area	Total oil consumption			Total primary energy consumption			Oil as % of total consumption 1982
	1973 million tonnes	1982 million tonnes	% change 1973–82	1973 mtoe	1982 mtoe	% change 1973–82	
U.S.A.	818.0	703.0	−14	1822.7	1728.4	− 5	41
Canada	83.7	73.0	−13	190.9	207.4	+ 9	35
Latin America	168.3	235.4	+40	244.6	362.2	+48	65
Austria	11.8	10.5	−11	23.8	24.8	+ 4	42
Belgium/Luxembourg	31.5	22.8	−27	50.8	44.8	−12	51
Denmark	17.9	11.7	−35	19.7	18.1	− 8	65
France	127.3	92.4	−27	186.1	180.6	− 3	51
Greece	10.0	11.9	+19	15.3	16.8	+10	71
Ireland	5.4	4.6	−15	7.4	8.0	+ 8	58
Italy	103.6	90.7	−12	137.9	141.4	+ 3	64
Netherlands	41.3	30.4	−26	76.9	63.4	−18	48
Norway	8.6	7.8	− 9	28.1	31.8	+13	25
Spain	39.1	46.5	+19	57.8	75.0	+30	62
Sweden	29.4	19.7	−33	46.2	37.6	−19	52
United Kingdom	113.2	75.6	−33	224.8	193.4	−14	39
West Germany	149.7	112.4	−25	264.9	250.4	− 5	45
Middle East	62.2	86.7	+39	87.1	125.7	+44	69
Africa	49.5	79.0	+60	98.2	181.3	+85	44
South Asia	31.3	45.1	+44	103.3	163.0	+58	28
South-east Asia	77.6	121.0	+56	127.4	200.2	+57	60
Japan	269.1	207.0	−23	347.7	340.2	− 2	61
Australasia	34.8	35.5	+ 2	66.6	89.5	+34	40
U.S.S.R.	325.7	448.5	+37	874.1	1242.0	+42	36
Eastern Europe	75.1	101.1	+35	357.2	455.4	+27	22
China	53.8	82.4	+53	362.5	522.1	+60	16
Total Eastern Hemisphere	1728.0	1807.4	+ 5	3665.0	4536.7	+24	40
Total Western Hemisphere	1070.0	1011.4	− 5	2258.2	2298.0	+ 2	44
WORLD	2798.0	2818.8	+ 1	5923.2	6834.7	+15	41

Source: Extracted from BP Statistical Review of World Energy 1983, BP, London.

and served to push prices even higher. The upward spiral was stimulated by some Western oil companies, which reacted to supply and price uncertainties by buying whatever oil they could, wherever it was to be sold, at whatever price. The OPEC producers, in turn, soon found themselves able to charge much more than the 'official' OPEC ceiling price. In June 1979 the official ceiling was U.S. $23.50 per barrel. By November the hard-liners, Libya, Algeria and Nigeria, were charging U.S. $25.27 for their light superior grade crude and selling it without difficulty. The Saudi Arabians were almost alone in trying to resist this upward pressure in prices. But they could only contain the official OPEC ceiling price until December 1979 when under hard-liner pressures, it was increased to U.S. $26 (Saudi light 34°). Similar pressures prevailed during 1980 when there were three further increases resulting in an official price of $32 by January 1981. By October the official price reached a ceiling of $34.

Growth of the Spot Market

The pressures leading to the leap-frogging of the OPEC 'official' ceiling price came from what is known as the spot market for oil. The spot market developed largely as a result of the changed status of oil after 1973, but began to make its effects felt only after the major oil crisis of 1979. Up to 1979, about 95 per cent of all OPEC oil was sold through long-term fixed price contracts to the large oil companies. The remaining 5 per cent of 'uncommitted' oil production sold outside these arrange- ments was available at short-term notice to refineries with temporary shortages. This was sold at spot prices somewhat higher than contract prices. However, with mounting uncertainties and the prospects of imminent shortages, the importance of the spot market grew. Producers were tempted to cancel long-term contracts and make higher profits by selling at spot rates. Consumers, faced with possible short- ages, were only too willing to pay the extra cash. The result was that the proportion of oil sold on spot markets rose from 5 per cent to 15 per cent by December 1979. In markets such as Rotterdam, Singapore and Rome, it was fetching prices as high as U.S. $40 per barrel. Some consumers whose long-term contracts had been cancelled found themselves buying the same oil on the spot market at considerably higher prices.

Under these pressures, it was inevitable that 'official' OPEC oil prices should continue to rise and that some OPEC countries should seek to exploit the market for their own advantage. Between December 1978 and December 1979, the weighted average price for OPEC crudes rose by no less than 117 per cent, and, even then, the price being charged by certain hard-line countries was still some U.S. $8–9 higher than the official level.

The immediate effect of these major price rises was to increase dramatically the revenues of the oil producers, including those from previously under-developed countries. By 1980 they had reached immense and unprecedented levels and were causing severe problems of financial management.

First the massive funds brought about major redistribution in the ownership of wealth in the world, from the industrialised countries to the oil producers. Second, they created considerable uncertainty on international financial markets since a large proportion of the oil revenues was lent on high-interest, short-term deposit in order to maximise profit. The only way the international financial institutions could relend this money was at medium or longer-term rates suitable for capital investments. Hence, there remained the inherent insecurity arising from the possibility that OPEC countries might suddenly decide to switch funds. Third, the massive inflow of funds to OPEC countries from the oil-importing Third World countries created enormous debts for these countries which they had no means of settling without harmful effects on their economies. Oil-importing industrialised countries also suffered growing balance of payments deficits as a result of oil payments. One victim was the U.S.A. The U.S. dollar had for some years been used both as the world's reserve currency and as the major currency through which oil transactions were conducted (the 'petro-dollar'). With huge balance of payments deficits, largely caused by higher energy costs and inflation since 1973, the U.S. dollar became much weaker in international markets. This, in turn, affected international trading in very many commodities other than oil.

Oil Prices and Inflation

The world economy was affected by these economic forces in two main ways. First, a major stimulus to inflation was created. Second, it generated a recession in world trade and through it, in internal economies. After 1973, and especially since 1979, for the first time in the post-Keynesian history of the industrialised world, inflation was accompanied by recession and redundancy.

There is however a direct and an indirect relationship between a rise in the price of oil and inflation. First, 3–10 per cent[6] of crude oil is actually used as a factor of production during the refining process and in the distribution of oil itself. A rise in the price of crude therefore directly affects the cost of the subsequent stages of petroleum manufacture. Furthermore, the price of refined oil may be affected indirectly by inflation insofar as rising capital, wage and other operating costs incurred will inevitably escalate the cost of the final product. These combined factors both influence the price of oil and are, in turn, influenced by it.

To illustrate, figure 2.3 compares the movement of the U.K. crude-oil price at current and constant (1970) values. Although in money terms oil prices have risen dramatically, in real terms they have barely kept pace with the cost of living except for the crisis year of 1973 and to a lesser extent in 1979. The reasons for this are partly that oil price rises triggered off rises in the general level of prices, partly because OPEC ministers raised oil prices in line with inflation expectations, and partly because initially there was no significant shift in consumer purchasing preferences away from oil (for example, towards conservation) despite the cost increases.

There is also a close association between crude-oil price rises and other prices in an economy. Oil is still today the life blood of modern industrial and rural economies of the world. Large conurbations depend on oil-powered transport for their livelihood. Industry depends on oil for everything from power generation to

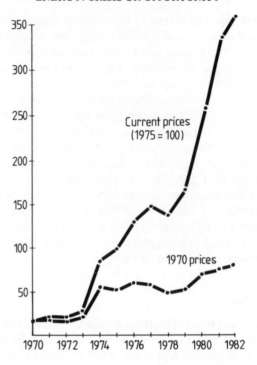

Figure 2.3   *U.K. crude-oil prices (imports and North Sea), 1970–1982*

obtaining raw material supplies, and transportation of the end products. Oil by-products are also used extensively in industry and agriculture. Even food production depends on oil to fuel agricultural machinery and to produce fertiliser and animal feedstock. Practically every sector of the industrialised world has come to rely on oil in one form or another for its survival. Consequently it is to be expected that major changes in oil prices and output would soon be felt in other sectors of the economy. This has in fact been the case. Table 2.2 compares the Wholesale Price Index (WPI) for imported petroleum to the U.S.A. with selected U.S. WPIs between 1973 and 1975, the period immediately after the first oil crisis. The table indicates that the price of petroleum imported into the U.S.A. rose by 320 per cent between early-1973 and late-1974 after which it stabilised again. Nearly half of this rise was in due course reflected in the price of refined petroleum, which in turn lifted the overall WPI by 40 per cent.

Oil Prices and the Gross National Product since 1973

Rising oil prices also affect economic growth. The model in figure 2.4 describes the main causal links. The model shows how higher OPEC prices reduce real GNP through cost-push inflation. This in turn lowers investment, employment and personal income and increases export prices. Consequently there is a fall in foreign import demand so that reductions in real GNP are 'exported' to overseas countries.

Table 2.2 *Wholesale Price Index (WPI) for imported and refined petroleum,*
USA, 1973–75

	WPI for imported petroleum (1972 ≡ 100)	WPI for refined petroleum (1967 ≡ 100)	WPI (1967 ≡ 100)
1973 (1st Quarter)	106.4	117.3	126.9
1974 (4th Quarter)	447.4	240.3	171.9
1975 (4th Quarter)	453.4	274.6	178.2

Source: Unpublished data adapted from Bureau of Labour statistics, U.S. Department of
Labour and quoted in A. Bradley Akin, *How Energy Affects the Economy,* Lexington Books,
Massachusetts, 1978, p.8.

The impact of an OPEC price increase, in short, is both to increase inflation and to
induce recessionary conditions in the world economy.

Several econometric studies have attempted to estimate the effects of rising
energy prices after 1973 on various major economic indicators such as GNP and
employment. This they have done by comparing existing historical behaviour with
simulations of how the economy would have performed without the energy crisis.
Two such studies[7,8] suggested that the 1973 oil crisis reduced GNP in the U.S.A.
by between U.S. $38 billion and U.S. $57 billion by 1975 during which period
unemployment increased by an additional 2 per cent over expected levels. There is
also evidence from Bureau of Census data to suggest that these figures would have
been even worse had manufacturers not responded to rising oil prices by devoting
greater efforts to inter-fuel substitutions and improvements in conservation and
energy efficiency. In the short run, higher energy prices act much like an excise tax,
raising the general price level and depressing aggregate demand. In the longer term,
the composition of GNP and the capital stock is affected in response to a movement away from energy-intensive modes of manufacture.

Whereas the 1973 oil crisis slowed down world rates of economic growth without
greatly affecting the demand for oil, the 1979 crisis created such repercussions that
the oil producers were themselves, in turn, affected. Resulting from the combination of recession and higher oil prices, oil-consuming countries were forced to
reduce their demand for oil, either through introducing more stringent fuel-
conservation measures as demonstrated by the previous examples, or by switching
from oil to other fuels, or even because one effect of the recession was to dampen
the demand for energy-intensive goods. An apparent 'oil glut' was consequently
created and the oil-producing OPEC countries were, this time, themselves forced to
cut back production. In the year March 1981/March 1982 OPEC states reduced
their output by one-third from 25 million barrels a day (mbd) to a production
ceiling of 17.5 mbd and to 15.0 mbd in July 1983. The fall represented just over
half of the 1979 peak output of 32 mbd, and was due this time to economic rather
than political forces.

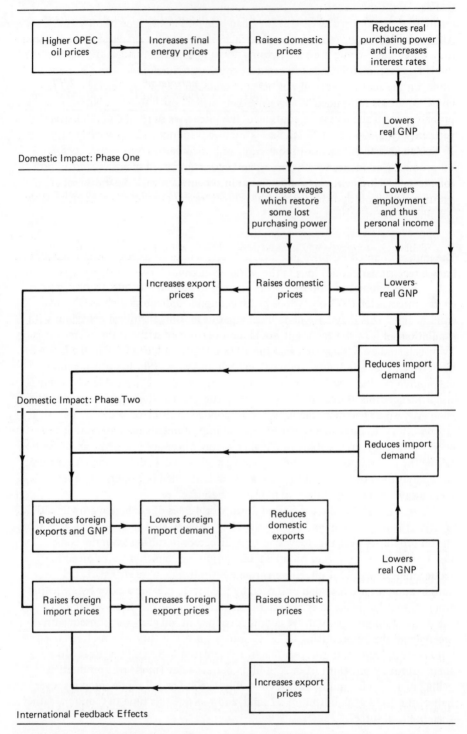

Figure 2.4 *Oil prices and their effect on GNP since 1973* (source: *The World Oil Market in the Years Ahead*, C.I.A. Report, Washington D.C., 1979)

Faced with this necessitated drop in output, OPEC states found it difficult to maintain the high price levels that they had achieved at the time of oil scarcity. Although OPEC ministers agreed in March 1982 to stabilise prices at $34 per barrel, they were under strong pressure to bring down price levels further. By March 1982 the U.K. government decided to reduce the price of North Sea oil to $31 per barrel while, on the spot market, oil was being obtained at $28.50. In March 1983 the OPEC countries responded by lowering their official price to $29 a barrel. It seemed probable that if the recession continued, the price for crude oil could fall even further. The 1982 cost of North Sea extraction was only $12 per barrel.

Such enforced cutbacks of production and prices have severe repercussions on oil-producing countries which during the 1970s committed themselves to ambitious long-term development plans and entered into contracts with the industrial countries on the assumption that their ever-increasing oil revenues would finance these. Table 2.3 illustrates this situation.

Table 2.3 *Example of financial pressures on OPEC producers, March 1982*

	Financial reserves (billion dollars)	Output levels (1000 barrels per day)	
		Needed to balance current accounts	Present output
Saudi Arabia	161.6	6,410	7,900
Libya	33.4	1,070	870
Kuwait	76.2	900	850
UAE	38.6	810	1,400
Qatar	16.1	60	360
Iran	3.0	3,610	950
Iraq	31.8	2,110	950
Nigeria	4.5	2,230	1,800
Algeria	3.8	1,200	700
Gabon	0.7	160	150
Venezuela	7.7	2,400	2,100
Ecuador	0.7	220	200
Indonesia	10.0	1,500	1,600
	388.1	22,680	19,830

Source: *Petroleum Intelligence Weekly*, February 1982.

The table shows that in early-1982 OPEC countries needed an output level of 22.68 mbd to meet the current balance of payments while actual output was running at 19.83 mbd. Of all the OPEC countries only Saudi Arabia, Qatar, the United Arab Emirates and Indonesia were still in surplus. It may therefore be assumed that, without external intervention, any cutback in oil revenues not only scales down OPEC's development but has direct repercussions on Western industry itself, such is the inter-relationship of oil-based economies.

On the other hand, it is probable that the post-1983 'oil glut' will be seen as a short-term divergence from trend as supply and demand adapt to new levels of equilibrium. In the longer term, prices are bound to continue to rise as oil supplies diminish in the absence of major alternative energy sources, unless the instability caused by the pressures on oil becomes so great that recession remains a permanent feature of the world economy. This view has been reinforced by a 1983 *Economist Intelligence Unit Report*[9] which argues that the oil glut is not merely due to a structural switch away from oil caused by increased conservation or a permanent move to alternative energies policies which could be laid aside if oil prices are to fall, but to the recession itself. It further argues that reductions in oil demand have a magnified downward effect on OPEC production owing to the 'OPEC multiplier', caused by consumers' tactical switching away from OPEC countries, and so on. Conversely, when demand rises again, there will be a renewed upward impetus to OPEC oil prices.

Steeply rising oil prices have also dramatically changed the ratio between oil consumption and economic activity in general. It used to be thought that there was a positive correlation between oil consumption and GDP which remained broadly constant over time. This has not proved to be the case. Up to 1973 oil consumption in the OECD industrialised countries was rising at a greater rate than GDP. In 1969 and 1970 the rate of increase in oil consumption was over double that of GDP since many industries and private consumers were continuing to switch to oil.

In the first seven years after the 1973 oil price rises, however, the ratio between the amount of energy, and even more so the amount of oil consumed per unit of output, declined rapidly. In 1980 GDP in OECD countries was 19 per cent higher than in 1973, but energy consumption grew by only 4 per cent and oil consumption was 3 per cent lower. The TPE/GDP ratio fell by 13 per cent.

Conclusions

The events and impact of the 1973 'oil crisis' may be seen as inevitable consequences of the preceding era of ever-increasing and often wasteful oil exploitation. However, two facts stand out as surprising. The first is that many countries took a very long time to adjust to the new energy scenario and to implement compensatory decisions. In some cases it was not until after the Iranian oil crisis of 1979 that significant energy policies came into effect. The second surprising fact was that the industrial nations should have been taken so unawares by the 1973 'coup'. To use an analogy from chess, it was as though the OECD's queen had been taken by OPEC's pawn. The OECD, whose very existence owed its origin to the assumption of economic growth, had in fact experienced 22 years of an annual average growth of 4.8 per cent in real terms. The assumption of continued growth came to an abrupt end with the recession that followed the 1973 crisis.

Undoubtedly, whatever the initial reactions and interpretations, the 1973 oil crisis was much more than a temporary interruption of energy supplies. It created worldwide inflation, recession, increased poverty in Third World countries and generated serious social consequences in developed countries, even affecting the distribution of jobs and patterns of industry. Politically as well as economically, the 1973 and 1979 oil crises exerted a major destabilising pressure on the oil-importing countries of the world, turning the Middle East into an even more highly sensitive and strategic area. Immense revenues, such as could alter the shape of the entire world's monetary system, were suddenly put into the hands of the OPEC countries. However, far-reaching and unexpected developments were to ensue, changing the world's patterns of energy usage. Although in 1973 the Middle East producers raised their prices and cut output of oil for political reasons and were regarded with hostility by the oil-importing countries, they also demonstrated to the industrially confident nations that their precarious existence was based on a finite and valuable commodity. Before 1973, most were aware that oil supplies would run out in the future, the future being always beyond the scope of the next demand forecast. 1973 acted as a catalyst and brought this 'future' into perspective. The time at which oil consumers would have to develop less energy-intensive lifestyles and alternative energy resources was telescoped to within the vision of the present generation.

References

1. Landsberg, H. H., 'Low Cost Abundant Energy: Paradise Lost?', *Resources for the Future,* Reprint No. 112, Washington D.C., December 1973
2. Tugendhat, C. and Hamilton, A., *Oil: the Biggest Business,* Eyre Methuen, London, 1975, p. 223
3. Gardener, F. J., '1973: The Year of Major Changes in Worldwide Oil', *Oil and Gas Journal,* 31 December 1973, pp. 83–8
4. *Coal for the Future. Progress with 'Plan for Coal' and Prospects to the Year 2000,* Department of Energy, London, 1978
5. Commission des Communautés Européennes, Direction Générale de l'Energie, *Évaluation des Programmes de Demonstration de la Communauté dans le Secteur de l'Energie,* EEC, Brussels, 1982
6. *Our Industry Petroleum,* British Petroleum, London, 1970, p. 191
7. U.S. Federal Agency, *'Project Independence' Report,* Washington D.C., 1974
8. Chapel, S. W., *The Impact of Higher Energy Prices on the U.S. Economy 1973–4,* Federal Energy Administration, Washington D.C., 1976
9. Mossavar-Rahmani, B. and Fesharaki, F., 'OPEC and the World Oil Outlook: rebound of the exporters?', *Economist Intelligence Unit Special Report No. 140,* EIU, London, 1983

Further Reading

Odell, P. R. and Vallenilla, L., *The Pressures of Oil,* Harper and Row, New York, 1978

Kirk, G., *Schumacher on Energy,* Jonathan Cape, London, 1983

Jones, A., *Oil: The Missed Opportunity or Naft and Shaft,* Deutsch, London, 1981

3

Oil and Natural Gas

'The oil sector, which is the leading sector, bears the responsibility of linking the present to the future'

Mara Saeed Al-Otaiba

A. OIL

Oil is found in the world's sedimentary basins and is widely believed to have been generated from countless millions of marine organisms which lived in the shallow waters surrounding pre-historic land masses. Through time, these drifted down to the sea-bed where the fatty acids contained in the organisms were transformed to a proto-petroleum product through bacteriological reduction. Probably through continuous and prolonged geothermal heating, this proto-petroleum was later transformed into the crude oil and natural gas we now know. In certain areas where porous rock was covered by impervious rocks such as salt or clay, the oil and gas gradually collected in these reservoirs.

Sedimentary conditions containing oil deposits are widespread and cover about half the world's land area and much of the adjacent sea-bed. There are about 600 known sedimentary basins in the world and about 400 of these have had some oil exploration. In every sedimentary area oil can be found in varying amounts. However, because formations of impervious rock are often absent from the world's sedimentary basins, so far under half of those explored have been shown to contain oil in usable quantities. Figure 3.1 shows the worldwide location of oil provinces, both proven and prospective.

Reservoirs containing crude oil almost invariably also contain saline water on which the oil floats, and natural gas dissolved in the oil, both of which have to be removed before transportation and refining. The amount of dissolved gas varies: little or sometimes none in heavy viscous crudes, frequently considerable quantities

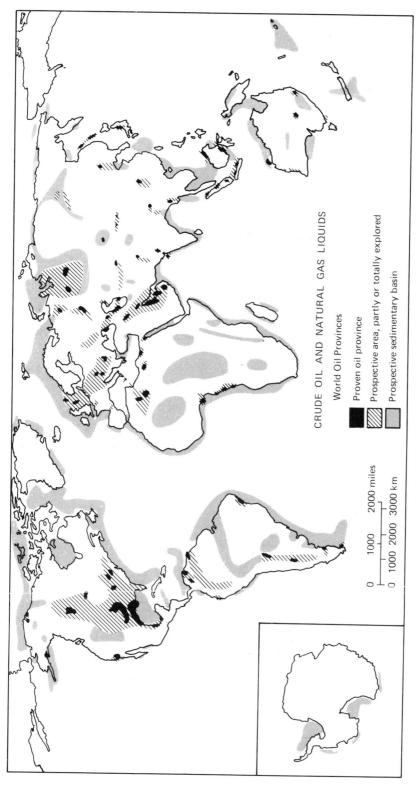

43

CRUDE OIL AND NATURAL GAS LIQUIDS

World Oil Provinces

Proven oil province

Prospective area, partly or totally explored

Prospective sedimentary basin

0   1000   2000 miles

0   1000   2000   3000 km

Figure 3.1 *World oil provinces* (based on Survey of Energy Resources, ©
11th World Energy Conference, London, 1980)

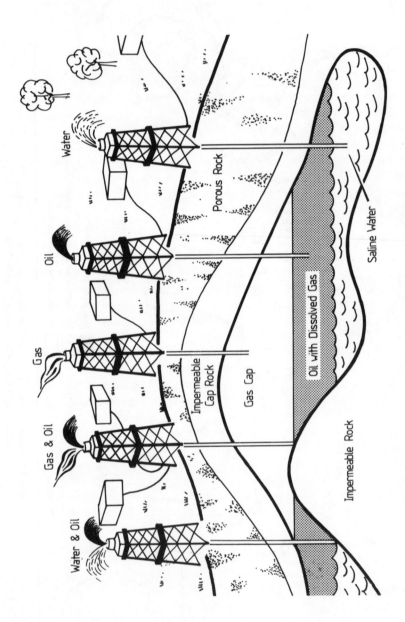

Figure 3.2   Cross-section of a typical oil reservoir in a sedimentary basin

in light thin oils which are easier to extract. Natural gas may also be present as a 'gas cap' above the gas-saturated oil. Figure 3.2 illustrates a cross-section of an oil reservoir.

Oil Extraction and Refining

Oil can normally be extracted from the ground by drilling a well into a reservoir. When a well is bored the pressure from the gas or water forces the oil spontaneously to the surface. This is called 'primary production'. As oil is removed, however, the pressure decreases and the rate of production declines. Normally, only between 20 and 30 per cent of the oil in a particular oil field is recoverable using primary production. It may then be necessary to employ 'secondary', artificial, production methods to extract the remaining oil. With secondary technology, usually involving the systematic injection of water, steam, gas or chemicals into the oil field for the purpose of drawing the dispersed oil in extractable quantities to the well, between about 35 and 50 per cent of oil is recoverable. Even more advanced technological methods (tertiary technology) have been tried under laboratory conditions, including alternative gas–water injection, thickened water injection, wettability reversal, miscible hydrocarbon injection, micellar water injection and thermal methods, but field applications are still under development and will largely depend on the scarcity and price of oil.

Oil from a number of wells is normally sent through 'flowlines' to a common 'gathering centre' where the water and gas are separated from the oil. Part of this gas is normally used as a fuel for heating and power purposes in the field, and may frequently be distributed by pipeline for other uses.

As reservoirs become depleted or when secondary production methods have to be employed, the amount of saline water and gases increases and separation becomes more difficult and costly. High separation costs also occur in crudes that contain unacceptable levels of hydrogen sulphide or other impurities. At high concentrations hydrogen sulphide can cause serious corrosion of equipment and the cracking activity of catalysts can also be adversely affected. Crude oil with low levels of sulphur compounds are often described as 'sweet' while high-sulphur oils are referred to as 'sour'. When the large, easily accessible and high-quality reservoirs have become exhausted, these less economic 'sour' deposits will increasingly have to be exploited.

Once the oil has been through the 'gathering centre' and has been purified in the separators and stripping towers, it is transported, usually by pipeline, to a coastal loading terminal where it is transferred into bulk oil carriers for shipment to overseas refineries. These are usually situated in the oil-importing countries where the requirements of indigenous consumer markets dictate the most appropriate refinery design. One of the longest pipelines in the Middle East is Aramco's Abquiq to Sidon crude line which consists of 1068 miles of 30-inch pipe. The Intercontinental Provincial line in Canada from Edmonton to Toronto and Buffalo, N.Y., is 3553 miles long.

Crude oil carriers can now be built in sizes of up to 500,000 dead weight tonnes (dwt). These 'supertankers', however, are too large to be berthed at most conventional jetties, so frequently purpose-built offshore unloading facilities have to be provided. Of the total world tanker tonnage in 1983, about 35 per cent was owned and operated by the oil companies themselves, about 58 per cent by independent operators mainly for charter to the oil companies; 4 per cent was owned by governments, and the remaining 3 per cent by other large consumers. The world oil fleet in 1983 (excluding combined carriers and all tankers under 10,000 dwt) totalled 283.2 million dwt.[1] However, owing to the drop in oil consumption and production cutbacks of 1983/4 a considerable number of these tankers was laid up.

At the refinery the oil is unloaded for distillation into a number of petroleum products. Refining is a highly complicated process owing to the extremely variable and differentiated nature of crude oil which is a complex mixture of compounds: no two oils, even from closely related reservoirs, are precisely the same in their chemical composition. All crudes, however, contain 85–90 per cent carbon and 10–14 per cent hydrogen and smaller amounts of sulphur (0.2–7 per cent), nitrogen (below 0.1–2 per cent) and oxygen (up to 1.5 per cent). Crudes can also contain trace elements such as vanadium, nickel, chrome, lead and arsenic, as well as other contaminating substances from the water and rocks of the host reservoir.

Crude oils may be classified according to: (i) their specific gravity (which ranges from about 10–50° API); (ii) the family of hydrocarbon compounds from which they come, notably the paraffin and naphthene; (iii) their boiling point (obtained by 'flash distillation'); and (iv) their wax and pour point (defined by the lowest temperature at which a petroleum oil will flow or pour when it is chilled without disturbance at a standard rate). Oils with different combinations of these properties lend themselves to different end-uses. For instance motor spirit has a low boiling point, a high naphthenic content, a low specific gravity and low wax content. It is the function of the oil refinery to distill the crude oil into forms that are appropriate for the various end-uses of different customers.

Distillation is carried out in fractionating towers or columns, as shown in figure 3.3. These have special perforated trays throughout their vertical section at intervals of about 2 ft. The crude oil is heated to a temperature of about 300°C or more, depending on the type of crude involved and the products required, and then pumped into the bottom of the tower as a heated oil and vapour mixture. As the mixture rises through the tower it cools progressively. With cooling, those compounds with higher boiling points condense near the bottom of the tower and the liquid oil is drawn off from the tray. The remaining vapour rises through the perforations in the tray until with further cooling another group of compounds condenses into liquid and is drawn off. The procedure is repeated until the light oils with the lowest boiling points are drawn off at the top of the tower. Normally, distillation columns separate crude oil into the 'fractions' shown in table 3.1. Typical proportions of each product obtainable from a given input of crude oil are shown in the right-hand column. As can be seen there is a wide range, deriving both from the many different types of crude oil in existence and the flexibility

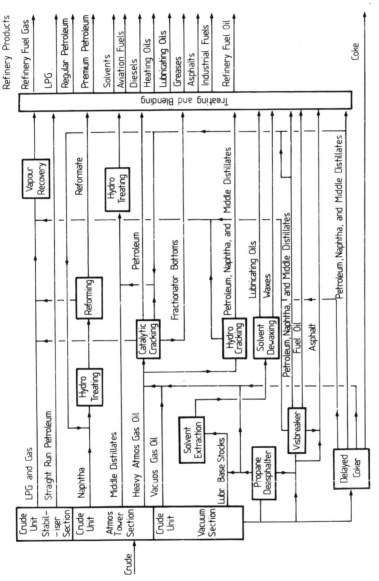

Figure 3.3 *Oil-refinery flow diagram* (source: Gary and Handwerk, *Petroleum Refining*, Marcel Dekker, New York, 1975)

Table 3.1 *Boiling ranges, various petroleum products*

Boiling range (°C)	Product	Initial crude (%)
−10–0	Propane/butane	2–5
25–80	Light gasoline	2–10
80–180	Naphtha	4–14
150–250	Kerosene	1–6
250–350	Gas oil (light)	16–28
	(heavy)	4–12
over 350	Fuel oil	35–50

Source: Our Industry Petroleum, BP, London, 1970.

with which modern refineries are designed. These ranges are important since there is no point in producing a refinery product unless there is a market for it.

In order to maximise flexibility and improve the quality of the product, various processes are carried out in addition to distillation. The main group of these is known as *cracking*, in which the large molecules that make up heavy oils are broken down into lighter molecules in order to produce oils with a lower boiling point such as motor spirit. Thermal cracking relies on temperature, pressure and time but the more recent development of catalytic cracking, as the term implies, involves the use of an appropriate catalyst. Other refining activities include conversion processes such as reforming, again both thermal and catalytic, alkylation and isomerisation, and extraction processes such as the dissolution out of aromatic hydrocarbons and dewaxing.

Main products of the total refining process include liquefied petroleum gases, motor spirit, aviation fuel, kerosene (paraffin), diesel fuel, lubricating oils, waxes, fuel oils and bitumen.

The proportions of petroleum products produced in each of these categories in the U.K. and in the world are given in table 3.2.

Since petroleum products are designed for specific end-uses it is possible to estimate, in some instances, oil consumption by end-use. For example, in 1982 nearly 50 per cent of all petroleum products was used directly for transportation (motor spirit, aviation fuel, some fuel oil and diesel oil) and up to another 35 per cent (fuel oil) for industrial purposes. A much smaller 3–5 per cent was used for the petrochemicals industry which is nevertheless important enough to warrant separate discussion.

The Petrochemicals Industry

In addition to the straight-line or blended fuels produced as a result of the refining process, petroleum, mainly in the form of naphtha or gas oil, is also used in many countries as the basic raw material for the petrochemicals industry. In some countries such as the U.S.A., natural gas liquids, ethane, propane and butane are

OIL AND NATURAL GAS 49

Table 3.2 *UK and world production, various petroleum products (1982 and 1979)*

Petroleum product class	% of total production	
	U.K. 1982	*World 1979*
Gasoline (motor spirit)	28.6	24.9
Aviation turbine fuel	6.6	3.8
Kerosene	2.6	4.5
Gas/diesel oil	24.2	26.4
Residual fuel oil	24.1	27.1
Lubricants	1.2	1.2
Butane and propane	2.0	
Naphtha	4.9	
Bitumen	2.9	
Other	2.9	12.1
TOTAL (million tonnes)	67.3	2956.5

Source: Derived from U.K. Department of Energy statistics; and *1980 World Energy Demand*, USDOE, Washington D.C.

the preferred feedstocks. There are four main routes by which petroleum hydrocarbons may be converted into the raw materials for the chemical industry: (i) by the removal of hydrogen to produce olefins, acetylene or aromatics; (ii) by oxidation; (iii) by chlorination; and (iv) by sulphuration.

The resulting substances can be further fractionated by cracking or catalytic reforming to produce ethylene, propylene, benzene, toluene, xylenes, as well as a large number of acids and alkalies, which in turn may be processed for use in a variety of industries. By cracking hydrocarbons at very high temperatures it is also possible to separate them into pure carbon and hydrogen. Carbon black is used for the reinforcement of tyres and has outlets in plastics, printing inks, lacquers and records. Hydrogen can be further reacted to produce ammonia (to produce fertilisers, for example), methanol and oxy-alcohols for solvents and plasticisers.

Other products of the petrochemicals industry include polymers such as the thermoplastics and thermosets which form the basis of the rubber, plastics and fibre industries. Polymers are produced by combining together chemicals made from molecules containing a small number of atoms and fusing them into huge macromolecules containing several thousands of atoms. These highly viscous synthetic 'networks' of molecules can then be used to make materials such as Perspex and other vinyl polymers, synthetic emulsions, cellulose, paints, moulded plastic products such as bottles, PVC, resins, films, plastic sheet, synthetic rubbers, polyesters and man-made fibres, nylons, and a multitude of other plastic products from refrigerator linings and electrical fittings to toys and plastic tea cups. By 1980 OECD countries were consuming about 350 million tonnes of oil a year in the petrochemicals industry alone, besides the far greater amounts being used for other purposes. Since reserves of oil are not unlimited the need was recognised to provide for some of these uses by other energy forms in order to safeguard oil supplies for the petrochemicals industry of the future.

Oil Reserves

The total amount of oil in the ground is termed *Initial Oil in Place* (OIP) and is determined by geological factors. The exact quantity of OIP is not known, since although seismic and exploration technology has been improving constantly there are still no reliable methods of anticipating where undiscovered oil fields actually lie and how much oil they are likely to yield. Despite increasingly sophisticated methods of assessing the age and structure of rocks beneath the earth from samples and sonic measurements, in the 1980s it is still only by drilling in an area that a commercially viable assessment can be made, and drilling is an increasingly costly process.

Even when drilling has been completed and an oil field located, there are formidable problems of assessing what proportion of the OIP is actually recoverable. This quantity, termed *Ultimately Recoverable Oil* (URO), depends on current technical and economic opportunities and constraints as well as an adequate know-ledge of the size, composition and yield of each reservoir. Although some 30,000 oil and gas fields have been found since the beginning of the oil era, much of the basic data on most of these fields has not yet been obtained or published. Nor do all existing sources state the economic assumptions on which their estimates are based. Grossling,[2] summarising the main criteria that have been used in analysing potential reserves, lists seven:

1. size of petroleum prospecting area;
2. density of drilling for petroleum;
3. drilling over past decades;
4. cumulative outcome of past exploration and development;
5. number of 'giant' fields discovered;
6. petroleum found per unit of prospective area;
7. reference to certain 'benchmark' areas such as the U.S.A. or the Middle East, and extrapolation to geologically similar unexplored Third World countries.

Widely differing estimates of URO exist, but in the 1970s there began to emerge a consensus among oil experts as to the ultimate amount of recoverable oil remaining in the earth. This convergence was partly based on the fact that the *rate* of discovery of new reserves is slowing down while the amount of exploration including unsuccessful drilling is rising. There is a statistically significant relationship between the increasing rate of unsuccessful exploration (as measured by the amount of petroleum found per unit of prospected area) over time, and the accumulated net addition to reserves consistent with a probable maximum amount of oil still to be discovered.

The declining rate of new discoveries is illustrated in table 3.3 which enumerates discoveries of oil in giant and supergiant fields by decade. Giant fields are those

Table 3.3 *Discoveries of giant and supergiant fields, by decade*

	Giant fields		Supergiant fields		Giants and supergiants: recoverable oil (billion barrels)
	Number	Recoverable oil (billion barrels)	Number	Recoverable oil (billion barrels)	
1848–1900	9	10	0	0	10
1901–10	6	7	0	0	7
1911–20	12	12	1	32	44
1921–30	21	21	3	37	58
1931–40	24	22	3	99	121
1941–50	19	21	3	104	125
1951–60	45	57	11	124	181
1961–70	69	80	12	117	197
1971–79	40	35	2*	15	50

*Assumes that Mexico has found one supergiant field onshore at the
A.J. Bermudez complex, and perhaps one offshore.

Source: The World Oil Market in the Years Ahead, issued July 1979 by C.I.A. Office of
Economic Research, Washington D.C.

with more than 500 million barrels of recoverable oil and supergiant fields are those
with more than 5 billion barrels.

As the table shows, in the 1970s only 40 giant and 2 supergiant fields were
located, with the lowest quantity of recoverable oil since 1920. By 1983 most of
the world's sedimentary basins had been explored to some extent and, apart from
offshore drilling, it seemed unlikely that many more major fields would be found.

A representative selection of attempts to fix a figure for Ultimately Recoverable
Oil is given in table 3.4. Despite a spread of over 30 years, these estimates, reviewed
by the 1980 World Energy Conference, converge on a figure of around 2000 billion
barrels. However, there are a few widely divergent opinions: for example, in 1978
Odell and Vallenilla argued that reserve figures could be as high as 4500 billion
barrels.[3] The differences arise mainly from differing views as to the proportion of
OIP that can be recovered by secondary and tertiary production methods, and also
because the larger estimates include shale oil and tar sands reserves.

While the amounts of oil in place (OIP) and ultimately recoverable (URO)
remain speculative, intensive drilling in specific areas has confirmed that certain
reserves are definitely in existence. These are called Proved Reserves, or more usually
'Published' Proved Reserves (PPR). The PPR are defined by the American Petroleum
Institute as the 'quantities of crude oil in the ground which geological and engineering data demonstrate with reasonable certainty to be recoverable from known
reservoirs under existing economic and technical operating conditions.'

Recent estimates of proved reserves (excluding shale and tar) suggest a figure of

Table 3.4   *Estimates of world ultimate recovery of oil*

Date of estimate	Estimator	Organisation	Estimate (billion barrels)
1959	L.G. Weeks	Consultant	2000
1965	T.A. Hendricks	USGS	2480
1967	W.P. Ryman	Esso (Exxon)	2090
1968	–	Shell	1800
1968	L.G. Weeks	Consultant	2200
1969	M. King Hubbert	National Academy of Sciences U.S.A.	1350–2100
1970	J.D. Moody	Mobil	1800
1971	H.R. Warman	BP	1200–2000
1972	Richard L. Jody	Sun	1952
1973	Adams and Kirby	BP	2000
1975	J.D. Moody and R.W. Esser	Mobil	2030
1975	–	Exxon	1945
1977	M. King Hubbert	Congressional Research Service U.S.A.	2000
1978	R. Nehring	Rand Corporation	2000
1979	J.D. Moody and M.T. Malbouty	10th World Petroleum Congress	2230
1979	A. Meyerhoff		2000
1980	–	World Energy Conference	2600

Sources: (1) M. Grenon, *On Fossil Fuel Reserves and Resources,* IIASA, Laxenburg, Austria, 1978.

(2) *Survey of Energy Resources,* © 11th World Energy Conference, London, 1980.

between 600 billion barrels and 750 billion barrels. For example, British Petroleum, in its annual statistical publication *BP Statistical Review of World Energy 1983,* estimated 677.7 billion barrels. Proved reserves are itemised by country in table 3.5.

This table shows that oil reserves are concentrated in relatively few areas. These are mainly in the Middle East (54.5 per cent); the U.S.S.R. (9.3 per cent); a few African countries (mainly Libya, Algeria and Nigeria) (8.4 per cent); the U.S.A. (5.1 per cent); some Latin American countries (mainly Mexico and Venezuela) (12.1 per cent); Western Europe (mainly in the North Sea) (3.5 per cent); and the Far East (mostly in China and Indonesia) (5.6 per cent).

These percentages may be compared with estimates of the regional distribution of Ultimately Recoverable Reserves, which puts 40 per cent in the Middle Eastern sedimentary basin, 25 per cent in the Communist Bloc and the remaining 35 per cent elsewhere, mainly in Africa and Latin America. The difference between the two sets of figures largely reflects the greater intensity of exploration and drilling in the Middle Eastern countries. About one-half of all the recoverable oil comes from only 33 supergiant fields, while an additional 250 giant fields account for a further 25 per cent.

Two other definitions of oil reserves that should be mentioned are *Extended Proved Reserves* (EPR) which are the 'Published' Proved Reserves plus an allowance

Table 3.5   *World 'published' proved reserves at end of 1982*

Country	Thousand million tonnes	% of total
U.S.A.	4.4	5.1
Canada	1.0	1.2
TOTAL NORTH AMERICA	5.4	6.3
Latin America	11.4	12.1
TOTAL WESTERN HEMISPHERE	16.8	18.4
Western Europe	3.2	3.5
Middle East	50.3	54.5
Africa	7.6	8.4
U.S.S.R.	8.6	9.3
Eastern Europe	0.3	0.3
China	2.6	2.8
Other Eastern Hemisphere	2.6	2.8
TOTAL EASTERN HEMISPHERE	75.2	81.6
WORLD	92.0	100.0

Source: BP Statistical Review of World Energy, 1982, BP, London.

for as yet undiscovered oil within existing known fields; and *Undiscovered Recoverable Oil*, which is the difference between Ultimately Recoverable Oil and Extended Proved Reserves. McKelvey has published a diagram outlining a petroleum resources classification system which is shown in figure 3.4.[4]

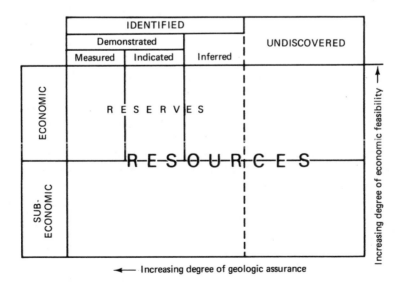

Figure 3.4   *Petroleum resource classification*

Oil Shale, Heavy Oil and Tar Sands

The quantities of oil found in shale deposits and tar sands are believed to far exceed current estimates of total world resources of conventional petroleum. The deposits located up to 1980 were mainly concentrated in the Western countries, the U.S.A., Canada, Venezuela and Australia.

Oil Shale

Shale oil or kerogen consisting of approximately 80 per cent carbon, 10 per cent hydrogen, 6 per cent oxygen, 3 per cent nitrogen and 1 per cent sulphur has been known to man for some time. In 1694 a patent was issued for 'distyllinge oyle from a kind of stone,' and in the nineteenth century Scotland had a successful shale industry. The total amount of oil potentially available is enormous. The *1980 World Energy Conference Report* estimated world resources of oil shales at 333,000 million tonnes, almost as much as its estimate of 354,000 million toness of recoverable conventional oil.[5] On the other hand, only 46.3 million tonnes of shale oil are deemed recoverable, of which the majority are to be found in the U.S.A., the U.S.S.R., Morocco and Thailand. Unfortunately, also, the oil content of oil shale varies considerably according to location and its costly commercial exploitation largely depends on the yield per tonne as compared with conventional oil prices. For example, one tonne of high-grade shale will produce over 120-125 litres of oil while low-grade shale will produce up to 75 litres. To extract even one million barrels of oil per day – about 7 per cent of U.S. requirements – would necessitate the annual mining of some 570 million tonnes of shale, which is nearly the entire annual output of the U.S. coal industry. A massive contribution to oil supply from shale is therefore difficult to foresee on resource, let alone on environmental grounds. When oil is extracted from shale by crushing and burning to liquefy the embedded kerogen, the residual rock has to be dumped. Not only are vast quantities involved, the volume of the waste being up to 30 per cent greater than the original solid rock, but as the rock has been 'burned' it is completely infertile. Also the dust-like residual mounds, when impregnated with snow or rain, can cause serious landslips. The dust created can be carried by the winds and dumped elsewhere creating other environmental hazards. *In situ* processes which create partial combustion underground are alternatives. For example, in the Oxy mining process the combustible gas is burned in a sealed underground chamber with injected air. The shale oil is drawn out through a trough in the floor. One problem with this method however, is that with all the pumps, transport of equipment and people to work the site, and water to find (which is often absent in oil shale regions), more energy may be expended in extracting oil than is produced by it. Finally, surface processing of oil from shale puts further extensive demands on water resources. For example, it has been estimated that to develop an oil shale industry of 3-5 million barrels per day in the Green River area of the U.S.A., all available water in the region would have to be devoted to the oil shale industry alone. Nevertheless attempts are being made to exploit oil shale where feasible. The Rundle deposit, Queensland, Australia, was the world's first large-scale oil shale project to be carried out commercially, since the shale was expected to yield 346 litres per tonne (31 per cent by weight) compared with 120–125 litres per tonne (9 per cent by weight) for the Colorado deposits. Work began in 1980[6] to extract oil from shale of 180 m thickness, although the ultimately recoverable reserves are not yet known. Because of the environmental problems, the difficulty and the expense involved, the future development of the extraction of oil from shale depends largely on the availability and price of conventional oil. However, by 1979/80 the exploitation of shale oil was thought to be competitive in certain regions.

Heavy Oil/Tar Sands

Tar sands, which should more properly be defined as bituminous sands or oil sands, constitute another significant alternative oil resource. These sands are reservoirs of 'heavy oil', an extremely heavy and viscous hydrocarbon liquid with a density of over 0.93 g/ml and a viscosity of below 20° API. The 1980 World Energy Conference estimate of world resources of oil from tar sands was 116,351 million tonnes, of which 40,051 million tonnes were proved recoverable resources – again nearly half the figure for proved reserves of conventional oil.[5] The proved reserves of tar sands are almost entirely concentrated in two areas: the Athabasca tar sands in Alberta, Canada, and the Orinoco belt in Venezuela. The first commercial-scale tar sand plant went into operation in 1967 in Alberta and has produced an average of 45,000 barrels a day since then; a second plant began production in 1978. Venezuela has been slower than Canada in developing its reserves.

In Canada, open cast extraction is extremely difficult since the sands lie mainly below ground in flat mosquito-infested muskeg swamps. For every barrel of bitumen extracted approximately 2 tonnes of sand and one tonne of overburden has to be removed and replaced. This is an even higher ratio of mineral waste than for average-grade shale. In summer, the viscous, sludgy nature of the heavy sands makes extraction very unpleasant and technically difficult since the equipment tends to sink into the swamps, the sands cling to it and working conditions are arduous. In winter when the sands are frozen and hard, temperatures are sometimes 45°C below freezing and digging teeth tend to shear off when they strike ice. Deep mining of tar sands presents other problems that are difficult to solve using conventional technology.[7] So far neither pumping nor underground burning techniques have been successful. Only about 1/20th of the oil in tar sands is currently extractable and the high capital costs of plant as well as the large amounts of energy used in extraction make it an extremely expensive energy option. Also long construction lead times, adverse environmental effects and difficulties of attracting labour to work in these extremely unpleasant areas make it unlikely that tar sands will achieve more than a marginal contribution to oil needs in the few countries endowed with reserves while cheaper oil is obtainable from other sources. In May 1982, Exxon, the world's largest oil company and a world leader in alternative-fuel technologies, announced that it was pulling out of the £2 billion Colony Shale Oil Development (losses were put at £508 million). The project was scheduled to produce 43,000 barrels a day of synthetic oil by the late-1980s. The announcement came soon after the decision by the Shell and Gulf Oil companies to drop out of the Alsands 'syncrude' project in Alberta, Western Canada (which would have produced 137,000 barrels a day).

Oil Production, Consumption and Trade

Production

Oil reserves may be known or thought to exist in the ground, but technical, economic and political factors will determine the extent and rate of production. Apart from the political considerations mentioned in chapter 2, the following are the main constraints affecting oil production from known reserves.

(1) *Technical factors.* Many technical constraints limit the feasibility of utilising a particular oil field. The porosity and permeability of the rocks and composition and size of the oil deposits have already been briefly mentioned. Another technical problem relates to oil under the sea-bed. The successful development of giant oil rigs to overcome adverse weather conditions and drilling requirements on shallow continental shelves was one of the more remarkable technical achievements of the 1970s. On the other hand, no technology yet exists to extract oil from locations under deep-sea conditions (over 2000 m deep) despite the fact that large potential reserves have been located in deep waters.

A further technical problem lies in separating oil from other natural elements which are sometimes found in combination in certain fields. The high sulphur and nickel-plus-vanadium content of oil in the Orinoco Heavy Oil belt in Venezuela is a case in point. The presence of these metals de-activates conventional catalysts used in hydrocracking. By contrast, the earlier technical problems involved in onshore deep-well drilling have been largely overcome since current drilling technology enables wells to be sunk to below the depth at which oil can be found, which is usually about 4500 m since oil cannot survive in geothermal temperatures of more than 260°C.

(2) *Economic factors.* The second major factor governing the volume of recoverable reserves that can be exploited is the price at which the extracted oil can be sold. The higher the price the more willing the oil companies will be to explore and extract oil from smaller fields and more inaccessible areas, such as the polar ice caps and the sea-beds. Against this, high conventional oil prices inhibit demand, make the alternative syncrude (synthetic) oils a more commercially attractive proposition and so reduce the propensity for difficult and costly exploration.

World Production and Consumption Patterns

Table 3.6 shows the major oil-producing regions with their 1983 production expressed as a percentage of the number of years' reserves (PPR) in that year. The table also shows (right-hand column) these countries' production expressed as a multiple of their consumption. Where this latter figure is less than 1, the country is a net

Table 3.6 *World oil reserves, production and consumption ratios, 1983*

Country	Number of years reserves at 1983 production rates	Production as multiple of consumption
U.S.A.	9.1	0.69
Canada	14.3	1.09
Latin America	35.3	1.49
Western Europe	18.7	0.29
Middle East	85.1	6.53
Africa	33.2	2.84
U.S.S.R.	13.9	1.37
Eastern Europe	15.6	0.17
China	24.6	1.25
Other Eastern Hemisphere	18.0	0.36
World	33.4	1.00

Source: BP Statistical Review of World Energy, 1984, BP, London.

importer. Obviously the countries with the lowest figures in each column are in the most vulnerable position in terms of their future oil security and hence their political and economic dependence.

Significantly, both the U.S.A. and the U.S.S.R. have the lowest number of years reserves remaining with published proved reserves likely to be exhausted before the end of the century in both countries. Western and Eastern Europe each import around three-quarters of total consumption, and are the world's most oil-vulnerable areas.

Offshore Oil

According to a 1983 Shell Report,[8] estimates of the world's proved oil reserves showed that nearly 50 per cent lay in offshore waters whereas only about 25 per cent lay in less than 200 m of water. Estimates of future potential crude oil discoveries indicate that one-third will be on land, one-third offshore on the Continental Shelf at depths of less than 200 m and one-third offshore in deep waters or polar regions. By 1983 some 20 per cent of world oil and over 5 per cent of natural gas were contributed by offshore fields.

In the late-nineteenth century, shallow offshore wells were drilled from piers off the coast of Southern California and around Lake Maracaibo in Venezuela. The first real offshore wells were drilled in the Gulf of Mexico after the Second World War. Figure 3.1 shows that half of the world's main sedimentary basins lies offshore. Since the world's onshore exploration and development up to the 1980s have been intensive, it is likely that there is considerable scope for future offshore developments despite the difficulty of obtaining data and the considerable development costs. By 1983 the most active offshore areas were the Gulf of Mexico, the North Sea, the Arabian Gulf, West Africa, Brazil and the South China Sea.

Oil Drilling Rigs

There are many types of offshore drilling rigs and production platforms and constant technical developments are under trial as the quest for oil extends to deeper waters. Whereas the early platforms could drill in depths of up to only 30 m, semisubmersible floating rigs can now drill in depths of over 300 m, the limitations being set by anchorage systems. By 1983, *dynamically positioned* rigs were capable of drilling in up to 1800 m, thus considerably extending exploration capabilities. These rigs, developed since 1970, with 20 functioning in 1983, receive monitored signals from an acoustic source in a known position on the sea-bed. The signals are passed to computers that control multiple propulsion units which automatically check and hold the vessel's position where the water is too deep for conventional anchorage systems.

Offshore Oil Production

There are also different types of oil drilling platforms ranging from *fixed production platforms* as commonly used in the North Sea in depths of up to 180 m or up to 300 m in calmer waters such as the Cognac field in the Gulf of Mexico, to the more recent *subsea development technology* for deeper waters. Four subsea systems are illustrated in figure 3.5. System 1 is suitable for depths of up to 130 m, Systems 2 and 3 up to 350 m, and System 4 is thought to be operable in depths of up to 1000 m or more.

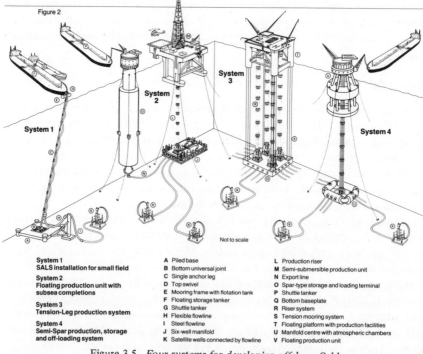

System 1 SALS installation for small field	A Piled base	L Production riser
	B Bottom universal joint	M Semi-submersible production unit
System 2 Floating production unit with subsea completions	C Single anchor leg	N Export line
	D Top swivel	O Spar-type storage and loading terminal
	E Mooring frame with flotation tank	P Shuttle tanker
	F Floating storage tanker	Q Bottom baseplate
System 3 Tension-Leg production system	G Shuttle tanker	R Riser system
	H Flexible flowline	S Tension mooring system
System 4 Semi-Spar production, storage and off-loading system	I Steel flowline	T Floating platform with production facilities
	J Six-well manifold	U Manifold centre with atmospheric chambers
	K Satellite wells connected by flowline	V Floating production unit

Figure 3.5 *Four systems for developing offshore fields*
(source: Shell Briefing Service, No. 1, *The Offshore Challenge*, 1983)

Trade in Oil

The world's largest net importers of oil include Western Europe, the U.S.A., and Japan. A significant proportion of the oil traded between these nations is supplied by OPEC. As can be seen from the map (figure 3.6), the largest flow (3.4 million

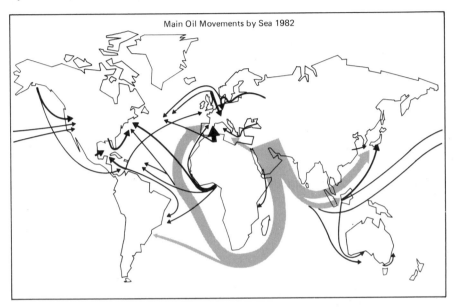

Figure 3.6 *Main oil movements by sea, 1982* (source: *BP Statistical Review of World Energy, 1983)*

barrels a day in 1983) is from the Middle East around the South African Cape and up the Atlantic coast to Europe, while 2.7 million barrels a day go from the Middle East to Japan. The extremely concentrated trading patterns pose serious strategic and political questions for the importing countries. Within the Gulf region itself, for example, virtually all the sea-borne oil for export has to pass through the narrow Straits of Hormuz. Continued free passage of oil through the straits is contingent upon political stability in these highly tense areas. This is, however, uncertain in view of simmering Syrian–Iranian, Iraqi–Kuwaiti and Iraqi–Iranian relations and by internal unrest within the countries themselves, for example, in Iran. The Gulf war of 1980 again demonstrated to Western societies the precarious position of their oil economies. Also as has been seen, the associated terminal facilities, refineries such as Abadan, and tankers sailing through the straits are particularly vulnerable to armed attacks and use of the 'oil weapon' by locally aggrieved states or possibly a world superpower using its strength to force a blockade in this exposed area.

The pattern of oil imports between 1972 and 1983 is shown in table 3.7. It can be seen that world oil trade reached a peak in 1979 after which it has been declining rapidly for reasons that were discussed in chapter 2. OPEC was hardest hit with sales falling from a 'high' of 31.8 mbd in 1979 to 17.5 mbd in 1983. Non-OPEC producers (including Mexico and the U.K.) increased their output during this period, overtaking OPEC production totals by early 1982.

Table 3.7   *World petroleum imports (million tonnes), 1972–83*

	1972	1973	1974	1975	1976	1977	1978	1979	1980	1981	1982	1983
U.S.A.	242	316	315	301	365	432	409	420	337	296	250	246
Western Europe	696	760	734	626	682	657	648	647	589	510	464	430
Japan	238	270	268	245	262	271	263	276	246	222	206	206
Rest of world	326	348	339	336	397	364	371	409	417	398	348	324
World total	1502	1694	1656	1508	1706	1724	1691	1752	1589	1426	1268	1206

Source: BP Statistical Reviews, 1980, 1984, BP, London.

North Sea Oil

The U.K. and Norway are fortunate in having access to oil in the Continental Shelf of the North Sea, first discovered in 1969. Significant production only commenced in 1975 when oil from the Forties and Argyll Fields came on stream. According to the Department of Energy, total U.K. reserves were estimated at 31 December 1983 at between 3.5 and 4.6 billion tonnes; these figures are broken down in table 3.8.

Table 3.8   *Proven UK oil reserves*

	Million tonnes
Proven reserves in present discoveries	1500
Probable* reserves in present discoveries	450
Possible* reserves in present discoveries	625
Total reserves in present discoveries	2575
Possible reserves in future discoveries	925-2000
Total reserves in known discoveries	3500-4575

*Probable reserves: estimated to have a better than 50 per cent chance of being technically and economically productible. Possible reserves: a significant but less than 50 per cent chance.
Source: Development of the Oil and Gas Resources of the United Kingdom, Department of Energy, HMSO, London, 1984.

The British National Oil Corporation and Offshore Oil Legislation

To provide the institutional framework for exploiting U.K. North Sea oil, the Petroleum and Submarine Pipelines Act came into force on 1 January 1976. The Act extended the government's control over oil exploration and exploitation in the North Sea. One of its main provisions was the setting up of the British National Oil Corporation (BNOC) with powers to explore and produce, transport, refine, store, distribute and sell petroleum and to take over the government's participation interest in U.K. licences. It could also carry out consultancy, research and training in petroleum matters and build, hire or operate refineries, pipelines and tankers. BNOC was given the right to acquire up to 51 per cent of the oil produced from any field in the U.K or its sector of the North Sea. In 1982, however, the Conservative government used enabling powers in the Oil and Gas (Enterprise) Act to split the BNOC into two separate parts. The trading arm remained state-controlled and retained its right to take up to 51 per cent of North Sea production, but the oil-producing and exploration side was set up as a separate corporation (Britoil). 51 per cent of the Britoil shares were subsequently sold by the government to the private sector.

Tables 3.9 and 3.10 show the rig activity and oil production over time of the

Table 3.9 *U.K. Continental Shelf rig activity 1972-83*

	Rig activity (in rig years) per year											
	1972	*1973*	*1974*	*1975*	*1976*	*1977*	*1978*	*1979*	*1980*	*1981*	*1982*	*1983*
Mobile rig	8.8	13.3	24.5	27.7	21.2	23.6	18.1	16.1	20.6	24.6	30.1	34.2
Fixed platform	3.8	3.2	2.8	2.6	9.3	14.9	18.6	21.5	25.2	27.0	25.0	24.1
Total rig time	12.6	16.5	27.3	30.3	30.5	38.5	36.7	37.6	45.8	51.6	55.1	58.3

Source: Development of the Oil and Gas Resources of the United Kingdom, Department of Energy, HMSO, London, 1984.

Table 3.10 *U.K. oil production: offshore fields (million tonnes/year)*

	Total from 1975 to end 1977	1978	1979	1980	1981	1982	1983	Cumulative total from 1975
Argyll	2.4	0.7	0.8	0.8	0.5	1.0	0.7	6.9
Auk	3.5	1.3	0.8	0.6	0.6	0.6	0.6	8.1
Beatrice	–	–	–	–	0.2	1.6	1.5	3.4
Beryl A	3.4	2.6	4.7	5.4	4.7	4.4	3.8	29.0
South Brae	–	–	–	–	–	–	1.0	1.0
Brent	1.4	3.8	8.8	6.8	11.1	15.2	18.7	65.6
Buchan	–	–	–	–	0.9	1.4	1.6	3.9
Claymore	0.3	3.0	4.0	4.4	4.5	4.8	4.7	25.9
North Cormorant	–	–	–	–	–	1.3	2.1	3.5
South Cormorant	–	–	0.04	1.1	0.7	0.9	1.5	4.2
Duncan	–	–	–	–	–	–	0.1	0.1
Dunlin	–	0.7	5.7	5.2	4.7	3.9	3.5	23.6
Forties	29.3	24.5	24.5	24.6	22.8	22.2	21.7	169.6
Fulmar	–	–	–	–	–	2.6	5.7	8.3
Heather	–	0.1	0.8	0.7	1.2	1.7	1.3	5.9
N W Hutton	–	–	–	–	–	–	1.9	1.9
Magnus	–	–	–	–	–	–	1.5	1.5
Maureen	–	–	–	–	–	–	0.8	0.8
Montrose	0.9	1.2	1.3	1.2	1.1	0.9	0.7	7.3
Murchison (U.K.)	–	–	–	0.4	3.1	4.4	4.5	12.4
Ninian	–	0.04	7.7	11.4	14.3	15.0	13.7	62.1
Piper	8.7	12.2	13.2	10.4	9.8	9.8	9.6	73.7
Statfjord (U.K.)	–	–	0.04	0.5	1.2	1.8	3.0	6.6
Tartan	–	–	–	–	0.7	0.6	1.1	2.4
Thistle	–	2.6	3.9	5.3	5.5	6.0	5.1	28.3
Total stabilised crude oil from offshore fields	49.9	52.8	76.5	78.7	87.7	100.1	110.5	556.1

Source: Appendix 8 in *Development of the Oil and Gas Resources of the United Kingdom,* Department of Energy, HMSO, London, 1984.

recovery of oil from the U.K. Continental Shelf. One immediate impression from these tables is the speed with which the enterprise got under way in the early-1970s, and also the extent of the activity that took place before any oil was actually produced. In the period 1971–74, before any of the oil fields came on stream, a total of 341 wells was drilled by the 17 rigs that on average were to be found in the North Sea at any one time. Once production did start, it built up very rapidly.

How long the oil from the Continental Shelf will last is difficult to forecast since it depends both on the sizes of future oil finds and the rate of extraction that prevails. In the early-1980s estimates were around 20–30 years, but in 1983 a Shell Report stated: 'If the United Kingdom alone is to remain self-sufficient (in oil) after the late 1980s, about 80 to 100 small oil fields will have to be appraised and developed during the next 20 years. These additional reserves will require another 500 exploration wells to be drilled at around $8 million apiece.'[8] In 1984 the Minister of State for Energy gave the following forecasts for U.K. petroleum production (in million tonnes) in the years 1984–88:

1984	110–130
1985	110–130
1986	110–125
1987	85–115
1988	85–115

The ranges, which make allowance for delays that may be caused by the technical and geological problems faced in the North Sea, show that production is expected to reach a peak in 1985 and decline thereafter.[9]

Effects of North Sea Oil on the U.K. Economy

The development of the North Sea oil industry has had far-reaching effects on the U.K. economy and on her energy policy. The amount of government revenue from the North Sea, in the form of royalties, Petroleum Revenue Tax (PRT), Corporation Tax, and for a period in 1981–2 a Supplementary Duty, is very considerable. In 1979–80 it was £2.3 billion and by 1983–4 it had risen to £8.9 billion. Oil from the Continental Shelf has helped to redress the balance of payments deficit, previously attributable in part to energy imports, and has also provided many new job opportunities in Scotland and other parts of the U.K. Sterling rose initially against other currencies on the strength of North Sea oil, although the fall in the price of Forties crude to below $30 per barrel in 1983 led in part to a substantial decline in the value of sterling, now regarded as a petrocurrency. Oil self-sufficiency was achieved during the latter half of 1980, partly on account of the increasing level of North Sea production and partly owing to a 14.8 per cent decrease in consumption over 1979 because of the economic recession. Self-sufficiency and a degree of net export capacity is expected to last until about 1990 when declining production is expected to fall short of consumption requirements. Thereafter the U.K. is likely to become a net importer of oil, although factors such as government depletion policies, fiscal measures, economic recession or rising production costs will ultimately affect consumption rates.

Economics and Technical Obstacles of U.K. North Sea Oil Production

Despite the enormous technical and economic successes, development of the North Sea oil and gas fields has been an extremely expensive enterprise. A total of £16.5 billion was invested up to 1981, and expenditure in 1982 was over £4 billion broken down as follows:

Exploration (oil and gas)	£860.9 million
Development (oil only)	£2348.4 million
Production (oil only)	£1175.2 million
TOTAL	£4384.5 million

According to Professor Fells of the University of Newcastle-upon-Tyne the Forties field alone cost over £1 billion to develop – the initial estimate had been £360 million but unexpected engineering problems, delays in construction times and inflation caused the increase. Similarly the cost of developing the Brent field rose from an estimated £800 million to over £1500 million. Since these debts, many of them contracted overseas, were to be serviced from oil revenues, substantial quantities of oil have to be sold just to cover them. This has entailed a very high rate of extraction which, in the opinion of some, is to the long-term disadvantage of the U.K., since the alternative would be to leave oil in the ground to ensure future supplies at times of scarcity or crisis.

After the 1983 General Election, Parliament passed the Petroleum Royalties (Relief) Act which gave tax relief for oil companies. The Act abolished royalties on new fields – a considerable incentive to the development of some of the smaller, less lucrative oil fields.

In addition to the enormous costs, other technical and social difficulties have been encountered in developing the North Sea oil industry. There were difficulties in locating reservoirs under the sea-bed and, even though seismic surveying is frequently cheaper offshore than on land, the cost of offshore drilling can be up to five times more expensive than on land. In the early days, many wells were drilled that were incapable of commercial exploitation. For example, the Ekofisk field, the first worthy of commercial exploitation by Norway, was eventually found after the 34th well had been sunk. More recent drilling techniques in the early 1980s and the increased use of computers to analyse complex seismic results have since greatly improved the speed and success of their interpretation. Shell claims that their worldwide exploration success rate improved from 1:2.9 in 1979 to 1:2.6 in 1981. Against these technological successes are the higher costs of employing engineers and divers with the necessary expertise to construct and maintain the rigs, and operators willing to work in extremely isolated and inhospitable locations. The collapse of the Alexander Keilland rig in 1980 was a vivid reminder of the dangers involved. There are further problems associated with the temporary employment and accommodation of a labour force of several thousand required to develop the oil fields, and the social dislocation caused by their sudden arrival in an area and their equally abrupt departure when the construction work has been completed.

The North Sea oil industry has also involved massive investments in the rapid development of a highly sophisticated infrastructure in these remote rural regions.

Examples include not only the expensive laying of oil pipelines but the building of roads, offices, accommodation and communication networks, and ensuring adequate transport and supply facilities. For example, Sumburgh civil airfield on Mainland, Shetland's largest island, by 1980 was already handling more air traffic movements than any U.K. airport other than Heathrow. With offshore extraction, the increased possibility of oil spillages involves additional capital investment. Dyce in Aberdeen has had to be equipped as the centre of emergency operations should there be a disaster or any oil spillages in the adjacent oil fields. The security and surveillance of offshore oil supplies is another problem requiring the maintenance of patrol forces.

As new discoveries are made there emerge other problems requiring increasingly advanced technologies. Although by 1983 several new offshore fields had been located, many of these lay in deeper waters such as at Porcupine Bank, 100 miles west of the Irish Republic at 1500 ft (450 m) and in the Goban Spur region south west of Kerry at a depth of 5500 ft (1700 m). Since conventional platforms cannot operate under such conditions, to extract this oil necessitates the use of a subsea system involving the linking of well-heads on the sea-bed to a central rising pipe by means of a flexible pipeline system. This pipe would feed a converted oil tanker which would separate the oil and the gas. These, in turn would be transported to the mainland by other tankers. (For example see System 4 in figure 3.5). In bad weather conditions, however, the oil flow would have to be shut off because of the much higher risk of spillage. The Advisory Committee on Oil Pollution of the Sea pointed out in its 1980 annual report the grave pollution dangers that would be incurred by developing such systems.

U.K. Onshore Oil

Whereas North Sea oil was only effectively developed during the 1970s, onshore exploration had been in progress for more than 60 years. Licences for exploration were issued under the provisions of the Petroleum (Production) Act of 1934, the Petroleum (Production) Regulations of 1976 and the amended Petroleum Regulations of 1978. By April 1982, 106 exploration licences and 63 production licences were in force and in 1982 total production had reached 240,000 tonnes. This was more than double the 1979 production level, owing in part to the development of the British Gas/BP fields at Wytch Farm in Dorset. Wytch Farm was the largest onshore field discovered in the U.K. by 1983 with confirmed reserves of 13.5 million tonnes (90 million barrels). After a period of reduced activity in the mid-1970s (no new production licences were issued between 1971 and July 1979 and on average less than three exploration wells were started in each of the years from 1975 to 1979), onshore exploration and production began to increase at the beginning of the 1980s. Nevertheless, no really significant onshore finds had been made to compare with the larger North Sea reservoirs.

The Future

In the short term, while the 'oil glut' continues under the impact of the world recession, the future of oil supplies is likely to remain reasonably secure. Trade patterns will continue to move away from OPEC countries as other countries' deposits are increasingly exploited. Hence, for a while, the future of OPEC is uncertain. Several members of the cartel are currently producing at less than 50 per cent of their maximum possible output and are suffering acute financial problems as a result. On the other hand, any upturn in demand could easily be met by existing production systems. By 1984 the decline in demand had levelled out, although OPEC members were still feeling continuing strains with production quotas having to be agreed, to continue the delicate balance of supply and demand. In the longer term, however, most authorities agree that oil is running out and that by the end of the century it will cease to be *the* major energy option for most countries, if only because it is a finite resource. Oil discoveries are becoming harder to make and continued expansion in output for many more years is impossible. As oil reserves in OPEC countries greatly exceed those in other countries, it will simply be a matter of time before the balance of economic power again moves in their direction. After that, undoubtedly one day oil supplies will fail to meet demand. When this day will arrive depends on how effectively and quickly oil-consuming countries cut back on their consumption and substitute other energy forms.

B. NATURAL GAS

Natural gas is derived from naturally occurring reservoirs below the earth's surface. Its main constituents are methane, butane, ethane and propane.

Seepages of this natural gas have been discovered from time to time throughout history, often giving rise to mysterious phenomena described by the ancients as 'eternal' fires, or 'burning springs'. Lamps that had been alight for centuries are mentioned by Plutarch (A.D. 60–140), St. Augustine (A.D. 354–430) and in early Chinese literature. In 1550 a vault was discovered on the Isle of Nesis near Naples with a lamp that had been burning from the beginning of the Christian era. The Chinese are thought to have put natural gas to good use some 3000 years ago by conveying it through bamboo pipes to heat pans of brine to obtain salt. When natural gas was first discovered and used in Fredonia in the U.S.A. in 1821 it was also piped through hollowed-out logs. Although, as has been shown in chapter 1, the use of manufactured gas increased in the U.K. and elsewhere during the first part of the twentieth century, it is only comparatively recently that natural gas has been developed as an important source of primary energy.

Natural gas has a higher calorific value (about 3.8×10^7 J/m^3) than manufactured gas (see chapter 4) and also possesses other economic and technical advantages. It can also be delivered at higher pressures. Thus when manufactured gas from coal faced increasing competition from cheap oil and electricity, natural gas, often found in association with petroleum, gradually replaced it, with similar distribution net-

works and appliances being used at extremely favourable costs. A major exception
to this pattern was the U.S.A. where natural gas and manufactured gas were
developed simultaneously. By the late-1960s, wherever possible, natural gas usage
had largely replaced manufactured gas and it had become one of the world's most
promising primary energy sources despite its limited reserves. The rapid expansion
of natural gas usage is reflected by the fact that in 1965 it supplied only 2 per cent
of total primary energy in Western Europe but by 1979 this figure had risen to over
14 per cent. In the period after 1973 the pressures to prospect for and develop
further natural gas supplies increased significantly.

Most oil reservoirs also contain gases dissolved in the oil (solution gas). Natural
gas can also exist in a reservoir above gas-saturated crude oil (gas–cap gas) but such
gas is almost never produced until most of the underlying economically recoverable
oil has been produced. Both these gases are known as 'associated natural gas'. For
various technical and economic reasons, their exploitation has been of secondary
importance to that of oil until comparatively recently. In the past most associated
gas was wasted owing to the practice of 'flaring' or 'venting' at the well-head. Even
gas in non-associated fields was often vented in the hope of producing the non-
existent oil underneath. In the U.S.A. legislation was gradually introduced to curb
these wasteful practices, but even in 1947 natural gas wasted or lost was estimated
at 3×10^{10} m^3 a year, while the total quantity wasted in the U.S.A. alone up until
the late 1940s was estimated by the geologist L. F. Terry at over 2×10^{12} m^3.
Other countries have followed suit, especially since 1973, despite the fact that to
curtail 'flaring' inhibits the rate of oil production. New pipelines and industrial
plant have been constructed to utilise this previously wasted energy source. By
1982 the amount of North Sea gas wasted by 'flaring' had almost halved from its
peak in the summer of 1979 (18 million m^3 of gas per day).

When gas reservoirs are located independently of oil fields, they are known as
'non-associated' gas. These have the commercial advantage over associated gas fields
since the gas can be extracted independently of oil demand. During the late-1960s
and 1970s industrialised countries such as the U.S.A., the U.S.S.R., Japan, the
Netherlands, the U.K. and Norway were spending considerable sums on prospecting for natural gas, developing indigenous supplies with the necessary infrastructure
and negotiating contracts with overseas suppliers. It had also become apparent that
considerable untapped reserves were located within the poorer Third World
countries, although they themselves rarely possessed the necessary funds and
expertise for prospecting and development.

While the exploitation and use of natural gas is relatively simple compared with
that of oil or gas from coal, its importance as an energy source in any particular
country depends on the extent of indigenous supplies and whether these merit the
capital investment required for its distribution. Obviously the nearer reserves are
located to available domestic and industrial markets the more economically attractive this fuel source becomes, a fact of particular relevance to natural gas development in Third World countries. However, as fossil fuels become scarcer it will
be increasingly attractive to transport the gas to distant markets either by long-
distance pipeline or via the gas liquefaction process. Against this must be weighed

the fact that local or indigenous supplies increase an area's energy autonomy whereas international pipelines, once laid down, place dependence for energy supplies on specific suppliers. This can have far-reaching economic and political repercussions. The U.S.S.R.'s supplies of gas to Western Europe which have been expanding since 1968, and which will further increase with the completion of the new pipeline agreed in 1982, are an illustration in point.

Natural Gas Reserves, Future Supplies and Consumption

It is almost impossible to arrive at an accurate assessment of world natural gas reserves which outside the U.S.A. were not studied in detail by major exploration companies until the late-1960s. On the one hand many geologists have estimated that most industrialised Western countries with indigenous reserves will find that output begins to decline before the 1990s, if the process has not already started. On the other hand in the view of others, such as Professor T. Gold,[10] vast reserves of methane lie further below the earth's surface and could be reached if exploration companies were persuaded to drill more deeply; and these could meet the present level of worldwide hydrocarbon demand for the next 10 million years! Unfortunately the economic realities of oil exploration and exploitation are such that the first view is more likely to be correct. Nevertheless it is true that much natural gas is likely to be added to estimated reserves, especially from more remote areas such as the North Sea north of the 62nd Parallel, the Middle East, Siberia, the Canadian Arctic and in several less-developed countries.

At the beginning of 1979 world natural gas reserves were as follows according to the 1980 World Energy Conference:[5]

Proved Reserves
   7.4×10^{13} m^3 or
   6.3×10^{10} tonnes of oil equivalent or
   2.9×10^{21} J
Estimated Ultimately Recoverable
   10.5×10^{21} J

Ratio of proved world reserves to 1978 consumption 51
Ratio of proved reserves to proved oil reserves 0.7
Ratio of ultimately recoverable reserves to 1978 consumption 184

The vast majority of the U.K.'s natural gas reserves so far discovered lie in the North Sea. Reserves in the Southern Basin, off the east coast of England, were the first to be discovered in the 1970s. Large deposits were found in the Frigg field which straddles the median line between U.K. and Norwegian waters, and subsequently further reserves were discovered in the northern Irish Sea.

In 1983 U.K. natural gas consumption totalled 39,529 x 10^6 m^3. Total U.K. reserves were estimated[9] at the end of 1982 to be between 1450 and 2125 billion m^3 (1.45-2.12 x 10^{12} m^3) equivalent to between 4.1 and 8.0 x 10^{19} J. The ratio of total reserves to 1982 U.K. gas consumption was between 27 and 45.

The Development of Natural Gas in the U.K. and other European Countries

The West Sole Gas Field was the first U.K. offshore field to be discovered in the North Sea by BP in 1965 and proved to be very rich. The gas was found to consist of 98.4 per cent hydrocarbons including 94.4 per cent methane. There followed the rapid discovery of other fields in the British, Norwegian, Dutch and Danish sectors of the North Sea and recently on a small scale in the Irish Sea. U.K. gas production from the Continental Shelf is shown in table 3.11.

Table 3.11 *Natural Gas production of U.K. Continental Shelf (million cubic metres)*

	Total to end 1976	1977	1978	1979	1980	1981	1982	Cumulative total to end 1982
West Sole Field	16,207	1,947	1,533	1,365	1,445	1,455	1,512	25,464
Leman Bank Field	96,730	15,581	14,719	13,831	9,482	13,207	11,675	175,225
Hewett Area	39,565	7,852	6,392	6,288	6,568	5,048	4,108	75,821
Indefatigable Field	27,377	6,779	6,450	6,006	6,878	5,613	5,720	64,823
Viking Area	21,311	6,330	5,238	4,397	4,689	3,307	4,381	49,653
Rough Field	522	1,063	931	1,005	467	99	101	4,188
Frigg Field		614	2,907	5,345	6,374	7,057	6,569	28,866
Piper Field			4	536	521	520	629	2,210
FLAG system							2,144	2,144
Other	10	138	323	455	866	1,098	1,437	4,327
Total production	201,722	40,304	38,497	39,228	37,290	37,404	38,276	432,721

Source: Table 24 in *Digest of United Kingdom Energy Statistics, 1983*, Department of Energy, HMSO, London.

Norwegian reserves of North Sea gas are many times greater than that country could possibly use over a long period of time owing to its sparse population and established use of renewable fuels such as biomass and hydropower, and further fields are still likely to be identified. However, there will be increasing competition between the U.K. and other European countries for a stake in Norwegian gas not yet committed to the market.

In the Netherlands the Groningen gas field, one of the largest non-associated gas fields in the world, was discovered in 1959. Its unique position near to many centres of population and industry including those of Germany, France and Belgium, made it one of the most favourable locations in the world and significantly affected patterns of gas usage in Western Europe. Besides taking over as the main supply for home markets, Dutch natural gas exports rapidly expanded between 1971 and 1978. By 1983, 6.83×10^{10} m^3 per year were being produced, with recoverable reserves estimated at around 1.4×10^{12} m^3.

World Natural Gas Consumption

The overall effect of the expanding interest in natural gas as a primary fuel has been that gas consumption has grown steadily in recent years. In 1983, world natural gas consumption, at 1328.9 million tonnes oil equivalent, was 25 per cent higher than

in 1973. The growth has been attributable mainly to large natural gas discoveries
and their exploitation by oil-rich nations such as the OPEC countries, the U.S.S.R.,
Norway and the U.K., and in some other countries such as the Netherlands with
non-associated gas resources. The world rate of increase in natural gas consumption
would have been considerably higher but for the fact that in the U.S.A. natural gas
fields are becoming depleted after having reached their peak production levels in
the early-1970s.

As a proportion of total world primary energy consumption, however, world
natural gas consumption has, perhaps surprisingly, not increased during the 1970s.
Table 3.12 shows the percentage of natural gas to total energy consumption in the
world and for selected countries from 1970 to 1983.

Table 3.12 *Natural gas as percentage of primary-energy consumption*

	World	U.S.A.	U.K.	Western Europe	U.S.S.R.
1970	17.9	34.0	5.2	6.6	21.2
1971	18.5	34.4	8.5	8.3	22.0
1972	18.6	33.2	11.7	9.7	21.9
1973	18.2	31.4	11.6	10.5	22.7
1974	18.5	31.4	14.8	12.2	22.8
1975	18.3	29.5	16.1	13.1	23.7
1976	18.2	28.3	16.7	13.2	24.9
1977	18.0	27.2	17.4	13.5	25.6
1978	18.0	26.6	17.9	13.7	26.2
1979	18.4	27.1	18.9	14.1	27.1
1980	18.9	27.9	20.3	14.4	28.1
1981	19.4	28.2	21.5	14.6	29.5
1982	20.0	26.8	21.5	14.3	30.6
1983	19.2	25.3	22.4	15.0	31.5

Source: BP Statistical Reviews of World Energy 1979–83, BP, London.

The stable world percentage has largely been due to the fact that countries such
as the U.K. have increased their consumption of natural gas at the same rate as its
reducing importance in the U.S.A. Gas production in the U.S.A. peaked in 1972
and by 1983 it had declined to 25.3 per cent of the total primary energy consumption in that country. By contrast, the U.S.S.R., with estimated reserves about seven
times greater than those remaining in the U.S.A., is still increasing both its home
consumption and its export trade to the West, as well as supplying natural gas to
its COMECON partners.

Pricing Policies

As fossil fuels become scarcer it is increasingly necessary for governments to
encourage restraint in their usage by exercising the pricing mechanism to reflect
their true value, defined in President Carter's National Energy Plan (1977) as the
cost of the fuel's replacement. Nowhere has this been more apparent than with
natural gas, which during the 1960s and early-1970s was slightly cheaper than coal

and in the U.S.A. cost about one-third of the price of oil. This led to very rapid growth in its use, both in industry and the domestic sector, and to lack of adequate attention to energy-conservation aspects. The U.K., Norway, the Netherlands and other natural gas producers have the opportunity of benefiting from the U.S. experience by pricing the fuel at levels that encourage conservation measures and so extend the life of reserves. For example, this was reflected in the U.K.'s gas price increases of 1980–83.

Natural Gas Markets and Patterns of Usage

The development of natural gas markets has varied considerably from country to country. In the U.S.A. about half the total is consumed by industry with domestic use accounting for about a third. In the U.S.S.R., on the other hand, 86 per cent is used by industry (including 34 per cent for power generation) and only 4 per cent for domestic purposes. In Japan, by far the largest consumer of gas is the power-generation industry; of the remainder, two-thirds is absorbed by the domestic market.[11] In the U.K., domestic use is greater than industrial: natural gas provides about half of households' total energy needs. Domestic consumption is mainly for space-heating, the load varying widely both between day and night and summer and winter. As gas can be readily stored either in its original reservoirs, in pipelines or in special storage vessels, these swings can be easily accommodated. The main industrial consumers in the U.K. are the chemical and steel industries.

Distribution

Most natural gas is distributed by pipeline from its original reservoirs to the final consumer. In the U.K. there is a major high-pressure national transmission system which brings gas from the North Sea to regional terminals, from which it is dispersed locally in subsidiary lower-pressure networks. The whole system is at present publically owned by the British Gas Corporation.

Japan, by contrast, has one of the world's most diversified and expanding gas markets of both manufactured and natural gas. The industry is mainly privately owned with very few controls and little standardisation. Fourteen different types of gas are distributed by some 250 companies of which 73 are municipally owned. As Japan has virtually no indigenous natural gas deposits, it relies mostly on LNG imports from Abu Dhabi, Alaska, Brunei, and Indonesia.

The gas industry in the Netherlands, as has already been mentioned, was during the 1970s and early-1980s the largest single supplier in Western Europe, and is owned by a mixture of private and governmental interests.

The U.S.S.R., with the world's largest reserves, has naturally become one of the most significant exporters of natural gas, mainly to Eastern European countries. In 1981 construction commenced on a pipeline from Siberia to West Germany. Another significant long-distance transmission project has been the COMECON-sponsored Crenburg project which involves the construction of a 2750 km pipeline from the Urals to the Czechoslovakian border.

Natural Gas Developments

Liquefied Natural Gas (LNG)

The impossibility of building long-distance pipelines across oceans has encouraged producers to seek alternative ways of transporting the gas. The most significant development has been the technique of liquefying natural gas at very low temperatures, at which point it reduces to about 1/600th of the volume of normal gas.

LNG consists mainly of methane which has been cooled to $-82.5°C$ and compressed to over 45 atmospheres to enable it to liquefy. The advantage of LNG is that the gas is reduced to a sufficiently manageable volume to be transported by ship, or stored until required. Originally LNG plants were developed to convert surplus pipeline gas for storage; the LNG could then be regasified during periods of high demand. However, since 1964 several base-load LNG projects have been developed and it is likely that this trend will continue.

Probably the French chemist Cailletet was the first to liquefy methane in 1877, and although the basic technology of liquefying gases was well established by 1914, the U.S. Bureau of Mines first liquefied natural gas on a practical scale in 1917. From 1941 to 1944 the world's first peak-shaving plant using LNG operated at Cleveland, Ohio. However, after an accident leading to massive conflagration (the Cleveland disaster) the plant was shut down and interest in LNG projects temporarily abated. In 1959 there was the first experimental transport of LNG from the U.S.A. to Canvey Island in the Thames Estuary, U.K., and in 1964 the first commercial LNG was delivered from the Arzew Plant in Algeria to Canvey Island, as part of a 15-year contract. A new high-pressure methane grid was built to supply 8 out of the 12 gas area boards with LNG which was subsequently reformed down to town-gas quality (approximately 19 MJ/m^3) before distribution to the end consumer. From this time international trade in LNG developed rapidly, despite the high capital costs and complexity of treatment, the necessity for transportation in specially constructed and insulated containment vessels and the political, commercial and technical risks involved. Reasons vary from a country's need to have access to this particular fuel to the need to diversify supply sources (as is the case with Japan). Despite its rapid expansion in the 1970s, international LNG trade remains small in proportion to total world consumption of natural gas: 36×10^9 m^3 out of 1450×10^9 m^3 in 1980, or 2.2 per cent. Japanese imports at 27×10^9 m^3 account for over 70 per cent of LNG trade.[12] However, LNG systems are likely to develop rapidly over the next decade especially in those parts of Africa, the Middle East, Central and South America, Asia, Australia and the U.S.S.R. where actual gas production greatly exceeds local market requirements.

Substitute or Synthetic Natural Gas (SNG)

This is being developed in many countries from the U.S.A. to Japan using oil as well as coal. However, as oil reserves are likely to run out before those of gas the main hopes for SNG are those projects using non-oil feedstocks.

Natural Gas Liquids (NGL)

These are liquid hydrocarbons which are extracted from 'wet' natural-gas fields together with other impurities. Those hydrocarbons heavier than methane, for example, ethane, propane, butane and isobutane, isopentane and natural gasolene, must be removed where they occur, to avoid technical problems with the transmission and distribution of natural gas. They can be a valuable by-product for a variety of uses such as transport fuels, refinery fuels or as feedstock for the petrochemicals industry. The U.S.A. has the world's leading NGL industry with over 750 processing plants in operation in 1976. Here the natural gas liquids, especially ethane, are used as the main feedstocks of the petrochemicals industry in contrast to most other countries where naphtha or gas oil is the main feedstock. Although NGL production in the U.S.A. will decline as gas production falls, various NGL export schemes are being developed in countries of the Middle and Far East, Africa and elsewhere.

The Potential for Natural Gas

Natural gas is a convenient and versatile fuel and like oil has many advantages over other energy forms. Its major ones are that: it is very cheap to extract and distribute; it is clean and does not require to be refined or transformed from a primary energy to secondary forms; it can be stored easily and so can cope with fluctuating consumption patterns; it is virtually pollution free and environmentally benign since pipes are laid underground; and it has a high calorific value of about 3.8×10^7 J/m^3. Disadvantages are that the gas fields are often located in remote areas; and the initial laying of gas pipelines and distribution networks is comparatively expensive, especially where the source of supply is far from the market. Nevertheless the development and judicious use of this limited fossil fuel should help to ease the transition stage from an oil-based economy to one dependent on more sustainable energy forms.

References

1. *BP Statistical Review of the World Oil Industry*, British Petroleum, London, 1984
2. Grossling, B., 'Window on Oil', *Financial Times Report 1976*, p. 71
3. Odell, P. R. and Vallenilla, L., *The Pressures of Oil*, Harper and Row, New York, 1978
4. Miller *et al.*, 'Geological Estimates of Undiscovered Recoverable Oil & Gas in the United States', *Geological Survey Circular 725*, Federal Energy Administration, Washington D.C., 1975
5. World Energy Conference, *Survey of Energy Resources 1980,* London, 1980
6. Tucker, A., 'Australia Looks Up', *The Guardian*, 24 April 1980
7. Bunyard, P., 'The Future of Energy in Society', *The Ecologist*, Vol. 6, No. 3, March/April 1976
8. Shell Briefing Service No. 1, *The Offshore Challenge: Exploration, Production, Subsea Developments, Costs*, Shell, London, 1983

9. Department of Energy, *Development of the Oil and Gas Resources of the United Kingdom 1981*, HMSO, London, 1984
10. *Financial Times*, 4 December 1980
11. Peebles, M. W. H., *Evolution of the Gas Industry*, Macmillan, London, 1980
12. Shell Briefing Service No. 3, *Oil and Gas in 1982*, Shell, London, pp. 8–9

Further Reading

Tugendhat, C. and Hamilton, A., *Oil the Biggest Business*, Eyre Methuen, London, 1975

Department of Energy, *Development of the oil and gas resources of the United Kingdom*, HMSO, London (annually)

Shell Briefing Service No. 3, *Oil and Gas in 1982*, Shell, London (updated annually)

Skinner, D. R., *Introduction to Petroleum Production, Petroleum engineering for non-engineers*, Gulf Publishing, Houston, Texas, 1982

Wheeler, R. R. and Whited, M., *Oil from Prospect to Pipeline*, 4th edn, Gulf Publishing, Houston, Texas, 1981

Berger, B. D. and Anderson, K. E., assisted by Farrar, G. L. and Pile, K. E., *Modern Petroleum: A Basic Preview of the Industry*, 2nd edn, Penn Well Publishing, Tulsa, Oklahoma, 1981

New Sources of Oil and Gas – Gases from Coal, Liquid Fuels from Coal, Shale, Tar Sands, and Heavy Oil Sources, Pergamon, Oxford, 1982

Al-Otaiba, M. S., *Essays on Petroleum*, Croom Helm, London, 1982

4

Coal

(by Israel Berkovitch)

'. . . used strategically . . . as the raw input for the production of synthetic liquid and gaseous fuels . . . coal would serve as a bridge to a future world built around non-depletable energy sources'

Energy in a finite world.
Report of a 7-year study by
the International Institute for
Applied Systems Analysis

The Origins of Coal

Though the fuel is now far less familiar a sight than it was a generation ago, coal is well known as a blackish rock hewn out of the depths of the earth or – if we are luckier – from near the surface. Rich in carbon, this rock is combustible. It is not a single mineral but a range of natural solids varying in composition. The complete range is broadly taken to extend from peat, through brown coal, lignite and bituminous coal to anthracite.

Coal is the fossilised product of decomposition of abundant tropical forest growths under marshy conditions. Within it is a proportion of material that will not burn. This is largely the residue of silt deposited between and on the rotting vegetable matter. It is this incombustible part that yields ash when coal is burnt.

In the swamps, plants such as mosses, sedges and trees grew thickly over long periods of time and their debris would accumulate. Then the area subsided and was flooded, putting an end to growth but also resulting in the vegetable remains becoming covered with sand and silt. Later the area was lifted by further earth movements and another generation of swamp forest began to grow. The process was repeated many times during periods measured in millions of years. Over these very long periods the plant components were partly decomposed and compressed. First, micro-organisms needing oxygen from the air (aerobic), then others that live without air (anaerobic) broke down much of the plant protoplasm, cellulose and other parts, though waxes and resins proved much more resistant. Then further weight of sediment deposited over these layers increased the pressure. The period when plants were being broken down is called the 'biochemical stage'; the later period of compression, which was also accompanied by mild heating is the 'geochemical stage'. The whole sequence of changes is known as 'coalification', and how far it has gone determines the 'rank' of the coal. The whole process is generally more advanced in the older seams, though coals are found of differing rank even at the same geological age, since earth movements, pressures and temperatures have all varied enormously. These variations in effects are quite marked, sometimes even within limited areas; consequently a coal seam may change appreciably in thickness and in rank as it is traced across a coal field. The successive layers of decaying plant residues in due course became a series of layers (seams) of coal; the intervening layers, also compressed and hardened, became rock strata such as sandstone (born of sand) and shale, resulting from sand and clay, respectively. Some researchers have suggested that the Dismal Swamp of Virginia and North Carolina, which is gradually being flooded by Lake Drummond, is such an area, and is currently undergoing active subsidence.

In what is now Britain, the forming of coal started much earlier in Central Scotland and the extreme north-east of England than elsewhere in these islands according to Dr W. A. Read of the Institute of Geological Sciences.[1] The earliest coals of workable thickness lie within groups of rock layers called the Calciferous Sandstone Measures of Scotland and the Scremerston Coal Group of Northumberland. Both are estimated to be 340 million years old. The geological division known as the Carboniferous (345 to 280 million years ago) is, as the name suggests, the chief coal-forming period, particularly in the Northern Hemisphere with brackish swamps covering much of Europe and North America. However there are also important deposits dating from the Cretaceous Period (from 140 to 65 million years ago) when some lower rank coals are thought to have been formed. All peat is reckoned to be geologically recent, that is the product of less than one million years. Peat is a low-grade fossil fuel with a calorific value lying between 8 and 12 GJ/tonne, according to its original composition and water content. At the upper end this is similar to the calorific value of dry wood or approximately one-third to one-half that of domestic coal. It is used as a fuel in countries such as Finland, Ireland, Sweden and the U.S.S.R. where it is locally available. Although it can never be a major fossil fuel, it can provide a small, but significant proportion of the energy demand of these countries. Brown coals and lignites are also of importance in some countries, notably in East and West Germany.

The Chemical Composition of Coal

Since coal is not a single chemical the chemical composition can only be given in ranges as shown in table 4.1. The second column shows the moisture of the fuel as

Table 4.1   *Composition of main types of humic coal*

Type of coal	Moisture as found (%)	Carbon	Hydrogen	Oxygen	Nitrogen	Volatile matter	Calorific value (Btu/lb)
			(%, all on dry, mineral-matter-free basis)				
Peat	90–70	45–60	6.8–3.5	45–20	0.75–3	75–45	7,500–9,600
Brown coals and lignites	50–30	60–75	5.5–4.5	35–17	0.75–2	60–45	12,000–13,000
Bituminous coals	20–1	75–92	5.6–4.0	20–3	0.75–2	50–11	12,600–16,000
Anthracites	1.5–3.5	92–95	4.0–2.9	3–2	0.5–2	10–3.5	16,000–15,400

Source: Slightly re-arranged from Dryden, article on 'Coal' in *Kirk–Othmer Encyclopedia of Chemical Technology*, 1965.

found; all subsequent columns give the properties corrected on the basis of the organic matter free of the associated moisture and mineral matter. The term 'humic' refers to the main range of common banded coal. There are other, less abundant, less important coals called 'sapropelic', based on different plant origins. 'Volatile matter' — as defined in the analysis of coal — refers to the loss in weight when the coal is heated in the absence of air under standard conditions. The determination of volatile matter is a practical, fast, 'empirical' test. It can be used for classifying coal and gives results closely related to those from chemical analysis.

In the U.K. coal is classified[2] by a system based on the percentage of volatile matter and the type of coke produced in a standard test under standard conditions (the Gray–King coke test). As rank increases coals tend to give ever stronger cokes down to volatile matter levels of around 30 to 20 per cent. Then with further increase in coalification, the coking property weakens until the anthracites with 9 per cent or less of volatile matter are found to yield only a non-coherent powder when heated in the standard test. Broadly these changes are related to the proportion of the carbon that is chemically combined in 'aromatic compounds'. This is the term used for compounds where the carbon is in a particular kind of six-membered ring — of which benzene is the simplest form.

Coal Production and Reserves

Production

In the U.K., coal production has fluctuated around an annual average of 122 million tonnes since 1973/4 as table 4.2 shows.

On a world scale coal accounted for 30.3 per cent of primary energy consumption in 1983 (2097.1 mtoe out of a total world consumption of all energies of 6925.5 mtoe). Only oil took a greater percentage (40.3 per cent). Even during the world recession of the early-1980s, world coal production increased (by 2.9 per cent in 1982 over 1981), reaching 2866 million tonnes. But within this increase were marked differences. The U.S. production was almost constant at 697.5 million tonnes, and so was Western Europe with outputs of 261.4 million tonnes. In Poland,

Table 4.2 U.K. coal production 1972/3–1982/3 (million tonnes)

Output (million tonnes)	1972/3	1973/4	1974/5	1975/6	1976/7	1977/8	1978/9	1979/80	1980/1	1981/2	1982/3
NCB mines (including tip and capital coal)	129.1	98.7	116.9	114.5	108.5	106.3	105.5	109.3	110.3	108.9	104.9
Open cast	10.1	9.0	9.2	10.4	11.4	13.6	13.5	13.0	15.3	14.3	14.7
Licensed	1.3	1.1	1.0	0.9	0.9	1.0	0.9	1.0	1.1	1.1	1.2
Total	140.5	108.8	127.2	125.8	120.8	120.9	119.9	123.3	126.6	124.3	120.9

Source: National Coal Board.

which was apparently recovering its strength after recent political upheavals (and severe falls in coal output), production rose 15.3 per cent to an output of 189.3 million tonnes, and the U.S.S.R. had a small increase to 558.0 million tonnes. China has announced very ambitious plans and has invited Western expertise and capital for a major expansion of its coal industry. In 1982 its production rose almost 5 per cent to 618.0 million tonnes, securing its position as second largest coal producer in the world. Future plans are on an astonishing scale. They include a surface mine with an intended capacity of 80 million tonnes per year, three others of 15 million tonnes and one of 10 million tonnes. In the U.K. coal production has been declining since 1980 and at the end of 1982 stood at 124.7 million tonnes.

The U.S.A., China and the U.S.S.R. alone account for no less than 65 per cent of world production of coal. They also have the biggest proven reserves. Estimates of the World Energy Conference published in 1976 were that the U.S.A. has 107×10^9 tonnes, the U.S.S.R. 104, and China 99. The U.S.A. and the U.S.S.R. also have roughly the same quantities again of sub-bituminous coal and lignites. Major trends of importance are that in the U.S.A. it is becoming necessary to transfer coal-mining operations from the eastern territories to the large deposits in the Midwest and the western part of the country. In the U.S.S.R. the opposite geographical movement is gradually being forced on producers. Deposits in the European part appear to have a relatively limited life. A gradual transfer of the principal coal-mining operations to the coal deposits east of the Urals and in Siberia will be needed during coming decades.

By far the largest exporters in the world have been the U.S.A. and Australia. Both send large quantities to Japan and to Western Europe. Other important exporters include South Africa, Canada and, now vigorously re-entering world markets, Poland. The U.K. is largely self-contained for coal trade, the amounts exported and imported being relatively small. The U.K. is therefore not an important element in the international market for coal. Major patterns of world trade in coal are shown in figure 4.1.

Worldwide during the continuing recession there has been in effect a surplus of coal which halted and slightly reversed recent improvements in U.K. exports. Total disposals of coal in the U.K. in 1982/83 were 113.5 million tonnes, with imports bringing the figure up to a supply of 120.6 million tonnes. At the end of the financial year there were 28 million tonnes of coal in distributed stocks, with a further 25 million tonnes undistributed, so that total stocks were at the record level of 53 million tonnes.

Reserves

Geological *resources* of coal include all untapped deposits that might be economically extractable some time in the future. Coal *reserves* consist of those deposits known to be technically and economically recoverable under today's conditions. According to the Department of Energy, it is probable that U.K. reserves of coal amount to some 45,000 million tons – enough to support the current rate of production for over 300 years and several times greater than all the oil and gas in the

Main Coal Movements by Sea 1982

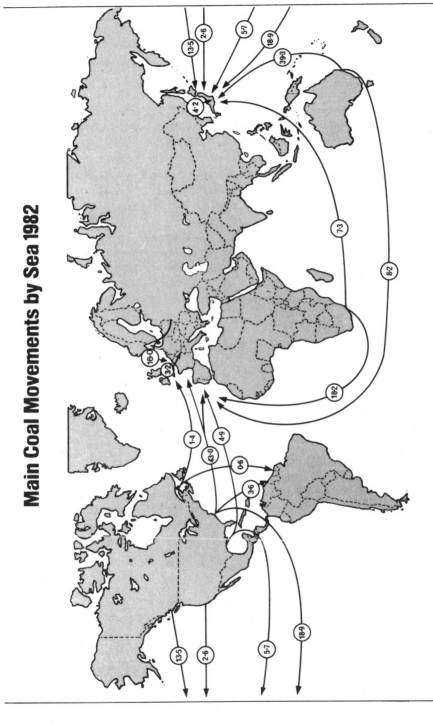

Figures are in Million Tonnes of Coal

North Sea fields.[3] Furthermore, though the distribution is uneven, a similar overall pattern applies for world supplies. Table 4.3 shows the relative strengths of reserves in major countries and confirms the dominating position of the U.S.S.R., the U.S.A. and China. Likewise, coal resources are abundant. The World Energy Conference in its 1978 appraisals of coal resources[4] stated: "The world possesses abundant coal occurrences. One may also assume, in addition to currently estimated resources and reserves, that there is a considerable 'potential behind the potential'."

There will certainly be new discoveries of coal. Even in such a well-surveyed country as the U.K. it proved possible to find the new coal field of Selby in 1972, the Vale of Belvoir in 1976 and the South Warwick coal field in 1978, and there are vast land areas, for instance in Africa and South-east Asia that have scarcely been prospected but are expected to contain coal deposits. Also, newer methods of extraction will unlock some of the more difficult deposits now listed as purely geological resources and make it reasonable to consider them as recoverable reserves.

In 1978, World geological coal resources were estimated at $10,000 \times 10^9$ tonnes of coal equivalent, of which technically and economically recoverable reserves were 640×10^9 tonnes. Peat, which covers about 1 per cent of the world's surface, has untapped resources estimated at between 200 and 330×10^9 tonnes on a dry-weight basis. The higher figure is equivalent to about 3.8×10^{21} J, very close to the 1981 figure for 'published' proved oil reserves.

Mining

Open Cast Mining

Very large tonnages are still obtained throughout the world by open cast (or strip) mining methods at various levels of mechanical sophistication ranging from shallow pits up to the use of giant draglines. Where surface working is possible it is by far the cheapest way of winning coal, and productivity is very high. In the U.K. in 1982/3, 14.7 million tonnes was produced by open cast mining with due attention to the environmental impact, both during operations and in later rehabilitation of agricultural land. These aspects are discussed further below. Open cast contributes about half of all U.K. anthracite and a large part of the prime coking coal particularly in Durham. There can be advantages in coal quality in working open cast because operations can be arranged to reduce contamination by dirt and water, without the need to resort to coal preparation. Peat is also obtained by the open cast method of draining the bogs in preparation for mechanical harvesting. In Ireland, the government organisation Bord na Mona has produced milled or powdered peat for electricity generation since the early 1950s and by 1980 had worked a total land area of over 700,000 hectares, with an estimated potential of over 6 million tonnes, or one-fifth of their total energy demand. Finland also has huge peat deposits, with about one-third of the land area being covered by bogs, and its Energy Policy Council envisages up to 10 per cent of energy demand being supplied by peat by the year 2000. The U.S.S.R. has the largest output in the

Table 4.3 World recoverable reserves of coal and peat (Gt ce)

Continent or economic–political groups	Proved recoverable reserves									
	Bituminous coal and anthracite		Sub-bituminous coal		Lignite		Peat		Total	
	(Gt ce)	(%)	(Gt ce)	(%)	(Gt ce)	(%)	(Gt ce)	(%)	(Gt ce)	(%)
Africa	32.5	6.7	0.1	0.1	0.0	–	–	–	32.6	4.7
America	111.4	22.8	75.3	67.4	13.2	15.0	0.3	5.2	200.2	28.9
Asia	113.9	23.4	0.8	0.7	1.4	1.6	–	–	116.1	16.7
U.S.S.R.	104.0	21.3	32.8	29.4	28.7	32.6	3.6	62.0	169.1	24.4
Europe	100.5	20.6	1.3	1.2	35.1	39.8	1.9	32.8	138.8	20.0
Oceania/Australia	25.4	5.2	1.3	1.2	9.7	11.0	–	–	36.4	5.3
Total	487.7	100.0	111.6	100.0	88.1	100.0	5.8	100.0	693.2	100.0
Common Market	70.0	14.4	0.0	–	10.6	12.0	0.5	8.6	81.1	11.7
OECD	205.9	42.2	74.6	66.8	34.9	39.6	2.1	36.2	317.5	45.8
COMECON	134.2	27.5	32.7	29.3	45.1	51.1	3.6	62.1	215.6	31.1
Developing countries	22.5	4.6	2.8	2.5	1.3	1.5	0.0	–	26.6	3.8
OPEC	0.4	0.1	0.2	0.2	0.2	0.2	–	–	0.8	0.1

Source: Survey of Energy Resources, World Energy Conference, London, 1980.

world, approximately 100 million tonnes per year. The U.S.A. also has considerable peat reserves but these still have to be assessed.

Deep Mining

As the continuing, growing need for coal in most coal-producing areas began to
demand ever-deeper burrowing into the earth, and ever-greater skills in the advance
prospecting, mines have become bigger and more complex. Sinking a pit is now a
major project that takes between 5 and 10 years, demands the investment of many
millions of pounds and involves high technology at all stages from the early prospecting, through the sinking of the shaft or the 'drift' (a tunnel driven in from the
surface down to the operating level), to the stage of full production. Though mining
engineers emphasise that every coal mine has its own individuality, there are general
features often in common. Production has become increasingly mechanised. In a
modern mine, coal is cut by a powerful cutting–loading machine using rotating
picks to shear it from a long wall face. Men work under hydraulic roof supports
which are advanced by power. The coal is thrown by the machine on to an armoured
conveyor that carries the coal to a transfer point where it is transferred to a main or
trunk conveyor. This carries it to a large storage bunker, that might be of, say,
1000 ton capacity. The bunker delivers coal at a controlled rate to a measuring
hopper that in turn loads a 'skip'. This is then raised in the shaft to the pithead.
Earlier systems of mining have included Bord and Pillar, where headings are driven
into a seam, connected by cross-headings. The pillars thus formed are then extracted
to the extent permitted by the need to support the surface. A diagram of a modern
mine is shown in figure 4.2.

Within mines, men, materials and coal all have to be moved, sometimes over
appreciable distances. At some pits, miners have to travel 11 km underground to
their work. Transport may be hauled by trolley locos drawing power from overhead
electric cables by means of a pantograph, or by diesel. Yet often the roadways
undulate and the engineers prefer a haulage system using a steel rope wound and
braked from a powerful electric engine at one end of the run. This depends on
signals from the train guard along signalling wires. New electronic controls have
now enabled this system to operate at higher speeds of up to 25 kph. The driver
rides on the train and controls it through a control stick in the cab sending radio
signals to what is called a cycloconverter.

Coal Washing (or Preparation) Plants

If the coal delivered from the pit is contaminated with too much stony material it
is sent to coal preparation plants. Coal and stone are separated by taking advantage
of the difference in specific gravity. There may be varying amounts of stone in
pieces of coal and vice versa. Consequently the separation is not perfect and some
of the dirt remains in the cleaned coal and reciprocally there is some coal in the
rejected dirt; material of intermediate specific gravity is called middlings and may

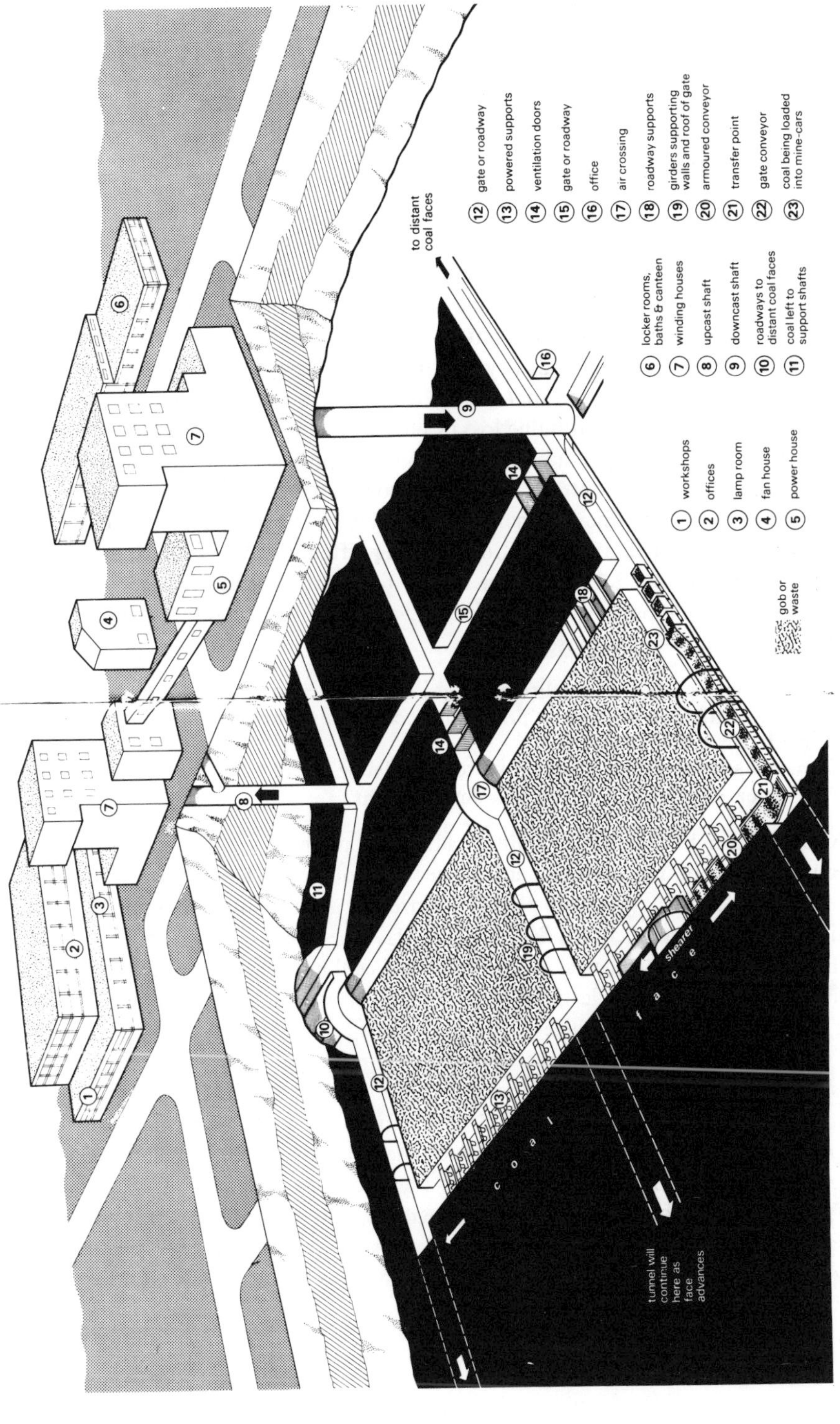

Figure 4.2 *Layout of a modern mine* (source: National Coal Board)

1 workshops
2 offices
3 lamp room
4 fan house
5 power house

6 locker rooms, baths & canteen
7 winding houses
8 upcast shaft
9 downcast shaft
10 roadways to distant coal faces
11 coal left to support shafts

12 gate or roadway
13 powered supports
14 ventilation doors
15 gate or roadway
16 office
17 air crossing
18 roadway supports
19 girders supporting walls and roof of gate
20 armoured conveyor
21 transfer point
22 gate conveyor
23 coal being loaded into mine-cars

gob or waste

to distant coal faces

tunnel will continue here as face advances

face

coal

be removed as an intermediate fraction, crushed to improve mechanical separation and sent for further 'washing'. The finest sizes of coal are separated by 'froth flotation'.

Exploration, Research and Recent Developments

The U.K. 'Plan for Coal'

In the wake of the fourfold increase in oil prices between 1973 and 1974, the government, Coal Board and mining trade unions came together to adopt a set of objectives to increase U.K. coal production. In essence this 'Plan for Coal' aimed to reverse the decline of the coal industry that had been allowed to take place during the era of cheap oil.[3] All agreed that the industry needed a substantial increase in investment to replace the continuing loss of output, by introducing new capacity and new technology. By September 1982, 238 major projects each costing over £0.5 million had been approved. 139 had been completed giving an estimated 16.5 million tonnes per year of new capacity to the industry. 60 collieries had been closed owing to exhaustion or exceptional mining difficulties. Overall output per manshift increased from 2.29 tonnes in 1974/5 to 2.44 tonnes in 1982/83, while new mines with the most advanced technology averaged 4.35 tonnes.[5] Investment is used for the following purposes:

(1) Expanding capacity at long-life collieries with sufficient reserves.
(2) Extending life at collieries by providing access to new reserves.
(3) Constructing new mines.
(4) A national exploration programme to locate new reserves.
(5) An increased level of research and development.

At September 1982 three new mines had been completed, capacity increased at 51 mines and major improvements introduced at another 85. The further 100 projects under construction were planned to provide an additional 26.3 million · tonnes per year. These included Selby, which was scheduled to cost £1 billion — the largest project ever undertaken by the British coal industry. It is claimed that its five linked collieries are destined to become the most extensive deep-mine complex in the world. It will produce 10 million tonnes a year brought out through a pair of sloping tunnels (drifts) on massive conveyor belts, into a reconstructed rail marshalling yard — therefore introducing no fresh despoliation of the area. All the coal is destined for power stations and is due to be transported by permanently coupled 'liner' trains. The associated 10 deep-shafts are used only for ventilation, materials handling and man-riding and their sites are being landscaped.

Exploration has continued by drilling programmes at existing mines and at new prospects. In addition, high-resolution seismic surveys have been carried out, many of which have been devoted to detecting faults ahead of existing mine workings.

Exploration involves a series of stages:

(1) to establish the existence of coal;
(2) to establish the boundaries of the coal field;

(3) to determine the main geological features;
(4) to assess the relative difficulty of exploiting the coal reserves.

There follows the development stage comprising discussions with local authorities and others, application for planning permission — all essential stages before starting the direct planning and construction of the new mine or the extension of an existing one.

In the mining operation itself, research has led to introducing ever more extensive use of remote and automatic control systems. The computer-based NCB system called MINOS (Mine Operating Systems) allows all the information to be monitored in a control room on the surface. To improve utilisation of equipment one programme is called FIDO (Face Information Digested On-line) which identifies and monitors causes of hold-ups or delays.[6]

At the coalface, the NCB are in 1983 concentrating on two programmes intended to improve the quality and durability of equipment.

(1) *Advance Technology Mining (ATM)* is a system being tested at selected collieries to automate support functions at the face and to relate the sequence control for supports to the operation of the power loader at the face.
(2) *Heavy Duty Mechanisation (HD)* is currently being tested in every coal field and aims to introduce, on the coalface, developed equipment of high durability, greater power and scope. It is planned to extend faces, have thicker coal extractions and to win a wider 'web' of coal with each pass. During these trials, daily output per face has been more than twice that from conventional faces.

By 1983 computers had been introduced in 16 out of 200 U.K. coal-preparation plants — dealing with controlling the washing, sizing and blending operation, and automatically analysing the coal at various stages for moisture and ash.

Summing up in his annual statement as Chairman of the NCB, Sir Norman Siddall refuted wild charges of 'butchery' of the industry by pointing to £3000 million invested under 'Plan for Coal' since 1975 and a continuing rate of investment of £700 m per year connected with all these developments.

In various countries besides the U.K. there are research projects for further reducing the need for underground labour in extracting either the coal or its energy. They include hydraulic systems for removing the coal from the face, often including pumping it to the surface, biochemical systems for breaking it down into a pumpable form, mechanical systems for remotely controlled (telechiric) coal cutting, and underground coal gasification. All these appear to be much longer-term prospects.

Transport

More than 95 per cent of all coal output is transported out of collieries by conveyors, and the rest in mine cars. In the remarkable Longannet project in Scotland, conveyors have so far in total carried over 20 million tonnes of coal from four collieries linked together and delivered it directly to Longannet Power station. Selby's output gathered from its 5 collieries will also travel on 2 conveyors each 14.7 km long.

They are monitored by electronic devices giving warning of excessive wear or over-heating, by sending signals to a control room.

Coal mining in fact, employs the U.K.'s biggest industrial transport system, moving each year 110 million tonnes of coal, 64 million tonnes of rock and waste and about 180,000 men underground on 70 million journeys. Since power stations take over 80 million tonnes a year, it is essential that this vast amount of coal is moved economically. Co-operation between the National Coal Board (NCB), British Rail and the Central Electricity Generating Board has resulted in the use of what are called 'merry-go-round' (MGR) trains. Special trains of permanently coupled wagons are both loaded and discharged without stopping. MGR services are scheduled by computer and run at speeds up to 72 kph. About 30 MGR trains a day are needed to supply the 6 million tonnes a year taken by a modern base-load station. Loading centres for these trains at collieries and open cast sites are highly automated with load cells weighing trains before and after loading; and results are recorded electronically. Four wagons are filled at the same time, and the whole process of loading a 1000-tonne train on the move is normally completed within an hour.[7]

This kind of pattern is matched at the power-station end where two trains can be handled simultaneously with equipment that opens bottom doors on the wagons, then closes them, resetting safety catches and discharging complete cargo in an hour. Other large users have also adopted the system including Associated Portland Cement, Bowater, and The British Steel Corporation.

Road deliveries have also been brought up-to-date by equipping specialised vehicles to off-load by conveyor or by pneumatic pipeline. Vehicles may carry their own compressed air unit and/or a range of pipes to connect on to a standpipe at the customer's premises. With this kind of aid, 15 tonnes can be delivered in half an hour up to heights of 30 m, without any human contact with the coal. In general, pneumatic conveying may be used in two possible systems. Dense-phase systems send the coal along as a series of 'plugs' several metres long; they use compressed air and are suitable for supplying small coal to boiler feed hoppers. Dilute-phase systems, using about half the ratio of coal to air, can use relatively low-pressure air or suction and are currently limited to the size range called 'singles'. In this system the coal is entrained in an air stream, then separated in a cyclone at the delivery end.

In North America coal is being moved over very long distances by pipeline systems but the most ambitious plans yet are reported from China.[8] For transporting coal as slurry from Shanxi province to the coast, a line is planned that will cover 535 miles (860 km). Fluor Mining of California signed a protocol agreement with the Chinese National Coal Development Corporation early in 1983. Two further such major pipelines are under study.

Uses for Coal

Overwhelmingly the largest current markets for coal in the U.K. are the power stations, which took 86.2 million tonnes in 1982/83 out of 120.6 million tonnes as the total for disposals. Large quantities used to go for carbonisation to gas, coke, ammoniacal liquor and tar with its further manifold products. But the introduction of natural gas killed the market for gas-making and the severe setbacks to the iron and steel industries have heavily cut the needs of the coke-ovens. Domestic space-heating requirements in the form of coal, anthracite or other coal-derived manu-factured fuels fell severely as users converted to more convenient forms of heating (electricity or gas) though coal is fighting back in this market. However, U.K. primary energy use as a whole in 1982/83 fell for the third year in succession and so did the use of electricity; all this inevitably hit coal consumption. Of the sales to power stations, 10.1 million tonnes were put into stock, creating the highest-ever stock levels amounting to 26.4 million tonnes. Coal supplied over 80 per cent of the primary fuel for the CEGB and there is what is officially called an 'Under-standing' to encourage use of coal. In its 1983 version, quantities over 65 million tonnes a year are sold at a significant discount and the CEGB undertakes to buy at least 95 per cent of its annual coal requirement — at least 70 million tonnes — from the NCB. This was scaled down from earlier years. Nevertheless the U.K. still imported some 4 million tonnes of coal in 1982. This was partly due to long-term contracts by CEGB but there are also actual shortages of special high-grade metallurgical coal and domestic anthracite.

Fluidised bed boilers — for many years largely laboratory curiosities — are being increasingly used in industry. In these, a 'bed' usually of coal ash, is kept in constant agitation by having air blown through it so that it looks as though it is boiling and behaves in many ways like a liquid. Coal is burnt on this at such a high rate that it forms only a very small proportion of the bed — about 1 or 2 per cent. This arrange-ment gives many advantages. It can burn a wide variety of fuels, gives high intensity of combustion and rapid heat transfer, can readily be adapted to trap oxides of sulphur (by adding limestone or dolomite to the bed) and is very readily linked with automatic controls and easy-feed systems including computer control. Indus-trial uses are for firing steam-raising boilers but also for drying grain, clay and stone.

If fluidised bed plants are operated under higher pressures, there are further gains. Thermal efficiency is improved and the size greatly reduced for any given rating. A power plant with a thermal capacity of 80 MW has been built at Grimethorpe as an international trial to assess the validity of the principle of pressurised fluid combustion.

Sales of domestic coal have dropped sharply from 30.5 million tonnes in 1956 to about 5 million tonnes in 1982/3, with an additional 3 million tonnes of manu-factured fuels. However, research and development by both the Board, at its establishment near Cheltenham, and by manufacturers has resulted in a number of highly efficient items of equipment, with the general title 'domestic solid fuel appliances'. These include both open fires and closed room heaters both equipped with back boilers and readily linked into central heating systems. Some need smoke-less or other special fuels but others, like Coalglo glass-fronted room heaters, burn house coal smokelessly, using ingenious designs that draw smoke through the fire and burn it. Sales of these appliances are increasing backed by a vigorous market-ing campaign, including a drive to ensure that houses are built with chimneys and to promote chimney systems that can be fitted to houses that do not have them.

In sharp contrast to the traditional open coal grate — generating smoke and draughts and having a low thermal efficiency — a typical modern smokeless fuel room heater with back boiler achieves an overall efficiency of 75 per cent; and a modern open fire with a wrap-round boiler obtains 70 per cent (figure 4.3).

Electricity Generation

The main difficulty in converting heat to energy is the low efficiency of the conversion. Steam is formed in the boiler at high thermal efficiency but in the following stage less than half its energy is converted into work in the steam turbine; the final stage, converting work into electricity is again highly efficient. But a large part of the primary heat energy of the coal is inescapably 'rejected' during the conversion to power. Many large factories, or groups of factories in industrial estates, have installed 'total energy' systems where they use this discharge heat. The main types of system in use are based on:

(1) *'Back-pressure' steam engines* or turbines generating power. For these steam is produced at a higher pressure (using coal or oil as fuel), used in the engine and then discharged at a pressure and temperature that still leaves it suitable for process heating in chemical plant, paper or textile mills or other processes:
(2) *Gas turbines*. These are used to generate the power. The exhaust gases, at say 500°C, are used for drying or heating or producing steam or hot water:
(3) *Reciprocating internal combustion engines*, which may be diesel or gas engines, for generating power. The reject heat from the cooling water and/or the exhaust gases may be used.

Combined Heat and Power

Studies sponsored by the Department of Energy[9] have led to estimates that harnessing the waste from public power generation, now discharged as heat to the atmosphere and rivers, could provide 30 per cent of the U.K.'s space-heat and hot-water needs. This system known as Combined Heat and Power (CHP) would save the equivalent of 30 million tonnes of coal per year and is discussed in more detail in chapter 10.

Improved Methods of Coal Usage

Since most coal is burned, the immediate way of improving its utilisation is to increase the efficiency of this operation. There have been improvements in well-tried methods of firing such as under-feed stokers by incorporating electric ignition controlled by time clocks and adding automatic de-ashing. For converting industrial oil-fired boilers to coal, methods have been developed for using pulverised coal, sometimes burning it with an oil flame to help stabilise it. A more interesting development is providing the fine-ground coal in an oil or water slurry, sometimes with stabilising agents. Some of the oil companies have been heavily involved in producing and extensively testing such mixtures, with the aim of producing a mixed

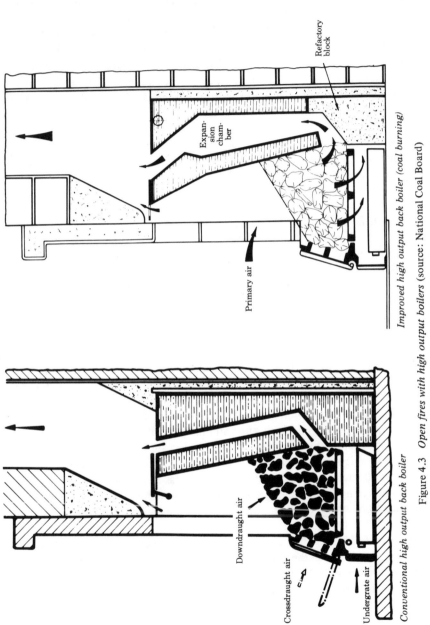

Figure 4.3   Open fires with high output boilers (source: National Coal Board)

fuel that can travel in oil tankers, be pumped like oil and then burn like oil with only minor adaptation of oil-fired equipment.

A considerable change of equipment, however, is required for fluid bed firing, briefly mentioned earlier. Though the CEGB still remains unenthusiastic in its attitude to this system, NCB researchers estimate that pressurised fluid bed operation will show gains in capital cost and should also lead to saving fuel. In the international project at Grimethorpe in South Yorkshire, experiments had been run with a total of over 1000 coal-burning hours by the end of 1982. One particular design of the tube bank in the fluid bed gave thermal efficiencies of 99 per cent but resulted in loss of metal, so that further research is still necessary in order to gain the high efficiencies while increasing the life of the equipment. By spring 1984 there had been over 3000 hours of coal-burning experience using a variety of coals.

Similar work in the U.S.A. on a 20 MW (electrical) prototype installation is expected to lead to the manufacture of a 200 MW (electrical) demonstration plant.

Chemicals from Coal

Coal may be converted into liquid or gaseous fuels and/or used as a chemical feed stock. Before 1939 most organic chemicals came from coal tar and coal was considered an important chemical feedstock. It was increasingly displaced by petroleum, so that it is now only a small contributor through the by-products from coke-ovens. There are new processes by which it can again be used for these purposes, though the conversions are currently generally uneconomic and in practice only one country, South Africa, now operates them. Here coal is exceptionally cheap, one of the major factors being the very low wages of its miners, and also it is national policy to substitute coal for oil as a chemical feedstock in order to overcome the attempted oil blockade in protest against South Africa's apartheid policy. In the coal-conversion plants, the coal is gasified in Lurgi plants by reaction with steam and oxygen under pressure, and the gases after purification are converted into a wide range of products by means of what are called the Fischer–Tropsch reactions. Products include petrol, oils of various grades including Diesel, waxes and chemicals such as alcohols and ketones. The gas emerging from the chemical plants is rich in hydrogen and is used for making ammonium sulphate, which is also made from what are called gas liquors from the process; ethylene and sulphur are among the other products.

Gas from Coal, an NCB report issued in 1983[10] reviews the Board's own work in this area but sets it in the perspective of information on other U.K. and international developments. It opens by referring to a source of gas from coal that does not involve manufacture but only drainage of the methane that is present in coal seams. At many mines this is removed by drilling into the strata as the coal seam advances, where it is collected and piped to the surface. Although the main purpose is to improve safety in mining, it also provides a clean industrial fuel of medium heating value. Quantities are appreciable and correspond to many hundreds of thousands of tonnes of coal per year. The gas is used in the collieries for heating and for generating electricity, or sold to industry either directly or blended into coke-oven gas. In South Wales there is even a limited gas grid connecting collieries and coke-ovens with consumers.

Gasification Processes

These are all based on the same set of chemical reactions whatever the gasifier used. Carbon is combined with oxygen to form both carbon monoxide and carbon dioxide, so giving out heat. Carbon reacts with steam forming carbon monoxide and hydrogen, thus absorbing heat; it also reacts with carbon dioxide, so absorbing heat. The fifth of the essential reactions is that of carbon monoxide with steam, which generates heat and produces carbon dioxide and hydrogen. The main products from the gasifier, as a result of these reactions, are hydrogen, carbon monoxide, carbon dioxide, steam and also nitrogen, if air rather than oxygen is used.

Mixtures of hydrogen and carbon monoxide are called 'synthesis gas' because of the great variety of materials that can be produced from them, depending on the reaction conditions and catalysts. Synthesis gas is manufactured from coal by gasifying it with steam and oxygen, which produces mainly carbon monoxide and hydrogen, and thereby reduces the complications that would otherwise be caused by the presence of nitrogen. There are three main commercially established processes that employ these reactions. These are: the Lurgi (mentioned above as the basis of the Sasol plant) in which a bed of coal slowly descends in a gasifier against an up-flowing stream of oxygen and steam at 20–30 bar pressure; the Koppers-Totzek process, where powdered coal is carried into a gasifier in suspension in a mixture of steam and oxygen at atmospheric pressure and is gasified at about 1500°C; and the Winkler process, in which a mixture of steam and oxygen is blown through a bed of coal particles kept in turbulent motion by the flow and maintained at a temperature of 800–1000°C. Each has its detailed advantages and drawbacks.

Finding that none of these processes met its own criteria, the NCB has developed its own process. The process needs to be met for both industry and power generation were seen as these:

(1) ability to use the wide range of U.K. coals;
(2) efficiency, with low capital and operating costs;
(3) quick response to fluctuating demands;
(4) a gas free from impurities;
(5) environmental acceptability.

This new NCB process is based on a fluid bed with a number of special design features; dolomite can be added to retain sulphur and it can be incorporated into a combined cycle to raise efficiency. In this cycle, the coal is partially gasified and the hot fuel gas produced, under pressure, is cleaned and burnt, driving a gas turbine connected to an alternator. Hot waste gas from this turbine is applied to produce steam. Extra steam is raised by burning the 'char' from the gasifier and is used to drive a conventional steam turbine–alternator set that generates more electricity. Discharge steam from this alternator goes to the gasifier and is used in the process. The process is estimated to be able to raise the efficiency of electricity generation from its present 38 per cent to about 45 per cent. The CEGB and NCB are working 'closely' together on this development.

The NCB is also working on the processes for making and using synthesis gas for chemical manufacture. Current economic assessments, made after discussions with the chemical industry, are that conditions are expected to favour this source as a chemical feedstock by about the year 2000, based on assumptions about trends in the real cost of oil.

Substitute Natural Gas (SNG)

Coal can also be converted to a substitute for natural gas (SNG). The steps are gasification by steam/oxygen, purification and a set of reactions called combined water gas shift and methanation. British Gas have been very active in developing a satisfactory process, an adaptation in which a Lurgi plant is operated at higher temperature so that the ash melts to a slag. The plant is called the British Gas/Lurgi Slagging Gasifier. A large part of the programme has been financed by U.S. companies and uses North American coals. But the NCB has also supplied test batches of British coals and collaborated in researches on these. In another project, British Gas are working on producing SNG by directly hydrogenating coal to methane. Yet a further process, developed in this case by the NCB for a U.S. consortium, is called COGAS; this overcomes some of the problems of using air as the oxidant by burning part of the char in a separate circuit to provide the heat for the steam–carbon reaction but keeps the products of combustion separate from the synthesis gas.

Methanol

About 300,000 tonnes a year of methanol — an important chemical intermediate — is now being produced in the U.K. from synthesis gas, and is becoming increasingly competitive. A process developed by Mobil produces a high yield of liquid products from methanol, notably petrol. Figure 4.4 shows the chemical process possibilities that are opened up by the use of synthesis gas directly and the further vast field offered by converting to the intermediate, methanol. Which, if any, of these possibilities is actually realised depends naturally on relative costs of producing by this route and by current processes.

The International Energy Agency (IEA) has also been studying[11] the economics of 'Coal as a petroleum substitute' (methanol and methanol-derived gasoline, Fischer–Tropsch liquids, direct liquefaction processes) and, more broadly, 'Advanced gasification and liquefaction processes'. Allied work has covered producing ethylene from feed stocks derived from oil, gas and coal.

On a pilot-plant scale, there are reports of processes being developed for direct hydrogenation of coal in the U.S.A., West Germany and Japan in varying quantities up to 150 tonnes per day. Though these all remain experimental, there are strong advocates of this route to light liquids from coal. It is estimated that they will call for less capital, be more efficient than the other routes examined and eventually produce cheaper products. Nevertheless, work continues on routes via synthesis gas and also other routes using liquid solvents.

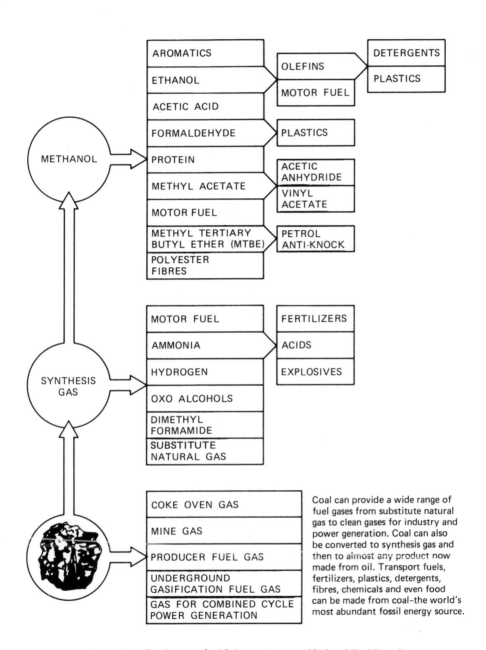

Coal can provide a wide range of fuel gases from substitute natural gas to clean gases for industry and power generation. Coal can also be converted to synthesis gas and then to almost any product now made from oil. Transport fuels, fertilizers, plastics, detergents, fibres, chemicals and even food can be made from coal-the world's most abundant fossil energy source.

Figure 4.4 *Gas from coal with its uses* (source: National Coal Board)

Syncrude

Coal may be dispersed in a liquid solvent that, with or without a catalyst, aids the reaction between coal and hydrogen. In these processes, the mineral matter is separated and the purified solution is concentrated until it resembles crude oil. This synthetic crude oil (syncrude) is then processed and, possibly after 'hydrocracking', refined in a petroleum refinery. The NCB has sufficient confidence in a two-stage liquid solvent extraction process for manufacturing transport fuels from coal to have started designing a 2.5 tonnes per day pilot industrial plant to be erected at Point of Ayr Colliery, North Wales. Financial support is being given by the ECSC and the U.K. Department of Energy.

Other Developments

There are also studies at various stages and in varying scales on such topics as underground gasification of coal — a concept that has tantalised many thinkers throughout this century. In this process, the reaction with steam and oxygen would take place with the coal *in situ* and only the product gases would emerge, there being no need for mining. A high-temperature nuclear reactor has been considered as the source of heat for gasification reactions. Alternatively off-peak nuclear electricity could be used to electrolyse water, yielding cheaper hydrogen for hydrogasifying coal directly to SNG.

The continuing drive to improve methods of using energy includes considering processes where the fuel may be derived from various sources, one of which may be coal. In this category are, for example, fuel cells. These are potentially highly efficient devices for converting the chemical energy of a fuel into electricity. There is an anode where fuel is oxidised, a cathode where an oxidising agent (usually oxygen) is chemically reduced and an electrolyte. This separates anode from cathode and conducts the electrically charged particles (ions) between them. So far they have proved expensive and far too short-lived, owing to poisoning of the electrodes.

Another direct conversion device where gases from coal have been considered is in magnetohydrodynamic systems (MHD). Hot fuel gas is ionised (electrically charged) and passed through a power channel surrounded by a magnetic field. Electrodes suitably placed in the gas stream can tap off electrical power. This is again a tempting prospect since it produces electricity without using any solid moving parts within itself but some 50 years of trials have so far achieved little success.

Environmental Issues

The NCB has given considerable attention to environmental issues in recent years and faithfully reported on them in a separate section of its Annual Report. Important stages in the development of attitudes have included the September 1972 report of the Commission on Mining and the Environment set up by 7 major mining companies under the chairmanship of Professor Lord Zuckerman OM FRS, former Chief

Scientific Adviser to the government. This distinguished body was set up to establish the way in which the two objectives of using mineral resources in Britain and conserving the environment could be 'harmonised'. In September 1981 the more official body, the Commission on Energy and the Environment, issued the directly relevant report *Coal and the Environment* — the Flowers Report[12] — named after its Chairman.

In response to this, the government issued a White Paper[13] in May 1983. This quoted with approval that the Commission saw 'no insuperable environmental obstacles to the role of coal as currently envisaged.' The government indicated that it wished the same environmental standards to apply to the coal industry as to other mineral operators. It recognised that, overall, standards have improved as new capacity replaces old and that the problems of the industry arise from much earlier operations. The wide-ranging scope of this document covers: energy policy and the future of coal; the environment, energy and coal; dereliction; spoil disposal; subsidence; open cast mining; transport and use of coal; and NCB/government relations on environmental matters.

Meanwhile the NCB continues to discuss with central and local government, the environmental implications of long-term mining proposals. It is running a major exercise, jointly with the local authorities, on spoil disposal in the Yorkshire coal field. Key undertakings have been given on a number of relevant issues including the strengthening of research effort on subsidence, underground disposal of colliery spoil and controlling noise at open cast sites. The NCB will improve tipping methods through progressive restoration and landscaping techniques — and can indeed point to a number of achievements in grassing over tips, transforming some into parks or golf courses, and working with the Ministry of Agriculture in restoring open cast sites to healthy, attractive farmland. A final commitment is to reduce the local environmental impact of coal handling and transport.

Acid Rain

On the much broader and internationally sensitive issue of acid rain wreaking widespread environmental damage in many parts of Europe (particularly Scandinavia), the NCB with the CEGB responded rather belatedly in 1983 by sponsoring research under the U.K.'s premier scientific organisation, the Royal Society, with the Norwegian Academy of Science and Letters and the Royal Swedish Academy of Sciences. The 5-year research programme will examine the causes of acidification of surface waters in Norway and Sweden and the implications for fisheries. A statement[14] issued by the Royal Society declares that it will work with its colleagues in the spirit of the UN Convention on Long Range Transport of Air Pollutants. This 'enjoins' signatories to promote research on 'natural eco-systems. . . with a view to establishing a scientific basis for dose–effect relationships designed to protect the environment.'

Carbon Dioxide Build-up

Assessment of the effects of increased carbon dioxide in the atmosphere is well summarised by a detailed IEA appraisal *Carbon dioxide and the 'greenhouse effect' – an unresolved problem* which concludes that the situation remains one of uncertainty. Although coal combustion is by no means the only contributing cause, the industry in the past has often been environmentally negligent. Recently, both the marked shift in public sentiment, and the changes in law have been reflected in what the NCB describe as 'their positive approach toward achieving environmental improvements.'

Provided that this 'positive approach' is adopted at the earliest planning stages, it is clear from both the Commission report and the White Paper that there is no basis for objecting to the further development of the modern coal industry on grounds of environment, irrespective of the grim and depressing past. The IEA findings after a major exercise on coal use and the environment included a similar conclusion.[15] The report urged countries to consider their environmental systems and energy systems as integrated wholes, and was confident that they could achieve reduction in net pollution loadings in a cost-effective manner.

The Coal Industry and the Economy

Throughout the current century, coal has been very much on a switchback reaching its highest production in 1913 when its output was 287 million tons; exports that year were effectively 98 million (direct exports, supplies to shipping and coal equivalent of exported coke and manufactured fuel). Demand fluctuated thereafter but the period between the two World Wars was generally one of decline character- ised at times by great hardship and misery for the miners. It also included the lockout of 1926 that led to the General Strike. Later in the 1930s, coal was the industry that suffered most from the world depression and the decline in British exports.

Though there have been a number of enquiries into the industry, probably the most pregnant was that of Charles Reid appointed in 1944.[16] Its remit was 'to examine the present technique of coal production from coalface to wagon, and to advise what technical changes are necessary to bring the industry to a state of full technical efficiency.' This group of current and former mining managers and directors of colliery companies surprised themselves. After analysing in great detail the general causes of the low productivity of British mines at that time, and making drastic and far-reaching recommendations for technical changes, they 'carefully considered' whether they had fulfilled their duty. They concluded that 'The Conditions of Success' (the title of their final chapter) went beyond these techni- cal recommendations and called for masters and men to work better together. They urged the merging together of collieries into 'unified commands'. 'An Authority must be established', they wrote 'which would have the duty of ensuring that the industry is merged into units of such sizes as would provide the maximum advantages of planned production, of stimulating . . . broadplans . . . A great pioneering task . . . nothing less than the rebuilding of the industry on the most modern lines'. Politically, it was evidently the return of the Labour government with its massive majority in 1945 that led to the nationalisation of the industry. But much of the heavy ammunition was supplied by this critical and honest group of mining engineers.

Before the coal industry was nationalised in 1947, there were 800 companies owning over 1400 collieries, 30 manufactured fuel plants, 55 coke-oven and by-product plants, 85 brick and pipeworks, 225,000 acres of farmland and 140,000 miners' houses. Some of their more surprising assets were shops, offices, hotels, cinemas, swimming baths, wharves, milk rounds, a holiday camp and a cycle track. Though these assets were taken over, many of the non-mining ones were quickly sold. 480 of the pits were called 'small mines', which employed fewer than 30 men underground. Generally these were granted licences by the NCB to continue in private operation. In 1982/83 they contributed 1 per cent (1.2 million tonnes) to the total of 124 million tonnes output.

The many early changes – in technical rehabilitation, mining conditions, coal preparation, scientific control, scientific research, miners' housing, baths and welfare services, training of new entrants – are all surveyed in detail in a *Colliery Guardian* publication called simply *National Coal Board. The First Ten Years.*

Technically, in terms of human relations, and in its concern for the environment, the industry was revolutionised. But the fluctuations in demand have meant that it stays on the switchback. Post-war inland consumption reached a peak in 1956 at a level of 218.4 million tonnes. Thereafter consumption of coal fell – despite vigorous marketing campaigns – as oil became cheaper. Later, as natural gas began to flow, programmes of building nuclear reactors were adopted and a *White Paper on Fuel Policy* in 1967 presented a new developing theme; we were moving from a two-fuel economy of coal and oil, to a four-fuel one, where nuclear power and natural gas would become major contributors.

Since the government accepted the advice of its fuel experts, that in addition oil would remain competitive with coal, the policy recommended was to run down manpower at a rate of 35,000 miners a year. The mood continued until 1973 when 'official' thinking began to turn. Tom Boardman (Conservative) Minister for Industry accepted that further contraction of the coal industry would pose an unacceptable danger to the country's supplies of energy. A Coal Industry Act was passed that gave substantial assistance to the industry. With the passing of the Act, the industry's mood changed and became more confident. The economy was improving and it was expected that there would be an increase in coal use in the country. The following (Labour) government promoted a Tripartite Examination (Government, NCB, Unions) which resulted in the 1974 *Plan for Coal.*

The Future

For the immediate future, the NCB and the Monopolies Commission – who were asked to sit in judgement on it – both see the major needs as bringing output into step with demand for coal and doing this by cutting down the numbers of high-cost pits. In June 1983, the Monopolies and Mergers Commission reported[17] that the major reason for the Board's financial problems had been the failure to reduce output by the broad average of 3–4 million tonnes per annum that was accepted by the NCB, the unions and the Government in the interim report of the Tripartite Examination of 1974. The Monopolies and Mergers Commission acknowledged that 'the other main element of that report – investment in new and modernised production facilities – has continued unabated.' Sir Norman Siddall in his valedictory message in August 1983 made much the same points: 'The most important task facing everyone in the coal industry is to move out of high-cost mining capacity. Thanks to our high rate of capital investment (about £800 million this year) we are bringing in modern, low-cost capacity much faster than we are getting out of the old places. Last year 12% of our output lost £275 million – almost three quarters of the £374 million deficit grant paid by the Government. That is a drain on the rest of the industry. It has to be remedied with as little hardship as possible to the people involved.' The bitterness of the 1984 miners' strike confirms the complexity of these problems.

In respect of long-term planning, the Monopolies and Mergers Commission made some play with the errors in the NCB's forecasts for coal demands compared with the actual demands, although many forecasts had proved erroneous in a decade when many trends had been upset. By contrast, in the White Paper on the Environment, the government specifically disclaimed any intention of drawing up an energy blueprint – an attitude very much in accord with a Parliamentary answer given earlier in 1983 by the Secretary of State for Energy who said: 'The NCB should aim at that share of the market which they can profitably sustain in competition with other fuels .' Nor did the remainder of the answer make any concession to the concept that coal should be financially supported on the grounds that it has large reserves and that other fossil-fuel sources will shortly decline.

Earlier we referred to the many studies on the economics of coal-conversion processes. The results understandably tend to emerge in the form of graphs showing the conditions for break-even prices. Results can only be presented in this way since firm statements about the dates when processes will compete imply an unrealistic certainty about future trends in supplies and prices. Yet at some stage within the coming decades it is evident that shortages of oil and gas will in fact drive their prices so high that converted fuels and chemical feed stocks from coal will become competitive. The policy question is whether the change in economic relationships will take place at such a rate that it will be possible to change the infrastructure of industry to move about and use the new feedstocks at a fast enough rate to prevent major crises arising from shortages. If this is not likely to be possible, then changes should be taking place in anticipation, with government financial support (which would later be withdrawn) to smooth the transition from a heavily oil and gas-based fuel and chemical economy to one largely based on coal, with initial support from nuclear-energy sources and later from renewable-energy systems.

'We are convinced,' write the authors of *Energy in a finite world*,[18] 'that a return to coal as a major energy source is not only necessary but also inevitable, both for the coal-rich nations and for the rest of the world.' But the scientists of the International Institute for Applied Systems Analysis add that 'this time round coal use will have to be managed differently.' The specific points are that it must be used "in such a way as to permit an orderly buildup of the coal industry and particularly of the coal supply industry to a level that will allow it to become a major source of secondary liquid fuels. Coal must become 'new coal'." This is not likely to happen by the blind workings of the market alone. Government and EEC guidance and subvention are going to be necessary.

References

1. Read, W. A., Institute of Geological Sciences, London, Personal Communication
2. *The Coal Classification System used by the National Coal Board*, NCB, London, 1964
3. *Plan for coal*, Reports of Coal Industry Examination, Department of Energy, 1974; *Coal for the Future*, Department of Energy, London, 1978
4. *World Energy Resources 1985-2020*, World Energy Conference, London, 1978
5. *Report and Accounts 1982/3* (and earlier *Annual Reports*), NCB, London
6. 'Mechanisation – the dream that became a reality', *Coal and Energy Quarterly*, No. 4, Spring 1975
7. *High Technology In Coal*, NCB, London
8. 'China encourages foreign investment', *World Coal*, June 1983, pp. 62-5
9. *CHP Feasibility Programme Interim Report* (known as the 'Atkins Report'), Department of Energy, London, 1980; *House of Commons Papers 314-I* and *II, Combined Heat and Power, Report and minutes of evidence, and appendices*, HMSO, London, 1982 and 1983
10. *Gas from Coal*, NCB, London, 1983
11. *IEA Coal Research Report, 1982* (with references to other IEA reports), NCB, London, 1983
12. *Coal and the Environment* (known as the 'Flowers Report'), Commission on Energy and the Environment, HMSO, London, September 1981
13. *Coal and the Environment*, White Paper, Cmnd 8877, HMSO, London, May 1983
14. *Acidification of Surface Waters in Norway and Sweden*, Statement issued by the Royal Society, London, 5 September 1983
15. *Coal use and the environment*, Report by the Coal Industry Advisory Board, IEA, Paris, 1983
16. *Coal mining. Report of the Technical Advisory Committee* (known as the 'Reid Report'), Cmnd 6610, HMSO, London, March 1945
17. National Coal Board, *A Report on the Efficiency and Costs in Development, Production and Supplies by the National Coal Board*, Cmnd 8920, HMSO, London, 1983
18. International Institute for Applied Systems Analysis, *Energy in a finite world; paths to a sustainable future*, Ballinger, Cambridge, Massachusetts, 1981

Further Reading

Merrick, D., *Coal Combustion and Conversion Technology*, Macmillan, London, 1984

Grainger, L. and Gibson, J., *Coal Utilisation: Technology, Economics and Policy*, Graham and Trotman, London, 1981

Hunt, V. D., *Synfuels Handbook*, Industrial Press, New York, 1983

Future Coal Prospects; country and regional assessments, Report of the World Coal Study (known as the 'WOCOL Report'), Ballinger, Cambridge, Massachusetts (distributed by Harper and Row, New York), 1980

IEA Coal Research Report, 1982, NCB, London, 1983

James, P., *The Future of Coal*, Macmillan, London, 1982

The Use of Coal in Industry, Report by the Coal Industry Advisory Board, IEA, Paris, May 1982

Environmental Aspects of Increased Coal Usage in the United Kingdom, Royal Society, London, February 1980

Steam Coal and Energy Needs in Europe beyond 1985, Special Report No. 134, Economist Intelligence Unit, London, 1982

Berkovitch, I., *Coal on the Switchback*, George Allen and Unwin, London, 1977

Berkovitch, I., *Coal: Energy and Chemical Storehouse*, Portcullis Press, Redhill, Surrey, 1978

5

Nuclear Energy

(by Ross Hesketh)

'Reducing the concentration of power will automatically reduce the size of corruption. This is not to say that it will eliminate corruption, only make it more visible and perhaps manageable'

Michael Zwerin

Nuclear energy is the most concentrated form of energy in use today. In the social sphere, the control of nuclear energy is no less concentrated. The two go together and any worthwhile discussion must cover both aspects.

Nuclear Energy – What is it?

An atom consists of a nucleus and a number of planetary electrons. The nucleus contains protons and neutrons which are held together by 'strong' forces. If we can make these forces do work nuclear energy is released. This can be achieved if the protons and neutrons are re-arranged into a more stable pattern. This means taking a less-stable nucleus and turning it into a more-stable nucleus. The most stable nucleus of all is that of iron; it is impossible to re-arrange the protons and neutrons of the iron nucleus in such a way that the re-arrangement leads to a release of energy.[1] On the other hand if we join together two lighter nuclei, so that the compound nucleus resembles that of iron, the re-arrangement leads to a release of energy. Equally, if we split a heavy nucleus so that the fragments resemble the nucleus of iron, the re-arrangement leads to a release of energy. Thus there are two ways of obtaining energy from the nuclei of atoms: the *fusion* of light nuclei and the *fission* of heavy nuclei. In real life the fusion products and the fission products are not iron nuclei; they are simply the nuclei of atoms lying somewhat closer to iron than the original nuclei.

All the hundred or so elements of which our earth is composed were themselves formed in a large but incomplete nuclear explosion. Had the explosion been complete, the earth would be solid iron.[2] Man-made nuclear fusion in particular is difficult because we are trying to create, on a small scale and in cold surroundings, the conditions of a stellar interior.

The protons and neutrons of the nucleus are controlled by two significant forces, the 'strong' attractive force between any two nucleons (that is, between proton and proton, neutron and neutron, or between proton and neutron) and the electromagnetic force, which acts between protons and is always repulsive. When any two nuclei are joined together the 'strong' force gives out energy. However, if large nuclei were to be joined together, all this energy would be used in doing work *against* the electromagnetic force, simply because the number of protons is so large. When small nuclei are joined together the work done against the electromagnetic force is small and there is a net release of energy, obtained from the 'strong' force. For the heavy nuclei the reverse is true: energy is obtained *from* the electromagnetic force, in the act of fission, and the work done *against* the strong force is in this case small. Nuclear fission is technologically possible because the nuclei of the very heavy metals, thorium, uranium and plutonium, are already so unstable (they already contain so many protons) that under reactor conditions they break up.

Fusion

Principles

The 'strong' nuclear force from which fusion energy is derived only operates when two nucleons are very close together, no more than 10^{-15} m apart; it is a short-range force. By contrast, the electromagnetic force that operates between two nucleons or two nuclei is a long-range force.

Thus, as we bring two nuclei together, with the object of fusing them, they at first repel each other, and we have to do work against this repulsion. Only when the two nuclei are close together does the 'strong' attractive force come into play and result in fusion. The problem of fusion therefore is like that of pushing a boulder up a hill so that it may run down into a much deeper valley on the far side. The hill is created by the repulsive electromagnetic force, and hence it is known as 'the Coulomb barrier'. The Coulomb barrier may be surmounted if two colliding nuclei have a kinetic energy greater than the height of the barrier. They are given this kinetic energy by being heated in a plasma, and they surmount the barrier in significant numbers when the temperature T is such that

$$kT \geqslant \text{height of barrier} \qquad (5.1)$$

where k is Boltzmann's constant. The height of the barrier is proportional to the product of the atomic numbers of the colliding nuclei (that is, proportional to the product of the number of protons in each). Hence the lowest plasma temperatures are obtained for collisions between the lightest nuclei, such as hydrogen and deuterium. The energy release per nucleon, in fusion, is also *greatest* for these light nuclei. Hence fusion reactors use the lightest nuclei possible.

To heat a plasma to, say, a hundred million degrees Celsius, requires an input of energy. A fusion reactor must therefore give at least the same output of energy, if it is to be of any use. This simple matching of output to the minimum input leads to 'the Lawson criterion'; the product of plasma density n and plasma lifetime τ must exceed a specific value which is typically 10^{21} m^{-3} s:

$$n\tau \geqslant 10^{21} \text{ m}^{-3} \text{ s} \qquad\qquad (5.2)$$

This expression is an absolute requirement and somewhat idealised. In reality a fusion reactor requires a larger energy input than the ideal minimum, and a correspondingly larger output is also required.[3]

Since 1955, the product $n\tau$ achieved has increased by a factor of 10 in each decade, as has the temperature attained. The conditions specified by expressions (5.1) and (5.2) are now being *approached*, but at the time of writing they have not been *attained*. If progress continues at the same rate it would appear that fusion power is unlikely to be available before the year 2020. Even then the cost to the community may well be prohibitive.

Practice

A plasma is a gas in which some or all of the electrons have ceased to be bound to particular atomic nuclei. A plasma may have a temperature of 50,000,000 degrees Celsius and obviously requires a container that does not melt. A magnetic field can be used. A charged particle, such as an ion of tritium, moving in a magnetic field, experiences a force which is at right angles both to its own direction of motion and to the magnetic field. Under this force all the charged particles in a plasma move in helices which have the lines of magnetic force for axes.[4] Therefore, if we have a tube of magnetic force, the charged particles may move freely along the length of the tube but they are restrained from getting out at the sides. If the tube has ends, particles may leak from the ends. A torus, illustrated in figure 5.1, is a tube without ends and is today the most popular shape of 'magnetic bottle'.

The magnetic field along the (circular) axis of the torus is produced by current flowing in the D-shaped coils shown on the right. Massive iron transformer cores link the torus. One of eight of these is shown on the left. This transformer induces a toroidal current in the plasma, and this current, in its turn, induces the poloidal magnetic field (the circular lines running around the torus tube). The toroidal and poloidal magnetic fields combined to produce a helical field around the torus. Only with such complexity is it possible to stop the plasma touching the cold, stainless steel wall of the vacuum chamber, also a torus.

The plasma is heated in two ways, first, by Ohmic heating (by electrical resistance to the current flowing in the plasma, round the torus), and second by a subtefuge called 'neutral injection'. This is necessary because the electrical resistance of the plasma decreases as it becomes hot, and the Ohmic heating therefore decreases. In the neutral injection method, hydrogen atoms are first ionised,

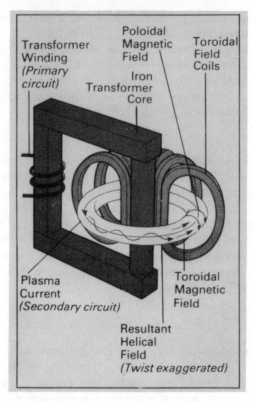

Figure 5.1 *Principal magnetic fields of a tokamak* (source: JET Joint Undertaking)

then accelerated to velocities characteristic of 100 million degrees, then allowed to pick up an electron and so to regain electrical neutrality. In this neutral form energetic atoms of hydrogen can penetrate the walls of the magnetic bottle, become part of the plasma and add energy to it, increasing the heat. Plasma temperatures of 60 million degrees have already been reached by this method.

There is a second method of bottling up the plasma which is called *inertial containment*. In essence this works like a rocket motor, or like the Crookes radiometers which rotate in the sunlight. Instead of sunlight, symmetrically placed lasers bombard the surface of a spherical pellet of hydrogen, vaporising its surface. The vaporising hydrogen compresses the remaining solid hydrogen, increasing its density by a factor of 100. Finally the whole pellet vaporises and flies apart. However, it remains together as a hot dense plasma for sufficiently long for a useful number of fusions to occur. Again, temperatures of 100 million degrees are required, together with a compression to 1000 times the normal solid density. Temperatures approaching 50 million degrees have been obtained and, independently, compressions to about 100 times normal density. To achieve an energy balance (that is, to get out as much useful energy as one puts in) the inertial confinement method will have to increase the achieved temperature by a factor of 2, and the achieved density by a factor of 10, and produce both increases, together, in one and the same pellet.

A third and totally different method of obtaining fusion energy has been canvassed, with apparent seriousness — namely, to set off a sequence of small hydrogen bombs in a large underground cavern and to use the resulting gas pressure to drive turbines at a power station situated on the earth's surface. There would I think be some small problems of containment, of contamination, and of refuelling in such a fusion reactor!

The machine illustrated in figure 5.1 uses two isotopes of hydrogen as its fuel, deuterium (D) and tritium (T), and the fusion of these gives helium (He), a neutron (n) and a release of energy:

$$D + T \rightarrow {}^4He + n + 28 \times 10^{-13} \text{ J} \qquad (5.3)$$

Since tritium is a radioactive element with a half-life of 12 years it is present in the environment only in negligible quantity, and has to be manufactured, either in an ancillary fission reactor or in the fusion reactor itself. The neutron shown in expression (5.3) can interact with either of the isotopes of lithium, to give tritium and helium:

$${}^7Li + n \rightarrow T + {}^4He + n - 4 \times 10^{-13} \text{ J} \qquad (5.4)$$

$${}^6Li + n \rightarrow T + {}^4He + 7.7 \times 10^{-13} \text{ J} \qquad (5.5)$$

Thus the fusion reactor — when there is one — will use as its fuel naturally occurring deuterium and naturally occurring lithium. Deuterium is comparatively plentiful; the world's oceans contain some 10^{12} tonnes of it. Lithium is less plentiful; the high-grade resources are some 5 million tonnes. To provide electricity for the U.K. at the present rate of use would require an initial supply of 10 thousand tonnes of lithium, and thereafter no more than 50 tonnes a year. All this presumes that the fusion reactor can be developed to give out much more energy than it consumes.

The 'fusion products', the waste from a fusion reactor, are formally only the helium atoms appearing in expressions (5.3), (5.4) and (5.5). However, the structure of the fusion reactor itself rapidly becomes radioactive (by the absorption of neutrons escaping from the plasma). This means that access to the reactor is restricted and that maintenance must be done remotely, which increases the difficulty and expense, not least because fusion reactors are so much more complex than fission reactors. Unlike fission reactors, fusion reactors contain much electrical equipment, and this deteriorates under the neutron irradiation. The whole machine therefore has a finite life, at the end of which it must be disposed of, since the radioactivity in the steel components will persist for 50–100 years. Despite equations (5.3), (5.4) and (5.5), the waste products of a fusion reactor are significantly more than helium; the toroidal vacuum chamber becomes highly radioactive.

The next stage, of extracting usable power from a fusion reactor, has not yet been approached, for the simple reason that there are other problems to be solved first. Diagrams of proposed fusion power stations show the fusion energy being extracted as heat from a lithium blanket surrounding the toroidal vacuum chamber which holds the plasma. The heat is then used to raise steam and drive turbines, as in any coal-fired power station. Especially for fusion, one would like to use the energy of the particles without having to use an intermediate thermodynamic engine, in which, of necessity, an appreciable and usually major part of the energy must be discarded as low-grade heat. A method for direct conversion into electricity of the energy released in fusion has not yet been discovered. By contrast, primitive technologies, such as hydropower and wave power avoid the thermodynamic cycle and the losses that go with it.

For a fusion reactor the greatest problem may well be that of economics. The present research machines are sophisticated, and if a doubling of plasma temperature has to be attained it seems unlikely that this advance will be attained by a method that is less expensive than the heating methods presently used. Similar remarks apply to the problem of increasing the time for which the plasma remains stable. Thereafter, today's research machines will have to be scaled upward in size. Design studies have been made for full-sized power stations (on the basis that a machine of the year 2020 will look much like one of today's), but the problems of size, maintenance and replacement lead to great complexity and consequently to major costs. The fusion power research project with which we gamble relies on the occurrence of technical developments that we do not yet see. The present scale of research requires international co-operation to build a machine of the type shown in figure 5.1, and the next generation of machines may number no more than 2 or 3 throughout the world.[5]

Fission

Principles

There are three nuclei in which the configuration of protons is very nearly stable but not quite: thorium-232 (half-life 14 billions years), uranium-235 (half-life 0.7 billion years) and uranium-238 (half-life 4.5 billion years). These are the heaviest nuclei to occur in significant quantities in the surface layers of the earth. They are available to us because their half-lives are comparable to the age of the earth, which is roughly 5 billion years. (The half-life of a radioactive element is the time in which half of any given number of nuclei decay.)

Evidently, when the earth was formed it contained approximately twice as much uranium-238 as it does today, and half of this primordial uranium-238 is now an isotope of lead, Pb-206. The thorium and the U-235 decay to Pb-208 and Pb-207, respectively.[6] Naturally occurring lead is not a 'fission product' in the accepted sense; the natural radioactive decay of thorium and uranium involves only the emission of alpha particles (the nuclei of helium atoms) and the emission of beta particles (electrons), it does not in general involve fission of the nucleus.

There is one interesting exception to this generality. In Gabon, in West Africa, there is a rich deposit of uranium.[7] About 1.7 billion years ago, in the presence of ground water as a moderator, this became a low-powered, natural reactor, though its total output was trivial by comparison with that of a man-made power station. When the earth was formed, the ratio of uranium-235 to uranium-238 was much greater than it is today; in today's language we should describe it as '30 per cent enriched'. Because uranium-235 decays faster than uranium-238, the enrichment had fallen to 3 per cent when the West African natural reactor came into operation. (Modern light water reactors employ a similar enrichment, but because the naturally occurring enrichment is today only 0.7 per cent we have to use gaseous diffusion plants or centrifuges to obtain a 3 per cent enrichment.)

In a *fission reactor* the nuclei of uranium or thorium are broken up into two approximately equal pieces, thus releasing energy. Uranium-235 can be induced to break up into two pieces of between 70 and 160 mass units. One break-up is as follows:

$$\ _{0}^{1}n + \ _{92}^{235}U \rightarrow \ _{92}^{236}U \rightarrow \ _{56}^{141}Ba + \ _{36}^{92}Kr + 3\ _{0}^{1}n + 2.8 \times 10^{-11}\ \text{J} \qquad (5.6)$$

The last term in this expression is the energy released by the fission. It is some 40 million times greater than the energy released in chemical combustion. The important fact attaching to expression (5.6) is that nuclear energy is released when fission products are created, in this case barium and krypton.

Expression (5.6) shows the uranium nucleus absorbing a neutron, and the two fission products being accompanied by three neutrons. In a fission reactor one of these three emitted neutrons is absorbed by another uranium-235 nucleus, and by a sequence of such absorptions energy is continually released. The energy itself appears as heat, first in the reactor fuel, then (by thermal conduction) in the less-hot reactor coolant, then, again by conduction, in the steam fed to the turbine.

Fission of the uranium-235 nucleus gives rise to *prompt neutrons* from the fission event itself. Their rate of emission is far too rapid to be followed by the machinery of a reactor. A nuclear reactor is controllable because the fission products emit *delayed neutrons*, on a timescale of seconds or minutes. The delay occurs when a fission product emits a beta particle, for example bromine-87 decays by beta emission to krypton-87, with a half-life of 55 seconds. The krypton then promptly emits a neutron to become krypton-86. Strictly, it is the beta particle that is delayed but the useful result is that a time delay is built into the neutron chain reaction. Control rods can be driven into a reactor, or removed from it in a time that is much less than this intrinsic delay. Hence it is possible to control the chain reaction. By contrast, the chain reaction of a nuclear bomb is sustained by the prompt neutrons alone, and for this reason the chain reaction is rapid; it is an explosion. Nuclear reactors avoid this condition of 'prompt criticality'.

Reactor Types and Systems

If one of the three neutrons shown on the right-hand side of expression (5.6) is absorbed by a U-235 nucleus, resulting in fission of that nucleus, a chain reaction can be sustained. There are two basic ways of ensuring the absorption of a neutron, and these lead to two distinct types of reactor. The first way is to make the neutron

travel at such a speed that the probability of absorption by a U-235 nucleus is very
high. This means reducing the speed of the fission neutron to a speed similar to that
of molecules of hydrogen in a gas at ordinary temperatures. The reduction is
achieved by bouncing the neutron, many times, against nuclei of similar mass, which
do not absorb the neutron. The reduction of speed is known as 'moderation' and
the three preferred moderators are the deuterium nuclei which occur in 'heavy
water', the hydrogen nuclei which occur in ordinary water, and the nuclei of carbon
atoms, of graphite. Reactors that moderate the fission neutrons are termed 'thermal
reactors', not because of the thermal energy released by the fission process but
because the neutrons are in thermal equilibrium with (that is, are in effect at the
same temperature as) the heavy water, light water or graphite.

Fast (Neutron) Reactors as Breeders

The second way of ensuring a chain reaction is to allow the neutrons to travel at
high speed but to place a large number of fissile nuclei in their path, so that the
necessary fissions occur. A reactor in which the fission neutrons are not moderated
but travel at high speed is a 'fast (neutron) reactor'. The term does *not* imply that
the mechanical parts of such a reactor operate more quickly than the mechanical
parts of a 'thermal reactor'. Both types of reactors utilise the 'delayed neutrons'.
Unlike the thermal reactor, a fast reactor must use highly enriched uranium in its
core (usually mixed with plutonium) simply so that the neutrons collide with U-235
nuclei (which are fissile) rather than with the U-238 nuclei (which are not). However, the core of a fast reactor is always surrounded by a blanket of U-238 and this
blanket, as the term implies, absorbs neutrons and breeds plutonium-239 which is
itself fissile:

$$\, _0^1 n + \, _{92}^{238}U \rightarrow \, _{92}^{239}U \rightarrow \, _{93}^{239}Np + \beta \tag{5.7}$$

$$\, _{93}^{239}Np \rightarrow \, _{94}^{239}Pu + \beta \tag{5.8}$$

Expression (5.6) shows three neutrons emitted by the fission process. One of
these is needed to sustain the chain reaction. If one of the remaining two is used to
produce the reaction shown in expressions (5.7) and (5.8) the reactor will 'breed';
for every fission of a U-235 nucleus (as in expression (5.6)) a Pu-239 will be created.
This forms the *fast breeder reactor*. Technologically it is difficult to ensure that
neutrons are not lost from the system by absorption in parasitic materials such as
the steel structure of the reactor or the steel cladding of the fuel pins. Nor does
every fission produce three fission neutrons, as in expression (5.6). The fission of
Pu-239 by fast neutrons yields, on average, 2.97 neutrons, a comparatively large
number, and this is why plutonium is so useful for a fast breeder reactor. The fission
of U-235 by fast neutrons yields only 2.51 fission neutrons on average and the
fission of U-235 by thermalised neutrons (in a thermal reactor) yields 2.47. In
practice it is impossible to build a thermal breeder reactor using only U-235 as fuel;
the figure of 2.47 is too low.

Sometimes, when a Pu-239 nucleus has absorbed a neutron it does not undergo
fission but simply becomes the heavier isotope Pu-240. This process is called

neutron capture and clearly it decreases the efficiency of the chain reaction. When Pu-239 is bombarded by fast neutrons fission fails to occur in some 15 per cent of the cases. When Pu-239 is bombarded by thermalised neutrons fission fails to occur in some 27 per cent of the cases and it is for this reason that a thermal reactor fuelled with Pu-239 does not breed more fuel than it consumes.

The breeding reaction is very important in the context of energy resources. Only 0.7 per cent of naturally occurring uranium consists of the fissile isotope U-235. In all reactors, the fuel elements are withdrawn before all the fissile material is consumed. Therefore, if the fuel is *not* reprocessed, but simply discarded as waste (this is Canadian and U.S. practice with the CANDU and Pressurised Water Reactor (PWR) systems) something less than this 0.7 per cent of fissile component will have been used. In the thermal reactor systems *some* U-238 is converted to Pu-239 and some of this Pu-239 undergoes fission. On a once-through system of operation, less than 1 per cent of naturally occurring uranium is consumed while 99 per cent is discarded. The breeding reaction increases the utilisation of naturally occurring uranium by a factor of 100, from, say, 0.6 to 60 per cent. Thus the breeding reaction is important given that the scarcity of naturally occurring uranium is a dominant consideration. If other considerations (such as the comparatively high cost of fast reactors or the considerable difficulties of reprocessing highly irradiated fast reactor fuels) weigh more heavily then, of course, the breeding reaction is less important.

Thermal reactor systems such as CANDU, PWR, the Magnox reactors and AGR (Advanced Gas Reactors) do not breed more fuel than they consume but they do 'convert' U-238 to the fissile isotope Pu-239. To take advantage of conversion, as of breeding, fuel must be reprocessed. Conversion occurs when the breeding ratio b is less than unity. During the first fuel cycle the destruction of N fissile U-235 atoms leads to the conversion of bN fertile U-238 atoms. After reprocessing, these bN (now fissile) atoms undergo a second irradiation in the reactor, so leading to the conversion of a further $b^2 N$ fertile atoms. By repeated fuel cycles the total consumption of fissile and fertile uranium atoms is

$$N + bN + b^2 N + b^3 N \ldots = N/(1 - b) \tag{5.9}$$

Hence if $b = 0.9$, 7 per cent of the uranium may be consumed, rather than the fissile fraction, the 0.7 per cent of U-235. Of course, if b exceeds unity the reactor can breed additional fuel. It is clear that, if uranium is a scarce and valuable resource, and reprocessing is desirable, 'conversion' is valuable and 'breeding' even more so.

Thus reactors may be divided into two categories: those in which the chain reaction is sustained by slowing the neutrons down so that they have a very high probability of being absorbed by a fissile nucleus (thermal reactors); and those in which the neutrons are not slowed down but in which, instead, the fissile nuclei are packed closely together and from the core of which non-fissile nuclei are substantially absent (fast reactors). The attraction of fast reactors is that they enable a large fraction of the naturally occurring U-238 to be consumed.

Fast Reactors

(a) *The core*. U-235 has a fission cross-section for fast neutrons of only 1.6 barn
(that is, the nucleus appears small to the fast neutron). Pu-239 has a similar fission
cross-section, 1.8 barn. By contrast, these two nuclei have cross-sections of 580
barn and 750 barn for thermalised neutrons. This difference means that the core of
a fast reactor must be made much smaller, more compact, than that of a thermal
reactor. A fast reactor core is typically a cylinder of 1 m diameter and 1 m high.
Within this cylinder the fuel is divided into many cylindrical pins, of small diameter.
This division provides a large surface area to the fuel and it is from the surface of
the pins that the fission heat is removed by a fluid, the reactor coolant. There have
been plans for gas-cooled fast reactors but the preferred reactor coolant is liquid
sodium. It has a high heat capacity and unlike a gas or a steam–water mixture it
does not have to be pressurised. It can remove heat from the core at a sufficient rate.
     The fuel pins themselves consist of stainless steel tubes a few mm in diameter,
filled with pellets of the mixed oxides of uranium and plutonium. The tubes are
sealed at each end. The outer surface of each tube is slightly ribbed to make the
flow of the sodium turbulent and so to improve the efficiency of heat removal. For
convenience of handling, the reactor core is sub-divided into sub-assemblies, each of
these being a bundle of fuel pins inside a larger steel tube of hexagonal cross-section
(open at each end to allow the free flow of sodium along the pins). The hexagonal
tubes are stacked together and form the reactor core.
     Because the flux of neutrons is very intense, the materials of the fast reactor
core behave in ways that are not encountered elsewhere, not even in a thermal
reactor (in which the neutron flux is less intense). The hexagonal tubes gently
deform under the pressure of the sodium being pushed through them. Under this
pressure they become circular rather than hexagonal and hence they jam together.
Within the steel of the hexagonal tubes and fuel pins microscopic voids are created,
that is to say the steel swells. The swelling is greatest where the neutron flux is
greatest, that is, at the centre of the core. The swelling is less at the sides of the
core. As a consequence the fuel pins and hexagonal tubes, all initially straight,
become banana-shaped. This disturbs the reactor physics of the core and the heat
transfer to the liquid sodium, and it can make the removal of a sub-assembly from
the core difficult. Some fast reactor cores are restrained with steel bands, to prevent
such deformation occurring during operation.
     Although the fuel pins are of small diameter they become very hot at their
centres. The gaseous fission products are released from the uranium–plutonium
oxide and a space is provided at the end of each pin for these gases to accumulate.
     The reactor core itself is surrounded by the 'blanket' of fertile material,
uranium (238) oxide similarly contained in fuel pins grouped into sub-assemblies.
     The power of the reactor is controlled (and the reactor is switched on and off)
by moving rods of boron steel into or out of the core. The isotope boron-10 is a
reactor poison: it has a high capture cross-section for neutrons and so can stop the
chain reaction.

(b) *The heat exchangers*. Liquid sodium can rapidly remove a large quantity of heat
from the small volume of the reactor core. Unfortunately, it is also a highly corro-

sive liquid metal, and as well as transporting heat around the reactor circuit it also transports some of the alloying elements specially put into the steels from which the reactor is built. This can lead to a degradation of various reactor components. Sodium must also be kept away from the steam of the turbines. For these reasons a fast reactor has two heat exchangers. The first is a sodium to sodium heat exchanger, the function of which is to isolate the sodium that flows through the reactor core. The second is a sodium to water heat exchanger, the function of which is to deliver energy to the turbines. A heat exchanger is in essence a membrane through which heat may flow but through which the heat-transporting fluids may not flow. Should the membrane in the first heat exchanger fail, sodium mixes with sodium and there is no damage to the reactor core. Should the second heat exchanger fail the sodium of the primary circuit is unaffected, the damage is restricted to a sodium–water reaction in a region remote from the reactor core.

Figure 5.2 shows the general layout of the British Prototype Fast Reactor at Dounreay in Caithness. The core and the primary sodium pump are immersed in a pool of sodium. Heat is removed by the intermediate (or first) heat exchanger. There is a convention, common to all power stations, which regards the reactor itself and the several pieces of new technology unique to nuclear systems as 'the nuclear island' and thus distinct from the older technology. In the fast reactor system the heat exchangers are properly regarded as part of the new technology. Their successful operation requires close control of the impurities in the sodium, particularly oxygen and carbon.

Thermal Reactors

To date (1984), all commercial reactor systems have been thermal reactors, that is, all contain a 'moderator' to slow the neutrons down to a velocity characteristic of the thermal motion of molecules. Within this general principle the particular designs are very diverse and, in their origins, owe as much to accidents of history as to conscious decision. During the 1939–45 war, heavy water from Norway was taken via France and Britain to the comparative safety of Canada, and all research using heavy water was conducted there. Forty years later Canadian nuclear power is based exclusively on heavy water moderation. At the conclusion of the war, scientists returning to Britain from their work in the U.S.A. on the atomic bombs (plural because the Manhattan project produced two distinct bombs, the uranium bomb and the plutonium bomb) were familiar with the graphite-moderated reactors first built by Fermi and later built on a larger scale at Hanford to produce plutonium for the bomb project. Forty years later British nuclear power is based exclusively on graphite moderation, though this will change if a PWR is built at Sizewell. Also after the war, the U.S. Navy developed a power reactor for its submarines using light water as a moderator. A few years later, as part of the race into civil nuclear power against the U.S.S.R., a submarine reactor was built on land, at Shippingport, the U.S.A.'s first nuclear power station. Today, 30 years later, the U.S. nuclear-power programme is almost exclusively based on light water reactors, the majority of them being of the pressurised type used in submarines. France initially followed Britain

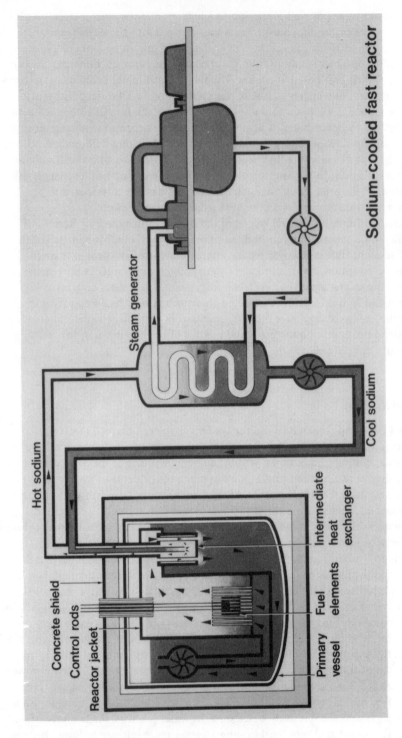

Figure 5.2 Schematic diagram of the prototype fast reactor (PFR)
(source: U.K. Atomic Energy Authority)

in choosing graphite-moderated systems but has recently switched to a large programme of light water reactors. Both France and West Germany have sought to cut corners, by devoting their efforts to second generation, rather than first generation, reactors. France and its collaborating nations now lead the world in fast reactor development. West Germany has developed a novel form of high-temperature, gas-cooled, graphite-moderated reactor, the 'pebble-bed' reactor. Britain and its Euratom partners have attempted to reach the same conditions as prevail in the pebble-bed reactor by developing the form of earlier British reactors. This has led to the 'Dragon' reactor at Winfrith in Dorset which, a few years ago, was thought of as the next commercial British system. The Americans have built a power station at Fort St Vrain, comparable to the Dragon reactor. The British have also built a prototype, the Steam Generating Heavy Water reactor, comparable to the Candian CANDU reactor.

In summary, when confidence was high, each of several thermal reactor types was separately developed with little regard to possible alternatives. As confidence has waned and with the exception of Canada which has frugally built only one type of reactor, the several nations have diversified their efforts.

Graphite-moderated Thermal Reactors

The first graphite-moderated reactors were cooled by air which was blown through cylindrical channels in the graphite bricks. These reactors operated close to ambient temperatures and did not produce electricity. To produce electricity, using a steam turbine and generator, a reactor must run as hot as possible. The higher the temperature the greater the thermodynamic efficiency, that is, the less the heat that must be rejected to the atmosphere, a nearby river, or the sea. However, air corrodes graphite at high temperatures, the graphite being converted to carbon dioxide. Carbon dioxide does not corrode graphite and it is comparatively cheap. Hence the first and second generations of graphite-moderated reactors (in the U.K., the 'Magnox' stations at Berkeley, Bradwell, Hunterston, Transfynnedd, Sizewell, Dungeness, Oldbury, Hinkley and Wylfa and the Advanced Gas Cooled Reactors at Dungeness, Hinkley, Hunterston, Heysham, Hartlepool and Torness) use carbon dioxide as a coolant.

The Magnox reactors take their name from the magnesium–aluminium alloy that is used to isolate the metallic uranium fuel from the coolant gas. An important property of this alloy is that it absorbs neutrons only slightly. It is therefore possible to build a reactor using only natural uranium as fuel, even though this contains only 0.7 per cent of the fissile isotope U-235. Because the fuel is clad in magnesium alloy rather than, say, stainless steel, it is unnecessary to artificially increase the proportion of the fissile isotope by either the expensive process of gaseous diffusion or, nowadays, by the less-expensive process of centrifuge enrichment. The magnesium cladding serves to contain the fission products, both solid and gaseous, which would otherwise escape from the metallic uranium fuel and contaminate the reactor. A great disadvantage of magnesium alloys is that they cannot operate at high temperatures. In consequence the outlet temperature of the carbon dioxide coolant cannot be more than about 400°C. This temperature is somewhat lower than that in a modern coal-fired power station and in consequence the Magnox power stations cannot take advantage of modern turbine engineering. Their thermodynamic efficiency is 30 per cent or less, that is to say between two-thirds and three-quarters of the nuclear energy generated in the reactor must of necessity be dumped, as low-grade heat, either to the atmosphere, to a river or to the sea — it cannot be used to generate electricity.

Because of this limitation on temperature (and hence on the usefulness of the energy released by the reactor) the second and third generations of graphite-moderated reactors have sought, above all, to raise the outlet temperature of the gas. Second-generation reactors (the AGRs) have raised the outlet temperature to 650°C. Third-generation reactors (the High Temperature Gas Cooled Reactors such as Dragon, Fort St Vrain and the West German AVR) have raised the outlet temperature to 750°C and above.

When the temperature is raised a cladding material other than magnox must be used. The metal beryllium would be an excellent cladding material if it were not so difficult to fabricate. Perforce, the AGRs use fuel clad in stainless steel. This absorbs more neutrons than does magnox and so it is necessary to enrich the uranium from 0.7 per cent U-235 to 2.5 per cent U-235.

Because of the temperature at which the Advanced Gas Cooled Reactors operate, a coolant gas that does not corrode the graphite tends to deposit soot on the stainless steel fuel cans, impeding the flow of heat from the fuel inside and making the fuel run hot. Conversely, a coolant gas that does not deposit soot on the fuel cans tends to corrode the graphite bricks which should be a permanent part of the reactor structure. Consequently the graphite takes on a very porous structure (similar to coke) and no longer fulfils its moderator role adequately. As a compromise the coolant gas of these reactors consists of a carefully proportioned mixture of carbon dioxide, carbon monoxide and methane.

Several of the metal components in this reactor system operate close to their natural limits and it is for this reason that the third generation of graphite-moderated reactors are very different from the first and second. In third-generation systems, the High Temperature Reactors (HTR), the carbon dioxide coolant is replaced by the more expensive but chemically inert helium. Metal components, so far as is possible, are avoided. The fuel itself is uranium dioxide (as in the second-generation reactors) but it is now clad in concentric layers of graphite and silicon carbide rather than in a metal tube. Because of this concentric construction each fuel element is now a short cylinder rather than a long one (in the British and American designs) and these short cylinders are stacked end to end. The West German design takes the concentric form to its logical conclusion and makes each fuel element a 2-inch diameter sphere, with the active uranium oxide at the centre.

The High Temperature Gas Cooled Reactor has suffered a sudden eclipse. In addition to its high thermodynamic efficiency for electricity generation (45 per cent rather than the 25 per cent of the Magnox stations) its high temperature of operation offered the prospect of using the heat directly in industrial processes. This would mean an efficiency approaching 100 per cent. However, there is now a wide choice of thermal reactor systems and, since the completion of the Magnox reactor programme, consideration has been given to the carbon dioxide-cooled AGR, the helium-cooled HTR, the light water-cooled, heavy water-moderated SGHWR, and the pressurised light water reactor, with some interest also being shown in the boiling light water reactor and the Canadian heavy water-cooled, heavy water-moderated reactor, CANDU.

The CANDU Reactor

Until recently[8] the Canadian system had the enviable reputation of the world's most reliable reactor. Like the Magnox reactors the CANDU system uses natural uranium and avoids the uranium-enrichment process. It uses the best moderator available: heavy water. Heavy water is better than light (or ordinary) water because it absorbs fewer neutrons. The mass of the deuterium nucleus is comparable to that of the neutron, as required for efficient moderation. The difficulty with the system is that heavy water is costly and requires a complex plant for its manufacture. The Canadians have two such plants.

Most of the reactors that require the coolant to operate under pressure put the entire core in a pressure vessel. The early Magnox reactors use steel pressure vessels, the later ones use pre-stressed concrete. The CANDU system is very different: the pressure vessel consists of several hundred tubes that thread the reactor core. Thus the diameter of the pressure vessel at any point is closer to 20 cm rather than to 20 m, and this smaller diameter means that the pressure is more easily, and safely, contained. Whereas the steel vessel of a Magnox reactor is some 10 cm thick, and that of a PWR is some 20 cm thick, the zirconium alloy of the CANDU pressure vessel is only 0.5 cm thick. If either of the larger vessels contains a crack several inches long the vessel may fail suddenly when pressurised. The virtue of the CANDU system is that such a crack will leak conspicuously before it breaks, which gives warning of failure. The *New Scientist* has reported the occurrence of a crack 2 m long in a CANDU pressure tube[8] – a sign of the intrinsic safety of the system is that the pressure tube did not explode!

The pressure tubes themselves are each concentric with and within a tube of the calandria, the vessel that holds the (unpressurised) heavy water moderator. The moderator is comparatively cold ($50°C$) and the coolant flowing through the pressure tubes is, of course, hot ($290°C$). The annular space between the pressure tube and the calandria tube therefore contains only an inert gas, a poor thermal conductor. All the calandria tubes are welded, at their ends, into two end-plates, to form the *calandria vessel*. This vessel contains the moderator, the heavy water, between the calandria tubes.

Within the pressure tubes, with coolant flowing over them, are the fuel rods: uranium dioxide clad in zirconium alloy. (Like magnesium, zirconium absorbs neutrons only slightly.) The coolant circulates through a heat exchanger, as in most reactors, and the heavy water coolant does not enter the turbines.

Like the civil Magnox reactors in the U.K., the CANDU reactors are refuelled while in operation and so offer high availability. New fuel is inserted into one end of a pressure tube while, simultaneously, old fuel is removed from the other end.

The machines that carry out this operation are of considerable sophistication and complexity. Of necessity they are operated by remote control. For some years the CANDU and Magnox systems jostled for first place in the yearly records for time spent at full power.

Light Water Reactors: the BWR and the PWR

Light water reactors avoid the use of costly heavy water. Instead they accept the penalty of using enriched uranium. They were developed in the U.S.A., which had a large enrichment capacity as a result of its wartime weapons programme, and so this type of reactor could be operated more cheaply than competing plant built in other countries in the post-war years.

There are two basic types of light water reactor: reactors in which the water is allowed to boil, thus forming a mixture of water and steam, and reactors in which the pressure is sufficiently great to prevent boiling, in which the coolant remains in the liquid phase. The first type is the boiling water reactor, BWR. The second type is the pressurised water reactor, PWR. A BWR operates at a pressure of 70 atmospheres (in the primary coolant) and a PWR at 150 atmospheres. By contrast the coolant pressure in a Magnox reactor is only 20 atmospheres. The power density in a PWR is also approximately twice that in a BWR, 100 kW per litre instead of 50. It is obvious that the physical conditions in a PWR are more severe than in a BWR, for example, a BWR is designed so that water may turn into steam whereas the PWR is designed on the assumption that the water will *not* turn into steam, in the reactor core. The formation of steam in the highly rated core of a PWR is therefore a fault condition, which must be avoided in operation.

If the pressure vessels of the two reactor types were of the same size, one would expect the wall of the pressure vessel of the PWR to be twice as thick as that of a BWR, in order to contain the larger pressure. However, because the power density in a PWR is twice that in a BWR then, power for power, the core volume of the PWR is half that of a BWR. The diameter of the containing vessel is typically two-thirds that of the BWR, and the pressure vessel wall need only be some 30 per cent thicker, 20 cm instead of 15 cm. The fabrication of a steel pressure vessel with walls 20 cm thick and to a standard of quality that requires the complete absence of any crack more than a fraction of an inch long is a severe test of metallurgical engineering and of crack definition. (It is evident that any welded joint must be sound to a depth of at least 4 inches from either surface, and that the strength of the weld-metal must not be less than the strength of the plates being welded together.) In the U.K. an inspection and validation centre (IVC) is to be built at Risley, Cheshire, as a joint project between the Central Electricity Generating Board and the United Kingdom Atomic Energy Authority for the purpose of monitoring the construction and testing of PWR pressure vessels.[9]

Both types of light water reactor suffer the penalty that light water (ordinary water) is an absorber of neutrons. In compensation it is necessary to enrich the uranium fuel from its natural content of 0.7 per cent of U-235 to a content of 2 per cent U-235 in the BWR and 3 per cent U-235 in the PWR. This may seem a very modest increase in comparison to the enrichment from 0.7 to 90 per cent that is required for nuclear weapons. It is therefore important to stress that by their nature the enrichment processes now in use (gaseous diffusion and centrifuge enrichment) consume a lot of energy in the initial stages of enrichment. It takes only twice the amount of energy to enrich from 0.7 to 90 per cent as from 0.7 to 3 per cent. The necessary enrichment of a large quantity of uranium fuel (100 tons per reactor charge in a PWR, 170 tons per reactor charge in a BWR) is therefore a significant factor in light water systems. Neither system can be refuelled while the reactor is generating electricity (that is, 'on load'). Thus a BWR or PWR is 'down', or out of use, for the time required to remove an old reactor core and instal a new one.

In a reactor designed to drive a submarine it is a prime requirement that the power is packed into a small volume. It is a secondary consideration that the several reactor components are consequently highly rated (that is, working close to the limit of their capability) and may require expensive replacement. Equally it is a secondary consideration that the reactor is replenished with fuel while the submarine is in harbour and the reactor is not working. The initial and basic criteria for submarine reactors are therefore noticeably different from those of a civil nuclear power station.

In both BWR and PWR the uranium fuel is in the ceramic form, uranium dioxide, and capable of sustaining, without difficulty, internal temperatures at the centre of each fuel pin of greater than $1000°C$. Cylindrical pellets of UO_2 are stacked inside a fuel can of zircaloy, the same alloy of zirconium as is used in the CANDU reactor. These fuel pins, several metres long, are stacked vertically in the conventional cylindrical array with coolant flowing between them.

In a BWR there is no heat exchanger. Steam from the reactor core is fed direct to the turbines. The absence of a heat exchanger reduces the capital cost of the system but allows any radioactivity released in the core (from a failed fuel element) to enter the turbine, with consequent limitations on turbine maintenance.

The PWR does have heat exchangers. The primary coolant, at 150 atmospheres pressure but at the comparatively modest outlet temperature of $320°C$ (compare the $400°C$ of the Magnox system, the $650°C$ of AGR and the $750°C$ of HTR) flows through a bank of small-diameter pipes immersed in the secondary coolant, water at a lower pressure, which feeds steam to the turbines. So that the heat may be efficiently transferred from the primary to the secondary coolant the pipes have a large surface area (there are many pipes of small diameter rather than a few pipes of large diameter) and the wall thickness of these pipes is as small as possible. The detailed design of PWR heat exchangers has been subject to revision throughout the history of the system. Many operating PWRs have had their heat exchangers replaced and it is known that replacement will be necessary in others.

It is evident that if the primary pressure circuit of a PWR were to be breached the pressure would rapidly fall from 150 atmospheres to a pressure at which (at a temperature of $320°C$) steam must be formed; the coolant would be a two-phase coolant rather than a single-phase coolant. The highly rated fuel pins would then overheat and perhaps melt. The problem of poor thermal conduction in the two-phase coolant would be compounded by the behaviour of the zircaloy fuel cans. These cans contain the fission gases, under pressure. Under normal operation all is well because, though the zircaloy is hot, and in consequence soft, the liquid coolant exerts a pressure on the outside of the can. If this external pressure is lost, the fuel cans may 'balloon' under the action of the internal gas pressure. The effect is very comparable to inflating, by mouth, a long toy balloon. If a fuel pin 'balloons', the flow of coolant is reduced, and the pin becomes hotter. The condition is clearly self-aggravating. It is thus imperative, as a safety measure, to provide an 'emergency core cooling system' (ECCS) to prevent an unwelcome rise in fuel temperature in the event of a 'loss of coolant accident' (LOCA). A LOCA is often envisaged as arising from a break in one of the pipes carrying the coolant from the core. It can also arise, as it did in the accident at Three Mile Island, by the failure of a valve to close.[10,11] The 1979 accident occurred five years *after* a safety assessment of light water reactors which cost $3 million and culminated in two long reports, WASH-1250[12] and WASH-1400.[13] This extensive study failed to identify the fault that actually occurred at Three Mile Island, perhaps because of a rational human tendency to set aside those possibilities that appear, from the committee room, to be irrational. It is irrational for a candle to set fire to the world's largest nuclear power station and put it out of action, yet this is what happened to the BWR at Brown's Ferry, Alabama, in 1975.[10]

Even the rational and admitted problems of emergency core cooling in PWRs continue to be a subject of dialogue in the U.K., between The Nuclear Installations Inspectorate and the designers of the proposed PWR at Sizewell, on the Suffolk coast. If the control rods can be driven into the reactor and if the fuel pins can be kept wet, all is well; it is possible to remove the 'decay heat' (the power delivered by the fission products and which cannot be shut down). If however the fuel rods become dry, it is difficult to wet them again; water striking them is immediately turned to steam and this, because of its large volume, blows out the remaining water. The prime function of the ECCS in a reactor is to ensure that the core stays wet. Patterson[14] calls the emergency core cooling systems 'easily the most controversial feature of the PWR' and the present technical debate in the U.K. seems to bear out that judgement.

Table 5.1 summarises the important physical parameters of the major types of thermal reactors that are in commercial use or that have been considered or prescribed for commercial use.

The Nuclear Fuel Cycle

The word 'cycle' implies an operation that returns to its point of departure. The word is therefore a misnomer in regard to nuclear energy; there is not and cannot be a 'cycle'. It is a *sine qua non* of fission reactors that there must be fission products; the nuclear energy is derived by re-arranging the protons and neutrons of one heavy nucleus into two less heavy nuclei. The process is non-cyclic. The same is true of fusion, though the intrinsic fusion products are less troublesome

Table 5.1 *Parameters of thermal reactor types*

Reactor	Magnox	AGR	HTR	CANDU	SGHWR	BWR	PWR
Country of origin	Britain, France	Britain	EEC, U.S.A.	Canada	Britain	U.S.A.	U.S.A.
Type of uranium fuel	metal	dioxide	dioxide or carbide	dioxide	dioxide	dioxide	dioxide
Enrichment	natural	2%	93%	natural	2%	2%	3%
Cladding	magnox	stainless steel	silicon carbide and graphite	zircaloy or zirconium/niobium	zircaloy	zircaloy	zircaloy
Moderator	graphite	graphite	graphite	D_2O	D_2O	H_2O	H_2O
Coolant	CO_2	$CO_2/CO/CH_4$	He	D_2O	H_2O	H_2O	H_2O
Pressure vessel	welded steel or concrete	pre-stressed concrete	steel or pre-stressed concrete	zircaloy and steel	zircaloy and steel	welded steel	welded steel
Pressure-vessel wall thickness	0.10 m (steel) 6 m (concrete)	5 m	0.05 m (steel)	0.005 m (zircaloy)	0.005 m (zircaloy)	0.16 m	0.20 m
Coolant pressure (atmospheres)	20	40	20/50	85	67	70	150
Coolant outlet temperatures	410°C	650°C	750°C	300°C	280°C	290°C	320°C
Power density (kW per litre)	1.1	4.5	8	16	11	50	100
Mass of fuel (tons)	240/600	114	2/16	93	22	170	100
Electrical output (MWe)	330/1200	660	42/342	500	100	1065	1050
Thermodynamic efficiency	25/30%	35%	40%	29%	28%	32%	32%
Core dimensions (height (m) × diameter (m))	7.4 × 13.1 to 9.1 × 17.4	8.3 × 9.1	1.6 × 1.0 to 4.75 × 5.9	5.9 × 6.4	3.66 × 3.12	3.7 × 4.8	3.6 × 3.35
Refuelling on or off load	on	?	off	on	off	off	off
Fuel burn-up (gigawatt-days per tonne)	5	22	32/100	7	21	30	33

than fission products. Fusion is non-cyclic in that two nuclei are re-arranged into one nucleus. It is at present only necessary to consider fission in any detail.

We may start with the fission products since these are unavoidable. Fission of the three heavy metals, uranium, thorium and plutonium produces a group of elements in the middle of the periodic table, which have mass numbers between 70 and 170. The fission fragments have two predominant masses, clustered around mass numbers 95 and 140.[6] These fission products are in general radioactive and many decay by emitting a succession of electrons, as in these two typical examples:

$$\underset{36}{^{92}}Kr \xrightarrow[3\,s]{} \underset{37}{^{92}}Rb \xrightarrow[5\,s]{} \underset{38}{^{92}}Sr \xrightarrow[2.7\,h]{} \underset{39}{^{92}}Y \xrightarrow[3.53\,h]{} \underset{40}{^{92}}Zr$$

$$\underset{56}{^{141}}Ba \xrightarrow[18\,min]{} \underset{57}{^{141}}La \xrightarrow[3.9\,h]{} \underset{58}{^{141}}Ce \xrightarrow[32\,day]{} \underset{59}{^{141}}Pr$$

The half-life of each transition is indicated. It is evident that 99 per cent of any krypton-92 that is created in a reactor (or that exists in any subsequent time, such as at the instant of release to the atmosphere in an accident) will have decayed to stable zirconium within a day. By contrast, 99 per cent of any barium-141 will take six months to decay to stable praseodymium. Strontium-90 and caesium-137 each have half-lives of about 30 years, and thus 99 per cent of any strontium-90 or caesium-137 that may be present at any time takes about 150 years to decay, leaving 1 per cent of the original activity. This 1 per cent may or may not be hazardous; the hazard is in proportion to the original intensity. The intensity of activity in the fission product wastes from a reactor is such that the wastes remain hazardous for roughly a thousand years, by which time their activity is comparable to the uranium ore that was mined to produce the reactor fuel.[1]

A reactor also produces radioactive actinides as a waste product. These are nuclei of the heavy elements uranium, thorium or plutonium that have *not* undergone fission (a small fraction do not break up) but which have simply added one or several neutrons to their nuclei. For example

$$\underset{94}{^{239}}Pu + 4\,\underset{0}{^{1}}n \rightarrow \underset{94}{^{243}}Pu \rightarrow \underset{95}{^{243}}Am + \beta$$

The americium-243 which is formed by the emission of a particle from plutonium-243 has a half-life of 7400 years and is an alpha particle emitter:

$$\underset{95}{^{243}}Am \xrightarrow[7400\,years]{} \underset{93}{^{239}}Np + \alpha$$

$$\underset{93}{^{239}}Np \xrightarrow[2.3\,days]{} \underset{94}{^{239}}Pu + \beta$$

Thus americium-243 will decay to 1 per cent of any given level in roughly 40,000 years. It is the presence of the actinides that makes reactor wastes hazardous for 'tens or hundreds of thousands of years' rather than for a few hundred.[1]

Part of the hazard from the actinides is that they emit particles and that these produce very intense damage over a short distance, in flesh. It is not an adequate compensation that the long half-life means that the rate of emission of alpha particles is low and that a human being can only absorb a fraction of a per cent of the alpha particles emitted. An actinide particle breathed into the lung, as dust, is in immediate contact with the tissue in which cancer may be induced. It is for this reason that the actinide dusts recently found in houses near Windscale (now re-named Sellafield) are a cause for concern. The timescale of the damage is reduced from a lifetime to a minute, by the act of inhalation.[12]

The term 'waste management' implies a system that does not allow actinides to blow around the landscape as dust, nor allow them to enter the food chain (where they are almost invariably increased in concentration by the species carrying them) and, after one or more steps of biological concentration, to enter the human tissue.

Waste Storage

When a reactor is in operation, up to 10 per cent of its heat output may come from the fission products, and this heat cannot be switched off by switching the reactor off. (It is the presence of fission product heating that makes it imperative to main-tain the flow of coolant through a reactor core in any accident.) This same fission heat poses a problem for spent-fuel transport and storage, and for the storing of reprocessed waste. The transport flasks must dissipate the heat and the storage pond or dry store must also dissipate the heat.

If the fission products are made into a solid, such as a glass block, the block must dissipate the fission heat without cracking or melting and it must be placed in an environment from which the heat can escape. Such a glass block must also resist the leaching action of any water in which it might be immersed over the next few thousand years. Blocks of borosilicate glass interred about 500 m deep in suitable geological structures and containing about 25 per cent of high-level waste that has already had a 10-year decay since removal from the reactor would still reach the unacceptable temperature of about $300°C$, at which they would disintegrate rapidly upon contact with ground water.[15] The crystal structure of naturally occurring minerals, on the other hand, is more stable than that of glass (glass is essentially a supercooled liquid, without a regular crystal lattice) and the crystal structures of minerals can accept radioactive elements into their lattices. Some effort is therefore being devoted to the production of synthetic rock, using the actinide and fission product wastes as one of the raw materials.[15] Such rock is between 1000 and 10,000 times more resistant to leaching than is borosilicate glass.

Prior to solidification, radioactive wastes are stored as liquids, in stainless steel tanks. The history of leakage from such tanks, at Hanford in the U.S.A. and Windscale in England makes plain that storage in liquid form is no more than a temporary expedient.

Liquid wastes arise by the act of reprocessing spent fuel elements, and here there are two schools of thought and practice. One holds that the present world price of uranium is so low that it makes more sense to buy new uranium to recharge a reac-tor than to reprocess (or recycle) the old fuel elements. It is easier and cheaper to store the *already* solid fuel elements than to go through an expensive recycling process at the end of which one still has to store the same waste products in a solid form. It is Canadian and U.S. commercial practice not to recycle spent fuel. The

repository for the spent fuel elements is known as a 'plutonium mine', since this
practice always allows the option of reprocessing at a later date.

A second school of thought advocates the reprocessing of fuel, particularly from
thermal reactor systems. In the words of the Flowers Report: 'The extraction of
plutonium still provides the main reason for reprocessing fuel: the element is
valuable as a source of energy either in fast reactors or as a substitute for uranium-
235 needed to enrich fuel in thermal reactors. There are other reasons.' One 'other
reason' is plutonium extraction for the manufacture of nuclear weapons. In the
U.K. the civil and military programmes were highly interdependent at the outset,
and the link between the two continues, in the reprocessing line and in the arrangements under which this reprocessing line operates. A relevant perspective is set by
the early history of civil nuclear power. In the words of the official history of the
Electricity Supply Industry,[16] 'a working party under a young Treasury Under-
Secretary, Burke Trend, considered whether there should be a civil nuclear power
programme to complement that required by the military for plutonium production.' The head of the electricity supply industry, Sir Walter Citrine, only received
a copy of the report of this committee some 3 months after it was written, and
'with the details of the military plutonium requirements excised.'[16] At that time
(1954) it was judged that civil nuclear power could be economic only 'on the
assumption that the plutonium could be sold for use in bombs or as fuel for a later
design of reactor.'[16] Britain's civil nuclear programme went ahead on this basis.
Quite arbitrarily, the value of this 'plutonium credit' was set at one-third of the
estimated cost of nuclear electricity;[16] it was anticipated that without such credit
nuclear electricity would be some 50 per cent more expensive than that from fossil-
fired plant. This genesis of the U.K. civil nuclear programme makes clear why
reprocessing was adopted in the U.K., and was not adopted, for example, for comparable reactors built in non-weapon states like Canada.[10,11] The second use of
plutonium, for a fast reactor programme, was, and remains, an unrealised possibility.
The slow progress being made with the reprocessing of oxide fuel from the AGR
programme (the THORP reprocessing facility which was the subject of the 1977
Windscale Inquiry) suggests that once the military pressure is removed the reprocessing option becomes less imperative.

Fuel Element Manufacture

Apart from the plutonium-bearing fuels, reactor fuel elements of either uranium
metal or uranium dioxide can be fabricated in comparatively unrestricted conditions.
High-quality control during manufacture is necessary to ensure that the fission
products are subsequently retained within the fuel cladding (the cans of magnox,
zircaloy or stainless steel) and do not contaminate the reactor circuit.

Enrichment Processes

The fuel for all reactors apart from the Magnox and CANDU must be 'enriched' and
the proportion of uranium-235 increased above the natural value of 0.7 per cent.
There are now many methods by which this may be achieved.[17] On a large scale the

technique of diffusing a gas through a fine membrane of sintered nickel powder (a method developed during the original atomic bomb project) and the technique of the high-speed gas centrifuge (a method abandoned during the original atomic bomb project[18]) are most common. A new technique, which promises to be of great importance in the present decade, is that of atomic vapour laser isotope separation, AVLIS.[17] It is important because of its comparative cheapness and for the ease with which it may be used to enrich plutonium. Plutonium presents difficulties to the conventional gas diffusion and centrifuge techniques in that its gaseous compounds are less stable than those of uranium. Plutonium hexafluoride decomposes while in the plant. It is at the enrichment stage that civil and military requirements once more become intertwined; the discarded fuel elements from the U.S. civil power programme of PWRs and BWRs contain some 70 tonnes of plutonium which is at present unsuitable for weapons. The AVLIS technique will enable this plutonium to be 'cleaned up' for weapons and the undesirable isotope Pu-240 to be extracted.

Uranium Mining

Uranium is mined from sandstone and similar rocks which typically contain 0.4 per cent of uranium metal. The ore is crushed to a powder and dissolved in chemical solvents from which an oxide, 'yellowcake', U_3O_8, is precipitated. Yellowcake is transported from mines in Canada, South Africa and Australia, by a variety of routes (some clandestine) to the plants that enrich it and convert it to fuel. Figure 5.3 illustrates the elements of the nuclear fuel cycle, showing the points of entry and exit, and the circulation.

Uranium mines are hazardous because of the presence of the radioactive gas radon-222. This is a decay product of uranium-238. Chemically, it is one of the noble gases (like helium). Being chemically inert it does not react with the rocks in which it is produced. Hence when these rocks are crushed it is liberated. Radon-222 is an alpha emitter and as such capable of severe biological damage to lung tissue, especially in conjunction with its daughter product polonium-218.[10] The U.S. Public Health Service has estimated that of some 6000 men who worked in the American underground uranium mines in the 1950s, between 600 and 1100 will die as a result of lung cancer contracted through radiation exposure in the mine.[10]

Safety and Risk Assessment

The high incident of lung cancer in uranium miners is a prominent safety aspect, and introduces two concepts, that of *assessment* and that of *evaluation*. Assessment involves giving risk figures (such as 1 in 10 workers in an uranium mine will probably die of lung cancer, as a result, before the age of, say, 50). Evaluation means putting a value on the numerical risk, in short, deciding whether or not to accept it. It is an aspect of choice. One may accept a risk even though it is large; or one may reject it even though it is small.

The distinction between assessment and evaluation lies at the root of debate

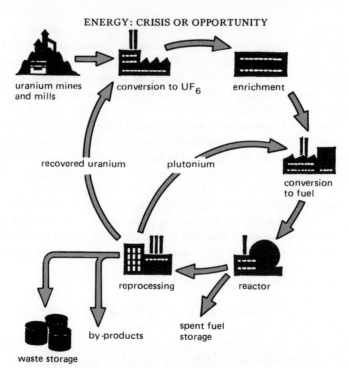

uranium mines
and mills

conversion to UF$_6$

enrichment

recovered uranium

plutonium

conversion
to fuel

reprocessing

reactor

by-products

spent fuel
storage

waste storage

Figure 5.3   *The nuclear fuel cycle* (source: Walter C. Patterson,
Nuclear Power (Pelican Books, 1976, p. 88).
Copyright © Walter C. Patterson, 1976.
Reprinted by permission of Penguin Books Ltd)

about the risk of nuclear power. The figures that can be produced as an assessment
of a risk often cover a broad range and are frequently arguable, but even when the
figure is low (by comparison with other known risks) it is a matter of evaluation as
to whether the risk should be run or avoided. The risk to a uranium miner is personal.
The risk from widespread contamination, say in the Irish Sea, is communal. Hence
nuclear power introduces questions of ethics as well as the particular numbers of
risk assessment.[19]

Historically, the trend of revisions in risk assessment is that a given risk proves to
be greater than was initially thought. For example, the uptake of plutonium by the
wall of the human gut is some five times greater for adults than previous estimates
have indicated, with an even higher factor for children during the first year of life.[20]
In consequence of this re-assessment, the discharge of plutonium and americium
into the Irish Sea from Windscale is to be reduced by a factor of 5.[21] The present
level of discharge of these isotopes (about 1000 curies per year) is a factor of 5 less
than the discharges that occurred in the mid-1970s.[21] Thus for several years in the
past decade the level of discharge from Windscale was some 25 times greater than
the level now considered safe. Since these isotopes are long lived (Flowers Report,[1]
Figures 9 and 10) it is clear that Windscale has released over a period of a few years
a quantity of isotopes that would more properly have been released over a period of
a century. Flowers under-estimated the risk; in only 7 years his cautious opinion has
been overtaken by events. For risks that are cumulative (such as exposure to

radiation) the young are more susceptible than the old for two reasons: the period of exposure is longer and, secondly, the very young organism is much more susceptible to damage than one that is fully grown. It is for this second reason that medical X-radiography of pregnant women is avoided. The effects of fall-out are not avoidable.[22,23]

Two decades ago it was common for textbooks to assert that an intensity of man-made radiation equal to the naturally occurring background radiation from cosmic rays and from radioactivity in the earth's crust produced no ill effect on the individual or the group. Graphs were drawn showing that 'radiation damage' began at a value of twice the natural background. This is the concept of the 'safe dose'. Experimentally, most of our information comes either from small populations irradiated to high dose or from large populations irradiated to low dose. ('High' in this context means a dose approaching that at which half of a population will rapidly die, or about 400 rem. 'Low' means either an intensity that is a few multiples of the background level of 0.1 rem per year, or a single dose of a few rem.)

The difficulty with the low-population/high-dose data is that one cannot sensibly interpolate to low doses without an understanding of the physics, chemistry and biology of the living organism. The difficulty with the high-population/low-dose data is that the magnitude of the dose is seldom well defined; the data themselves seldom arise from a designed and controlled experiment but more often from unplanned circumstances, such as the irradiation of servicemen close to weapons tests in the Nevada desert, in the Pacific or, of course, the incineration of Hiroshima and Nagasaki. The Japanese data show that a low dose of about 10 rem doubles the incidence of leukaemia in the population.[24] The most reasonable attitude to radiation damage, both genetic and somatic, would seem to be to assume that it is proportional to absorbed dose, until there is convincing evidence to the contrary.

The assessment of risk of failure of a reactor should be simpler. The difficulty is to define all the modes of failure. The method used is that of 'fault tree analysis'.[12,13] A postulated fault leads to the imposition of more severe conditions elsewhere in the system, with a recognised possibility of failure at these other points. One thus has a sequence of possibilities, often branching into two or more possibilities, like a tree. For each limb of this tree a risk assessment can be made and a number assigned. The probability of a sequence of events can therefore, in principle, be stated. It is difficult to recognise all the possibilities, rational and irrational. The history of reactor accidents[14,18] shows that events frequently mock the rational basis of forward calculation. The best estimate of risk to a reactor system is perhaps to be found in the premium an insurance company would charge to insure against a specified failure, of identifiable cost. Hannah remarks that Sir Christopher Hinton was pleased to see how low the British insurance firms set the premiums for the Magnox stations[16] and that nonetheless the insurance companies have found the risk a profitable one to cover. On the other hand, the Price–Anderson Act in the U.S.A. and similar legislation in the U.K. limits the maximum liability for an accident to a reactor to $560 million, and it is generally recognised that there are highly improbable accidents that could exceed this maximum.[14,18,19]

Within a reactor itself elementary steps are taken to ensure that failure of a component does not lead to an accident. For example, if the control rod drive mechanism fails, the control rods drop into the reactor and close it down (known as the 'fail-safe' technique.) Electric and electronic circuits to monitor reactor performance are often triplicated (the 'two out of three' system) so that if one circuit develops an instrumental fault the existence of the fault is immediately evident in the concordance of the two intact circuits. Further, there are now computer programmes to check the performance of the reactor as a whole, to ensure the consistency of one set of instruments with another.

The Nuclear-power Industry and the Economy

Each country has its own pattern, determined by governmental aspiration and the response of large-scale industry. For example, in the U.S.A. government research and development funding was disassociated from the commercial power programme at an early stage. In Britain both the CEGB and the UKAEA devote resources to research and development of current and future reactor systems, rather than leaving this wholly with the reactor vendor.

In the U.K. there were initially four reactor vendors, four groups of companies, the consortia, formed with government encouragement. In the initial flush of enthusiasm industry was over-stretched[17] and subsequently the number of consortia has been reduced until, in 1983, only the National Nuclear Corporation remains.

In the U.S.A. the General Electric Company was initially regarded as a greater industrial force than Westinghouse, but, via the submarine reactors and the subsequent land-based PWRs, Westinghouse has eclipsed General Electric and GE has withdrawn from the reactor business. Perhaps the dominant force in the U.S. nuclear industry is now the Bechtel Corporation (Mr Caspar Weinberger and Mr George Shultz are both Bechtel executives).

The number of organisations continues to decrease and a reactor system may now be built by a group of governments or by companies from several nations. Thus the fast reactors Phénix and Super Phénix, sited in France, are built by a group of several governments, with France and West Germany the major contributors. In the U.K. the proposed PWR at Sizewell is being handled jointly by Bechtel, the National Nuclear Corporation and the CEGB. A major component, the steel pressure vessel, will probably be made in France. It is evident from this brief description that the nuclear industry is moving from a national to a multi-national basis.

Regulating Authorities

Reactor systems are promoted by Governments, by utilities and by reactor vendors. The regulation of reactor systems falls to a government-established agency, in the U.K. the Nuclear Installations Inspectorate (NII) and in the U.S.A. the Nuclear Regulatory Commission (NRC). Their prime task is to separate the promotional and regulatory aspects. This is difficult in so far as a government regards itself as a promotor of nuclear energy. In the U.S.A. the NRC was formed from the previous Atomic Energy Commission so that the two functions should be more separate.

In the international sphere, the International Atomic Energy Agency (IAEA), with headquarters in Vienna, is charged with promoting the use of nuclear power, particularly in developing countries, and also with regulating its use, by ensuring that the military use of nuclear power, so evident in the U.S.A., the U.S.S.R., the U.K., France and China, is not imitated elsewhere. The Agency is funded by the larger governments, notably the two superpowers. The regulatory function of the IAEA depends on the willingness of any nation to be regulated; regulation is by consent rather than by constraint. Thus small nations with no aspirations to either nuclear weapons or nuclear power readily consent, larger nations less willingly. India and Brazil do not consent.

In similar fashion, the nuclear weapons states write their own rules for how they will be regulated by the IAEA. In the words of the IAEA annual report: 'the level of assurance associated with the Secretariat's findings for a particular installation or state depends – *inter alia* – on the content of the safeguards agreement concluded with the state in question. . .'[25] The rules written by the U.K., for itself, are contained in the U.K.-Euratom–IAEA Treaty[26] which is largely concerned with limiting the power of the IAEA to inspect U.K. facilities or receive information. It is because of these written limitations that U.K. Ministers and spokesmen can confidently declare that 'the IAEA has said it has no evidence to indicate the diversion of material from civil to military purposes in the U.K.'. By the terms of the treaty the significant information is withheld from the IAEA.[27]

The activities of terrorist groups are often cited as one of the putative dangers of nuclear power. In the writer's view this danger is commonly over-rated. The smaller the group the more formidable become the technical difficulties it faces, either of obtaining fissile material (enriched uranium or plutonium) or of fabricating it into an explosive device. A large group, such as a government-run industry (taking the worldwide context) presents a correspondingly greater danger, in that the technical facilities at its disposal are very large and immediately available, generally under conditions of considerable secrecy. The social control over these facilities remains extremely concentrated, in all countries.

Economics

The civil power programmes of the U.S.A., the U.K. and France were all initially undertaken without regard to economics. In the U.S.A. the Shippingport reactor was built as part of the cold war contest with the U.S.S.R. In the U.K. the Magnox reactors were conceived as an adjunct of the military system.[16] One may start a nuclear-power programme without regard to the economics, and perhaps one has to do so, but the programme cannot continue (without subsidy) if it is not economic. Two main questions thus become apparent: is the programme continuing or is it running down, and is it subsidised? The answer to the second question differs from country to country, but the answer to the first is more uniform, on a global scale. The projected world capacity of nuclear electricity, at the year 2000, will have fallen in an almost hyperbolic fashion during the past two decades, from 3500 GW[28] or more to around 700 GW. Zaleski anticipates that world nuclear generating capacity will reach its maximum value before the year 2000, with roughly the following distribution.[29]

Europe, Japan, U.S.A.	400–500 GW
Europe, centrally planned economies	150 GW
Developing countries	70 GW

Each programme grew, following the familiar sigmoidal learning curve, from the mid-1940s but the year 1975 saw a dramatic halt to the U.S. civil programme. Since then the orders have been negligible and the cancellations have run at roughly 10 per year. This reversal of ordering policy antedates the accidents at Brown's Ferry (1978) and Three Mile Island (1979).[14] In the U.K. building has continued, with stations commenced at Heysham and Torness in 1980 and 1981. In France the construction programme has gone ahead forcefully and against the worldwide trend: in the two years 1979 and 1980 some 44 GW of capacity were constructed throughout the world, 30 GW of this being in France. The French programme, provoked by a lack of indigenous fossil fuel, has required a massive borrowing of capital and a large construction industry which, having completed its programme, now has its order books empty, apart from such sales as can be made to China or the Third World.

The nuclear programmes of the U.K. and France show strong governmental intervention. Market forces are more evident in the U.S.A., and it is in the U.S. programme that the economics is least perturbed and therefore best assessed. The overall picture in the U.S.A. is that nuclear power has passed through a phase in which the utilities judged it to be competitive with coal-fired electricity generation and conservation, and into a phase in which the reverse is true. This is seen both in the reduction of the projected capacity at the year 2000 and in the cessation of orders since 1975. The reasons for the reversal appear to be:

1. A steady increase in the real capital cost of nuclear stations through the decade, by a factor of approximately 2.
2. An increase in construction times by a factor approaching 2, from 50 to 60 months to over 100.
3. An actual capacity factor of 60 per cent rather than the predicted value in excess of 70 per cent.
4. A slower rise in coal prices than anticipated.
5. A falling demand for electricity as a result of increased conservation measures.

For a nuclear station with a 20-year life the capital cost and running cost divide in the approximate ratio 2:1. Conversely, for a coal-fired station the capital cost and fuel cost divide in the approximate ratio 1:2. Thus the first two of the above factors increase the larger cost of the nuclear station and the fourth decreases the larger cost of the fossil-fuelled station. Nuclear stations are most economic when used to supply the 'base-load' of the electricity supply system. The rapidly changing loads are best supplied by fossil-fuelled stations, simply because nuclear fuel elements do not like a rapid change of power level ('power ramping'). Secondly, it is uneconomic to have a nuclear power station standing idle for an appreciable part of the day, because of its capital cost; interest on capital has to be paid for the inactive time as well as for the active time. Thus the economic performance of a nuclear station is dependent on the load it supplies. A coal-fired station is also most economic on base-load, but the difference between base-load and peak-load cost is less than for its nuclear counterpart. Because of its lower capital cost the coal station has a more flexible use.

Beyond these major factors lies controversy. What overheads should be added to the cost of nuclear and of fossil-fired stations? What of reprocessing costs, of waste storage? Is the cost of decommissioning a nuclear station negligible or appreciable? What is meant by decommissioning? Is the decommissioned reactor to be a permanent feature of the landscape or is the site to become a green field? In the latter case what is to happen to the radioactive structure of the reactor? Is the interest charged on capital, for either system, a market value, or does its value represent a government subsidy? What is the social cost of comparative energy systems? What is the cost to the community of sulphur dioxide deposited across northern Europe and the north-eastern U.S.A. from coal and oil-burning industrial plant? What is the cost of removing this at the power station, and is the community willing to bear the cost? The U.S. experience seems best to illustrate the economic realities since it is there that market forces are strongest. Elsewhere, civil nuclear power is more a matter of government policy, even to the extent that to some countries a nuclear programme is a status symbol. It is unlikely that a Third World country, in choosing to have a nuclear-power programme, is responding either to internal market forces or to the needs of its own community. As with other industrial products, if the First World has spare manufacturing capacity, the Third World is invited to buy.

Advantages and Disadvantages

An obvious advantage of nuclear power, by comparison with its present major competitors in the field of electricity production, is that it does not chemically pollute the environment. The corrosion, health risk and damage to forest and lake that results from the burning of fossil fuels is absent. Secondly, the use of nuclear power preserves our coal and oil resources for petrochemical feed stock.

The main disadvantages of nuclear power are that it is centralised, capital-intensive and generates long-term health and safety hazards which have so far not been surmounted. It encourages fantasy at national and international level. In the civil sphere one may look back to the widespread fantasy that it would provide 'power too cheap to meter.' This was never true of even the most rudimentary reactor, and is less likely to be true of more sophisticated reactors, either fission or fusion. In the military sphere, nuclear power, through its by-product plutonium, was to provide each nation with 'weapons with which no-one can argue.' This fantasy is openly contradicted by the present state of the world.

References

1. *Sixth Report of the Royal Commission on Environmental Pollution* (The Flowers Report), Cmnd 6618, HMSO, London, 1976
2. Hoyle, F., *Astronomy*, Macdonald, London, 1962
3. Pease, R. S., 'Progress and Prospects of Nuclear Fusion', *Royal Society of Arts Journal*, August, 1981
4. *The JET Undertaking*, Culham Laboratory, Oxford
5. Pease, R. S., 'Fusion Research 25 years after Zeta', *New Scientist*, 20 January 1983
6. Semat, H. and Albright, J. R., *'Introduction to Atomic and Nuclear Physics'*, Chapman and Hall, London, 1972
7. Cowan, G. A., 'A natural fission reactor', *Scientific American*, July 1986, p. 36
8. 'Canadians close in on CANDU's cracked reputation', *New Scientist*, 6 October 1983, p. 4
9. 'Cottrell will head PWR safety panel', *New Scientist*, 16 June 1983, p. 61
10. Harris, J. E. and Crosland, I. G., 'Fin waving, bolt strain and the railings round St. Pauls', *Endeavour*, Vol. 3, No. 1, 1979, p. 15
11. Nuclear Engineering International, *Charts of individual reactors*, in *The World's Reactor Series*, Full colour sectional drawings, IPC, Sutton, Surrey
12. United States Atomic Energy Commission, *The Safety of Nuclear Power Reactors (Light Water Cooled) and Related Facilities*, WASH-1250, USAEC, Washington D.C., 1973
13. United States Atomic Energy Commission, *An Assessment of Accident Risks in US Commercial Nuclear Power Plants*, WASH-1400, USAEC, Washington D.C., 1974
14. Patterson, W. C., *Nuclear Power*, 2nd edn, Penguin, London, 1983
15. 'A rocky graveyard for nuclear waste', *New Scientist*, 15 September 1983, p. 756
16. Hannah, L., *Engineers, Managers and Politicians: The first fifteen years of nationalised electricity supply in Britain*, Macmillan, London, 1982
17. Krass, A. S., Boskma, P., Elzen, B. and Smit, W. A., *Uranium enrichment and nuclear weapon proliferation*, Taylor and Francis, London, 1983
18. Pringle, P. and Spigelman, J., *The Nuclear Barons*, Michael Joseph, London, 1982
19. Shrader-Frechette, K. S., *Nuclear Power and Public Policy: the social and ethical questions of fission technology*, Reidel, Dordrecht, 1980
20. National Radiological Protection Board, *Gut uptake of plutonium, americium and curium*, R-129, HMSO, London, 1983
21. 'Sellafield cuts plutonium discharges', *New Scientist*, 13 October 1983, p. 73
22. 'Secrets of the Windscale fire revealed', *New Scientist*, 29 September 1983, p. 911
23. 'Atom bomb test leaves infamous legacy', *Science*, Vol. 218, 15 October 1982, p. 266; 'Scientists implicated in atom test deception, *Science*, Vol. 218, 5 November 1982, p. 545
24. Japan National Preparatory Committee, *International Symposium on the Damage and After Effects of the Atomic Bombings of Hiroshima and Nagasaki*, Pergamon, London, 1978
25. International Atomic Energy Agency, *Annual Report for 1981*, IAEA, Vienna, 1982 (GC(XXVD1664)), p. 57
26. *Agreement between the United Kingdom of Great Britain and Northern Ireland, the European Atomic Energy Community and the International Atomic Energy Agency for the Application of Safeguards in the United Kingdom of Great Britain and Northern Ireland in connection with the Treaty*

on the *Non-Proliferation of Nuclear Weapons (with Protocol)*, Cmnd 6730,
   HMSO, London, 1977
27. Hesketh, R. V., 'Social pressures on the use of plutonium from the civil nuclear
   programme', in *Issues in the Sizewell 'B' Inquiry*, Vol. 5, Centre for Energy
   Studies, Polytechnic of the South Bank, London, 1982
28. Zaleski, C. P., *Nuclear Energy in Today's World: Paradise Lost, Hell, or just
   Purgatory?*, British Nuclear Energy Society, London, 1982
29. Marrone, J., 'The Price–Anderson Act: The Insurance Industry's View', *Forum*,
   XII(2), Winter 1977, p. 607

Further Reading

Sixth Report of the Royal Commission on Environmental Pollution (The Flowers
   Report), Cmnd 6618, HMSO, London, 1976
Flowers, Sir Brian, 'Nuclear Power and Public Policy', *Journal of the British Nuclear
   Energy Society*, No. 16, 1977
The Ranger Uranium Environmental Enquiry, First Report (The Fox Report),
   Australian Government Publishing Service, Canberra, 1976
Patterson, W. C., *Nuclear Power*, 2nd edn, Penguin, London, 1983
The Fissile Society, Earth Resources Research, London, 1977
Pringle, P. and Spigelman, J., *The Nuclear Barons*, Michael Joseph, London, 1982
Gowing, M., *Britain and Atomic Energy 1939-45*, Macmillan, London, 1964
Gowing, M., *Independence and Deterrence*, Macmillan, London, 1974
Hannah, L., *Engineers, Managers and Politicians*, Macmillan, London, 1982
Lord Hinton, *Heavy Electric Current Engineering in the United Kingdom*, Pergamon,
   London, 1979
Hewlett, R. G. and Anderson, O. E., *The New World 1939-1946*, Pennsylvania
   State University Press, Pittsburg, 1962
Hewlett, R. G. and Duncan, F., *Atomic Shield 1947-1952*, Pennsylvania State
   University Press, Pittsburg, 1969
Pocock, R. F., *Nuclear Power: its Development in the United Kingdom*, George
   Allen and Unwin, London, 1977
Williams, R., *The Nuclear Power Decisions*, Croom Helm, London, 1980
Burn, D., *Nuclear Power and the Energy Crisis*, Macmillan, London, 1978
Sweet, S. (ed.), *The Fast Breeder Reactor: Need? Cost? Risk?*, Macmillan, London,
   1980
Shrader-Frechette, K. S., *Nuclear Power and Public Policy*, Reidel, Dordrecht, 1980

6

Solar Energy

(by Judith Stammers)

'Two roads diverged in a wood, and I —
I took the one less travelled by,
And that has made all the difference'

Robert Frost from 'The road not taken'

Solar Radiation[1]

Current world energy consumption is around 3.5×10^{20} J per year. The solar radiation reaching the earth's upper atmosphere has an intensity of around 1300–1400 W/m^2. Not all of this reaches the earth's surface but even so the energy from the sun received at the earth amounts to some tens of thousands of the world's energy requirements for a given period. Clearly there is the potential for solar energy to supply some, if not all, of our energy demand. However, the low power density and intermittent nature of solar radiation makes collection, conversion, and storage a challenge.

In 1974, prompted by the sharp rise in prices of fossil fuels, the disruption of oil supplies from the Middle East and awareness that supplies of conventional fuels were finite, many governments, universities and industrial companies throughout the world began work on the effective harnessing of solar energy.

The sun is one of the main sequence of stars, supposed to be formed from a gas cloud having an earlier existence. Random movements within this cloud have given rise to regions of density greater than average, so causing concentration of matter round these regions of high density under the action of gravity. The particles interact, moving fast in relation to each other, and this general state of agitation causes a temperature rise. The temperature near the centre of the sun is thought to be about 10 million °C.

The nuclear reactions responsible for the solar radiation occur in the central core,

132

which occupies about 3 per cent of its volume. The radiation begins as gamma radiation (with a very short wavelength of the order of a hundred-millionth of a millimetre) and is modified by the 500,000 km of material surrounding this central core. For instance, the gamma-radiation photons collide with nuclei and electrons or are scattered in near collisions and thereby lose some of their energy.

The radiation leaving the sun has a wide range of wavelengths. The energy distribution is fairly close to that of the classical 'black body' at a temperature of 5500°C. About half of the sun's energy is radiated with wavelengths between 0.35 and 0.75 microns or micrometres (a micron is a thousandth of a millimetre) and is in the visible band. There is very little radiation in the ultra-violet band, below 0.35 microns wavelength. There is more in the infra-red, above the visible band, which contributes to our warmth.

By the time the solar radiation reaches the earth the spectrum has been substantially modified and considerable energy has been lost.

At the highest level of the earth's atmosphere virtually all the ultra-violet radiation is removed: it is used up by a process in which molecular oxygen is dissociated into atomic oxygen. Some of these atoms recombine into molecules but most react with other oxygen molecules to form ozone, itself a strong absorber of radiation.

The visible and infra-red radiation are scattered by gas molecules, water droplets and dust particles in the air so that some radiation is redirected away from the earth and back into space again. The water droplets occurring in thick cloud may cause as much as 80 per cent of the incident radiation to be turned back into space. Absorption by gas molecules and its re-emission is another way in which the incoming solar radiation is modified.

By the time the solar radiation has reached the earth's surface its intensity is little more than half the value at the top of the atmosphere. Prediction of the value at a particular location is difficult. It will depend on local conditions such as pollution, amount of cloud cover and the length of path the solar radiation takes through the atmosphere. The latter varies with the time of day, season of the year and position on the earth's surface. The geometric effect of the apparent motion of the sun with respect to the earth has the largest effect on this. The earth's rotation about its north–south axis produces the obvious night–day variation and, because the earth's rotational axis is not perpendicular to the plane of its orbit around the sun, we have the seasons.

The intensity of solar radiation at a particular location will have two components – the direct radiation and diffuse radiation. The intensity of the direct component will be influenced by the angle of the rays relative to the plane of the surface and the length of their path through the atmosphere. The diffuse component consists of radiation which has been scattered by constituents of the atmosphere and redirected on to the surface. This diffuse radiation represents a significant fraction of the total. On a completely overcast day all the radiation will be diffuse.

At mid-summer in the tropics the sun's path is almost directly overhead and the hours of daylight are, for instance, greater than at the equator. The day lengthens with latitude and this compensates for the decrease in solar intensity. Therefore, in mid-summer, under clear conditions, the amount of solar radiation reaching a surface will be approximately the same from the tropics right up to the polar circles. In the mid-winter, however, the values fall with increasing latitude. Thus the yearly totals at different latitudes differ considerably.

In the tropics, about 2300 kWh of solar radiation fall on 1 m² horizontal surface over a year: about 17 per cent of this is diffuse radiation, the rest direct. In the south of England, about 900 kWh fall annually on 1 horizontal m² : about 60 per cent of this is diffuse. By tilting the surface to an optimum angle, the amount of radiation falling on it will be increased to about 1100 kWh/m² /year on average.

Converting Solar Radiation into Heat[1,2]

The warming effect of the sun is well known. When radiant energy strikes the surface of an object a proportion is reflected, part is absorbed and part may be transmitted through the object. The proportion that is reflected, absorbed or transmitted will depend on the wavelength of the radiation, its direction and the condition of the surface of the object. With a very smooth surface the reflectivity will be high; if the surface is rough, the absorptivity will be high because the photons will have a greater chance of being trapped.

When an object absorbs radiation it is raised to an excited state with electrons at high energy levels and the temperature is raised. The object endeavours to return to its original state by re-radiating the extra energy. In a solid body or dense gas the energy is transmitted to neighbouring atoms and the temperature is evened out. Thus the emitted radiation may have a different wavelength distribution from that of the absorbed radiation.

An 'ideal body' (a 'black body') is able to absorb all the radiation falling on it. The radiation emitted by a black body by virtue of its temperature has a particular distribution of energy density with respect to wavelength.

For real bodies the radiation emitted is not distributed quite like that of a black body but, for simplicity, can be expressed in terms of the latter. Thus a real body might be assigned an overall emissivity E, such that at a temperature T it emits a fraction E of the energy emitted by a black body at that temperature.

Similarly, we can assign to the body the properties reflectivity (ρ), absorptivity (α) and transmissivity (τ) such that if radiation of intensity P falls on it, the rates at which energy is reflected, absorbed and transmitted are ρP, αP and τP, respectively.

Polished metals have low emissivity at all temperatures. Matt black paints have high absorptivity and emissivity at 20–100°C and high absorptivity at higher temperatures. Glass is virtually transparent to the short-wave radiation we see by but is virtually opaque to longer wavelengths.

If we consider a body left lying in the sun in such a way that it exchanges no radiation with its surroundings, and if the solar radiation falling on the body is of intensity P and the body has absorptivity α for this radiation, then the body will heat up until it reaches an equilibrium temperature T. At this temperature the rate of emission of radiation will just balance the input. The situation can be expressed by the equation

$$\alpha P = E\sigma T^4$$

where σ is a constant of proportionality called the Stefan–Boltzmann constant. Thus the equilibrium temperature T is given by

$$T^4 = \frac{\alpha}{E} \cdot \frac{P}{\sigma}$$

Therefore, to obtain the highest temperature, we need a high ratio of absorptivity to emissivity. If we are to use the body to collect solar energy we also need a high value of absorptivity.

The simplest solar-collecting device is a flat-plate solar collector. Here the collecting body is a thin plate sandwiched between an insulated backing and a glazed front. The performance of such a collector will be improved if the ratio of absorptivity of solar radiation to emissivity of long-wave radiation can be increased without decreasing the absorptivity by much. This can be achieved by making the collector of polished metal and coating it with, for instance, a thin deposit of black metal oxide. This has a high absorptivity for short-wave radiation but is transparent to radiation longer than the coating thickness. Thus for longer wavelengths the emissivity is close to that of the metal beneath. The insulated backing minimises heat loss from the back of the plate. The glazed front makes use of the 'greenhouse effect': the glass allows the solar radiation to pass in but absorbs the long-wave radiation emitted by the surfaces of the plate and does not allow much to pass out again.

The Romans used transparent window coverings to help heat their public baths and sought a sheltered spot for such buildings, which faced the winter sunset.

Probably the simplest practical uses of solar heat are in the cooking of food, distillation of contaminated or brackish water to obtain potable water and in crop drying.

The Greeks, Romans and Chinese developed curved mirrors to concentrate the sun's rays on to an object to make it ignite. The glazed flat-plate collector is today most commonly used in systems for heating domestic hot water. Active space-heating systems, where the water heated in a large area of flat-plate collector is used to make a contribution to the domestic hot water and space-heating of the house, represents another application. Solar swimming-pool heating is, of course, widely used nowadays, too.

The Greeks were aware that the sun in winter travelled in a low arc across the southern sky and that in summer it passed high overhead. They built their houses so that the solar radiation could easily enter the house in winter through a south-facing portico. The main rooms of the house were sheltered from the north to keep out the cold winds. During summer overhanging eaves shaded the rooms of the house from the high sun during much of the day.

Passive Solar Design of Buildings[3]

This building design that makes the most effective use of the solar energy falling on it is known today as passive solar design. In buildings designed in this way all the parts — walls, floor, roof, windows — are arranged so that the benefits of solar heating in winter are maximised and the risks of over-heating in summer are avoided. When this is done, the consumption of fossil fuels for heating and cooling is brought to a minimum.

The materials from which the building is constructed are used to capture, store and distribute the solar heat. Materials with a high thermal capacity — such as brick, concrete, stone, or water — absorb the solar radiation as it enters the building and store it in the form of heat. The rooms are arranged so that they are in direct thermal contact with this store. The rooms are thus heated directly without the expense of special plumbing or forced hot-air distribution systems.

Careful consideration has to be given to solar and heat flow. The floor plan layout, circulation patterns, window location and building materials all affect how well the passive design works. The building itself is a solar system. Windows not only let in light but collect heat. Dividing walls also store and radiate heat.

A number of special measures are commonly used in passive solar design. These can be categorised as direct gain, indirect gain, isolated gain, and cooling measures.

The simplest and most commonly applied passive solar measure is the direct gain method. Here a south-facing window is the solar collector. The sun enters the living space and its energy is stored in thermal mass within this space, externally insulated from the outside atmosphere. Moveable insulation may be used on the south-facing windows to prevent winter and night-time heat loss. Shading devices may well be needed in the summer to minimise over-heating and glare.

The indirect gain methods are those using the Trombe wall, named after Dr Felix Trombe's and Jacques Michel's work in Odeillo, France. Here a wall, dark coloured on its south face and massive, is placed behind glazing. Behind both the glazing and the Trombe wall is the living space. The wall stores solar energy during the day and transmits it to the living space several hours later. The time taken for this energy to travel through the wall is related to the thickness of the wall. For immediate transfer of heat it is possible to open the space between the wall and glass. Care must be taken in the building design to minimise the possibility of over-heating. A water Trombe wall can be used in place of a mass Trombe wall. Here contained water replaces the solar wall. A roof pond is the same as a water Trombe wall but is placed on the roof. The heat is radiated through a metal ceiling. Moveable insulation reduces unwanted heat loss in winter and heat gain in summer. Care must be taken that the water load can be carried by the building.

Isolated gain methods are epitomised by the atrium, greenhouse or conservatory. Here a south-facing glazed space is isolated from the living areas. It is used as an extension to the living space but must be used only seasonally. The storage mass can be a Trombe wall or may be placed in the floor or ceiling of the living space and linked to this by means of a small fan.

A thermosyphon system is another form of isolated gain. This employs the change in density of a collection fluid to create circulation through heat-storage mass. This thermal mass may be placed in the living room or ceiling: it must be below the storage mass.

Cooling methods make use of radiant cooling, dehumidification and underground building.

Site investigation is a major factor in passive solar design. The building must respond to its site and micro-climate. Shelter and shade should be provided from surrounding land and vegetation. The path that the sun will take throughout the year should be determined to see if trees or buildings will block out the sun. Protection from cold prevailing winds can be provided by adjacent evergreen trees and earth beams or the shape of the building itself.

Good insulation must be provided to give thermal comfort and yet the building must be protected from over-heating by good ventilation, additional storage mass or reduced glazing.

Recently, many buildings incorporating the elements of passive solar design have been built throughout the world. In the U.K., one of the earliest buildings to use such measures was St George's School, Wallasey, which was built in 1961. During more recent years a number of dwellings have been constructed at Milton Keynes. The Pennylands estate, which was designed by Milton Keynes Development Corporation, was built in 1980 and consists of 177 houses owned by the Development Corporation and rented by their occupiers. Of these, 94 are well insulated and 83 are very highly insulated. Most of the houses are oriented within 45° of south, with minimal over-shadowing. Two house types have been built: one with dual aspect and the other with single, south-facing aspect. The latter has a small amount of glazing on the north side and large windows on the south, to provide direct gain passive solar heating. Both types occur in the well-insulated and very highly insulated areas. The U.K. Department of Energy has funded some monitoring aimed at determining the space-heating consumption of the different house types, as well as the social acceptability of this type of estate layout and these types of houses, and the costs–benefits of the energy-saving measures. The performance figures are still being collected and analysed but preliminary results appear to be in line with the original predictions and show that the energy consumption of the very highly insulated houses is consistently below that of the well-insulated houses and that the single aspect south-facing, mid-terraced houses have the lowest annual energy consumption.

Pink Cottage, designed by Ralph Lebens Associates of London, is an extension, completed in 1981, which makes use of a very small and restricted site within a corner of an L-shaped cluster. The cluster is made up of two semi-detached houses in an English country setting. The south and east faces of the extension are built right up to the boundaries. The west side of the extension attaches to the existing cottage. The location of the north wall was dictated by the existing roof. The only possible collection area was the south slope of the roof, the south wall being a party wall. The cottage is attached to a dovecot which now forms the main reception area to the neighbour's cottage and it was necessary that the experimental conservatory blended in with this. The system used is direct gain through double-glazed sealed units on the south slope of the roof with the possible use of the remainder of the cottage as remote storage (fan forced). Louvres vent off excess heat at the roof top. In summer, if shading and cooling are required, an insulated blind is pulled down to below most of the collection surface and the ridge vents can be winch-opened to form a solar chimney. During winter a reflective blind is drawn down the inside of the north slope of the roof, and directs most of the solar energy on to the gallery. Strapped to the underside of the gallery supports are packs of Glauber's salts phase-change thermal storage, as an experiment. At night in winter the insulating blinds can be drawn fully to prevent further heat losses. It is predicted that the percentage of the heating load met by direct solar gain is 51 per cent. Monitoring is being carried out to determine the effects of the Glauber's salts storage and of the insulation blinds.

Solar Water-heating[4] and Related Applications

We have already referred to the use of solar energy to heat water. The most frequently used system incorporates a glazed flat-plate collector. The absorber in such a collector consists of a thin plate which can be made of a variety of materials – copper, aluminium, mild steel, stainless steel, or plastic. All have advantages and disadvantages, both economic and technical. This absorber is sandwiched between an insulated backing and a transparent front made of glass or a suitable plastic, securely fixed with adequate provision for thermal movement.

The most commonly found solar system is that for heating solar domestic hot water. There are nearly a quarter of a million such systems in Western Europe. Large numbers are found in the Mediterranean countries. In Greece, for instance, 1 in 60 households has a system and in both Italy and France, 1 in 300. In the U.K. and West Germany the penetration has been lower to date (1 in a 1000), nevertheless a solar water-heating system can make a useful contribution to household requirements even in northern climates. In Israel, where legislation requires all new buildings to be fitted with solar water-heating, 1 in 3 houses has such systems. In Japan there is a strong government support scheme and there are hopes that solar water-heating systems will have been installed in 1 in every 5 private houses by 1990. In parts of the U.S.A. and Australia, too, a high penetration of the housing stock has been achieved.

A conventional solar domestic hot-water system consists of the solar collector, through which water is circulated and where it is heated, and some form of well-insulated storage tank. Commonly it means installing an additional tank although this is not necessarily the most efficient system. The solar-heated water is piped to this storage tank, circulation being achieved by natural convection or by using a pump. To bring the water to the temperature at which the householder will want to use it some additional heating by conventional means will be required for much of the year in countries such as the U.K. A thermal differential controller can be used to switch the pump on and off according to the availability of heat in the panel. This simple direct system can be modified in several ways. For instance, the water passed through the collectors may not be used directly but may itself be used to heat the water used in the domestic hot-water system. Alternatively, oil rather than water can be used as the heat-transfer medium. In all methods adequate measures must be taken to prevent freezing of the fluid passing through the panels in cold weather. The most general method is to use a non-toxic antifreeze additive. The formulation also incorporates a corrosion inhibitor since special care must be taken to overcome corrosion problems, which could occur by interaction of materials from which the collector is made with other metals in the system. An example of an indirect system for providing domestic hot water is shown in figure 6.1.

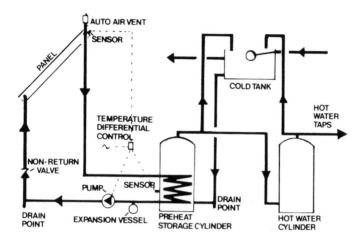

Figure 6.1   *An indirect system for providing solar domestic hot water*
(source: Solar Trade Association Ltd)

In the U.K. most systems for heating household water are able to use about half the radiation falling on the collector. The area of collector that might optimally be installed in a particular dwelling will depend on water usage and the alternative fuel options. A conventional domestic system in the U.K. for a 3–4 person household, where the hot-water usage is about 45 litres per person per day, is generally a 3–5 m^2 collector area; for 5–6 people, 4–8 m^2 might be optimal. For a particular case, the optimum might differ from this by as much as 25 per cent. Usually, panels are sited on the unshaded slope of a pitched roof that faces between south-east and south-west. The optimum tilt of the panels depends on the latitude and on the relative amount of direct and diffuse radiation at different times of year. In general, the pitch of the roof is a good compromise.

Solar systems have to comply with any building regulations that might apply, such as those dealing with structural stability or structural fire precautions. The responsibility for administering and interpreting the building regulations rests with the local authority. Similarly, water by-laws must be respected. The carrying out of operations that materially affect the external appearance of a building constitutes development that requires planning permission. In the U.K., planning permission for existing dwellings may be granted in one of two ways. First, certain types of minor development are given general permission by Article 3 of the Town and Country Planning General Development Order 1977. Most solar-heating installations come into this category and a formal application for planning permission to a local authority is not therefore usually necessary for the installation of solar panels in an existing dwelling. Local planning authorities have, however, the power to withdraw the permitted development rights given them by the General Development Order. In this case formal application to the local planning authority must be made. This must also be done for new buildings and listed buildings. An efficient conventional system in the U.K. could provide perhaps two-thirds of a family's hot-water requirements over good summer months and rather less over the rest of the year, averaging out at 40–50 per cent over the whole year. A good contribution can be obtained by using hot water at periods of high solar gain and minimising the time hot water has to be stored in the tank. In addition, setting the hot-water supply thermostat to, say, 45°C so that the water from the hot tap does not require diluting with cold will enable the best use to be made of any solar system.

Economics of Solar Water-heating Systems in the U.K.

In the U.K. the price of conventional fuels increased by 20 per cent a year on average from 1973 to 1979. General inflation was 14 per cent a year over the same period. Average incomes were related more closely to general inflation than to fuel price rises. Thus, as the years went by, the householder spent an ever-increasing proportion of his income on buying energy for the home. If, in 1973, capital had been invested in a building society, by 1979 it would have yielded a return of 17 per cent a year on average, and on the stock exchange it would have given an average return of 1.6 per cent a year over the same period (7.9 per cent with the dividend payments re-invested). Both these investments failed to keep pace with the general inflation rate of 14 per cent over the same period. If a solar system had been purchased in 1973, with this capital, by 1979 it would have yielded a return better than inflation. This continues to be true.

Let us look in detail at a good case where we have a domestic hot-water system designed to provide 200 litres of water a day in the U.K. at a temperature of 50°C and which has a solar collector area of 5 m². Provided that the system is well installed, the gain from the solar collectors might be 400 kWh/m²/year giving a total of 2000 kWh for the whole system. In 1982 a typical price for such a system was £1400. To examine whether this is a good investment we need to know the lifetime of the system, the rate at which fuel prices will increase in the future, likely maintenance costs, and the method of financing. It is sensible to ensure that systems are well constructed and durable. Central heating systems are generally given 25 years of life with the exception of the boiler and so we shall use the same value for the solar system in this example. The increase of fuel prices is difficult to predict and probably each fuel will fluctuate in a different fashion. However, compared with the historical figure, an average of 14 per cent a year seems reasonable, assuming a general inflation rate of 10 per cent a year.

The monetary savings from the solar system will differ according to the fossil fuel replaced and the conversion efficiency of the boiler. Boilers perform at low efficiency in the summer when they are heating domestic hot water only. Taking a boiler efficiency of 60 per cent with gas or oil, the value of savings from a 5 m² solar system would be £35 a year with a gas system at 1982 gas prices, £59 for an oil system, and £98 compared with electricity.

The solar system with a gas back-up will give an average annual return on investment of 13.3 per cent. With an oil back-up the solar system would give a 15.5 per cent annual return on investment averaged out over the lifetime of the system. Compared with electricity, the solar system would give 19.5 per cent annual average return on investment. All these returns on investment assume a future general inflation rate of 10 per cent a year.

Space-heating

Although the most common application of solar water-heating is for domestic hot water, there has been a fair amount of work on active space-heating, where the water heated in a large area of flat-plate collector is used to make a contribution to both the domestic hot-water and space-heating requirements of the building. Good solar systems of this type contribute as much as 40 per cent of the total heat requirements of the household. In some cases a phase-change salt store, rather than the traditional water or pebbles, is used. This store has applications in industry and agriculture as well as in dwellings.

Swimming Pools

Swimming-pool heating provides an attractive and cost-effective outlet for solar heating. Conventional glazed collectors may be used. The unglazed panels and slabs specially designed for heating large volumes of water at relatively low temperatures — say, 2–6°C above what it would be otherwise — are particularly appropriate for this application. By circulating a large volume of water rapidly through the solar collector, advantage can be taken of the fact that solar collectors are more efficient working at low temperatures. In most cases in the U.K. it is recommended that the area of collector equals one-third to three-quarters of the pool surface area depending on the nature of the site. Covering the pool when it is not in use, particularly at night, will help prevent heat loss and enable a smaller area of collector to be used than would otherwise be the case. The panels are installed on a supporting frame in an unshaded position on the side of the pool or on a nearby roof or, in the case of slabs, as a pool surround.

Solar Ponds

The solar pond is a still body of shallow water that collects solar radiation and stores it in the water as thermal energy.[5] It has a dense salt layer on the bottom and increasingly fresher layers of water on top. The dense solution is heavier and is held at the bottom by gravity. Solar radiation will heat the blackened bottom of the pond and, because of its high relative density, the hot water there cannot rise into the low-salinity layers at the top. Temperatures of the order of 60–90°C can be built up in this fashion. The solar-heated water can be used for industrial process applications or to operate heat engines whose output flows into turbines to generate electrical power. The efficiencies of solar pond systems are low but they have the advantage of being capable of operation throughout the year, day and night, in cloudy and clear weather. Naturally occurring salt-water bodies can be used to create salt ponds. There is a 7000 m² pond at Ein Bokek on the bank of the Dead Sea which began generating some 150 kW of electricity in the late-1970s.

Solar Stills

The *per capita* consumption of water[6] varies throughout the world from 900 tonnes a year in an agricultural country such as Iran to about 2700 tonnes a year in a highly industrialised country such as the U.S.A. Of the 2700 tonnes needed by the average American 0.3 tonnes is for drinking, 200 are used in the house, and all the rest is used in agriculture and industrial applications. All of the agricultural and domestic water and most of the industrial water is used in its fresh form. In many parts of the world solar distillation can be used to produce potable water from either sea or brackish water. A layer of such water is introduced into an airtight and watertight basin exposed to the sun. This still is covered by a material such as glass which allows the solar radiation to pass through to the water surface but which prevents most of the infra-red re-radiation and decreases convective heat loss. The solar energy trapped in the still evaporates the water, filling the air with water vapour and leaving the salt behind. As the humidity in the still increases water vapour condenses on the inside of the cover, which is sloped to allow fresh water to trickle into a catchment. Solar stills have been built throughout the world, mostly of the basin single-effect type described above. It is possible to increase the efficiency of such stills by modifying the design. Examples of advanced-design single-effect stills are inclined tray, stepped, multi-level, double-tube stills and stills with external condensation. More recently a new generation of stills called multiple-effect stills have been developed, examples of which are diffusion and diminishing type stills.

Solar Air-heating[7] and Related Applications

Flat-plate collectors have been available for some years where air and not water or an oil is used as the heat-transfer medium. It is difficult to combine such a system with a water-heating device although it can be done; its most appropriate use is for space-heating with a warm-air system. Thus such systems are not found very frequently in the U.K. where water-filled radiators are the most common method of house-heating.

The principle of operation is that heat is transferred from elements heated by radiation to the air in the collector. The warm air is then used directly or put into storage. This store must have a high thermal capacity and a large transfer surface. Crushed rock can be used or phase-change salts.

Solar Cooking and Refrigeration

In many developing countries a high proportion of domestic energy is supplied by firewood or charcoal (see chapter 11), and since the late-1950s much research has been carried out on solar cookers in order to reduce unnecessary use of firewood. There are a number of designs available: the hot box, the solar oven, parabolic direct-focusing cookers, the Tabor cooker, the Fresnel direct focusing cooker, the combination cooker, the sun basket, and the solar steam cooker.[8] These are describ-ed briefly below.

In the simplest hot box cooker, solar radiation enters a well-insulated box through glass panes and is absorbed by the dull black inside surfaces. Two reflecting surfaces can be put in the box at an angle to increase the efficiency. Cooking must be done with very little water since the box has to be put at an angle to the sun. A triangular box, where the longest side consists of double glazing placed so that it is perpendicular to the solar radiation, is more suitable for cooking with large volumes of water. An insulated lid with a reflecting inside surface, adjustable to the angle of the sun, can be closed to keep food hot after sunset.

Telke's solar oven is similar to the hot box but radiation entering the oven is augmented by four flat reflectors. The cooker, which needs to be oriented towards the sun, incorporates a moveable food platform which can be kept horizontal. Heat-storage materials allow heat to be released during cooking or after sunset.

In the paraboloidal direct-focusing cooker the cooking pot is suspended at the focal point over a parabolic dish.

In Tabor's cooker the reflector consists of 12 glass mirrors mounted in parabal-oidal array. The geometrical arrangement of the mirrors keeps the reflected beams to the bottom of the pan and its low pivot point makes the cooker more stable and easy to use because the reflector is closer to the ground at low sun angles.

The concentration ratio that can be achieved by a single lens is limited. The Fresnel lens combines the advantages of a multi-lens system within a single unit as each segment is designed to concentrate the incident radiation on to a centrally positioned receiver – in the case of the Fresnel cooker, the cooking pot.

In the combination cooker, the best features of the direct-focusing cooker are combined with those of the oven type. The design consists of an oven with a small window area on to which rays are reflected from a fairly large reflecting area so that the net energy delivery rate to the cooking pan is relatively high.

The sun basket is a recent design. It consists of a parabolic mirror at the focal point of which the cooking pot is suspended. A cheap automatic tracking device has been developed to increase efficiency: this operates on the principle of a weight floating on a gradually descending water level which pulls a string and in turn repositions the basket.

In the solar steam cooker, which can be used indoors, a flat-plate collector is used to provide steam to an insulated cooking vessel.

To be used for refrigeration and air conditioning, solar radiation has to be converted either into mechanical power or to electricity. This can be achieved by two different methods: solar thermodynamic and photovoltaic conversion. Photovoltaic generation of electricity is described below. In the remainder of this section we shall examine the solar thermodynamic routes.[9]

There are two main thermodynamic procedures. In the first, solar radiation is converted into heat which drives a heat engine to produce shaft power. This is coupled to a conventional vapour-compression chiller. The second approach to cooling is to use solar heat to drive an absorption-cycle chiller. In addition, evaporative cooling and radiative cooling can be used. The heat engines most commonly used today are Rankine cycle engines which use low-temperature boiling organic liquids as the working fluid. The organic liquid is pumped into a boiler and evaporated by solar thermal energy. The vapour generated then provides shaft power by driving an expander such as a turbine. The vapour is then condensed and cooled to complete the cycle. In the refrigeration cycle the shaft power drives a compressor, directly or indirectly, in a vapour-compression refrigeration cycle.

The more favoured route for solar thermodynamic refrigeration is the use of an absorption-cycle chiller. This uses heat as the energy input to the generator of the chiller. Thus a solar collector can be linked directly to the cooling unit without the need for an engine.

Absorption cycles are based on evaporative cooling where the latent heat of vaporisation of the refrigerant produces the cooling effect.

In the intermittent absorption cycle the usual refrigerant/absorbent combination is ammonia/water. The process takes place in two phases. In the refrigeration phase the refrigerant is evaporated in one vessel, called in this instance the evaporator, and the vapour is passed to a second vessel, the absorber, where it is absorbed by the absorbent. The heat of absorption must be removed by air or cooling water. In the regeneration phase, the second vessel becomes the generator and solar heat is applied to this to drive off the refrigerant vapour, which is condensed back in the first vessel, now the condenser. The simplicity of construction and operation of the intermittent system is offset to some extent by its poor thermal performance, need for manual operation, and inability to provide continuous cooling.

In the continuous absorption cycle, the common refrigerant/absorbent combinations are water/lithium bromide and ammonia/water. The evaporator, absorber, generator and condenser are now four separate vessels and the system operates in a continuous closed cycle. A solution pump is usually required to circulate the absorbent between the generator and absorber and a heat exchanger may be used between the absorber and generator to recover some of the sensible heat of the absorbent solution.

Water/lithium bromide operates with a high coefficient of performance but cannot produce ice and, to prevent salt crystallisation, the condenser must be water cooled. A number of solar air-conditioning systems are commercially available which use flat-plate collectors and lithium bromide/water-absorption chillers. Ammonia/water systems operate with a lower coefficient of performance but the refrigerant vapour must be rectified to remove water. Some of its advantages are the capability to make ice and a greater flexibility of operating conditions, including the possibility of air-cooled condensation. Variations of the continuous cycle exist. Multi-stage cycles can be used at operator temperatures above 140°C to give an improved coefficient of performance. Improved heat-exchange arrangements are also possible. Several other cooling processes also exist. Evaporative systems achieve cooling through evaporation of water. In the simplest method for house-cooling, water from a storage tank is trickled down an unglazed north roof.

A recent development is based on the use of natural zeolites which have the capacity of adsorbing large quantities of a variety of refrigerant gases. The zeolite collector is heated and water vapour starts desorbing from the zeolite. The vapour cools in the condenser and the liquid water is stored in a receiver. In the night, the zeolite container cools by convection and radiation and the zeolite begins to adsorb the water vapour produced by the evaporator.

Radiative cooling is particularly suitable at night with clear atmospheres and this technique can be used by pumping water into roof-mounted collectors.

Solar Concentrators

As we saw at the beginning of the previous section, solar radiation can be concentrated on to a point by means of mirror reflectors and parabolic focusing devices. The amount of heat collected can be increased by allowing the parabolic device to track the sun. At its simplest this technique can be used in solar cooking where the cooking pot is placed at the focal point of a parabolic reflector. Domestic hot water and swimming-pool water, too, can be heated by passing it through tubes at the focal point of a parabolic trough made of some highly reflective material. These collectors are widely used in countries with a high proportion of direct sunlight and their performance compares well with that of flat-plate collectors under similar conditions. Large-scale concentration is used in solar furnaces and solar thermal power plants.

As there are difficulties in making a large parabolic mirror track the sun, it is common to mount the parabolic mirror in such systems in a fixed position with its axis horizontal and facing north, in the northern hemisphere. Opposite the parabola is a heliostat which tracks the sun. This method was used by Trombe for his solar furnace in the Pyrenees in the 1950s. Later, the French government research organisation, Centre National de la Recherche Scientifique, built a 100 kW furnace at Odeillo. The parabolic mirror at Odeillo contains 9500 individual mirrors with a total reflecting area of 1920 m^2. It faces a field of 63 heliostats, with a total mirror area of 2839 m^2.

Solar Thermal Power Plants[10]

Solar thermal power plants generate electricity from direct sunlight by reflective focusing to achieve a concentration of solar energy. The latter heats a working fluid which, in turn, drives a turbo generator. The main elements of solar power plants are the collector/reflector system, the tracking system, a boiler at the focus of the collected energy and the turbo-generator.

There are three main types of solar thermal power systems: point-focus systems, linear-focus systems, and central-receiver systems. Point-focus systems are devices in which the solar energy is concentrated on to a receiver at the focus of a single collector/mirror unit which is composed of spherical or compound flat-mirror facets. The whole mirror/receiver unit may track the sun, or the receiver/mirror only may be moveable. Working temperatures are between 350°C and 900°C. The French government programme includes two types of point-focus system. The West German company MBB built a 100 kW(e)/500 kW(th) power plant for Kuwait which came on stream in 1981; it consists of 56 parabolic collectors of 5 m diameter each with a total reflecting area of 100 m². The tracked collectors reflect the solar radiation on to a spherical surface absorber placed at the collectors' focal point.

Linear-focus systems are similar to point-focus systems but with a linear focus in a parabolic trough collector. Like point-focus systems, they can be assembled into collector fields to form large-scale power plants. Working temperatures range from 180 to 350°C. A large example of this type is the 500 kW(el) power plant of the International Energy Agency at Almeria, Spain. This project is located adjacent to a 500 kW(e) central-receiver station.

Central-receiver systems consist of mirrors mounted on tracking heliostats, focusing on one large central receiver/boiler. They are the largest type of single system. Working temperatures range from 400 to 900°C. Probably the best-known example of a central-receiver system is the Eurelios 1 MW(e) power plant (see figure 6.2) built as part of the Commission of the European Communities' solar research and development programme at Adrano, Sicily, by Ansaldo, Cethel, and MBB.

Figure 6.2   *The 1 MW(e) central receiver system 'Eurelios'*

Solar Energy Storage

Because the availability of solar energy fluctuates according to the time of day and season, storage of the heat from the sun plays an important part in its usage. We have seen how the materials from which a building is constructed are used to store and distribute the solar energy coming into the building. Materials with a high thermal capacity — brick, concrete, stone, or water — absorb the solar radiation as it enters and store it in the form of heat. In good passive design the rooms are arranged so that they are in direct contact with this store. They are therefore heated directly without special plumbing systems.

We have also demonstrated how, in a conventional solar domestic hot-water system, water is circulated through flat-plate collectors and some of the heat collected is stored in a well-insulated storage tank. The water passed through the collectors may not be stored directly but may be recirculated through the collectors after passing in a coil through the storage tank. To prevent freezing of such an indirect system a non-toxic antifreeze additive, usually propylene glycol, may be used. Oil, rather than water, may also be used as the heat-transfer medium. In the U.K. any chemical added to a domestic hot-water system must have the approval of the local water undertaking. In practice those chemicals that have been tested and approved for this purpose by the National Water Council are generally permitted.

Because solar energy is not available at times when space-heating is needed, provision of an adequate store is particularly necessary in active solar space-heating systems. Traditionally, buildings using active solar space-heating have been custom designed and the store has been an important part of the design. Frequently the store consists of a large tank at the base of the house filled with water, pebbles or rocks.

Because of the labour, cost and inconvenience of providing such stores, work has been done in the U.K. and other parts of the world to develop stores made of phase-change materials. The heat of fusion (latent heat) which is involved when a substance changes from a solid to a liquid phase provides a method of storing a given amount of heat within a much smaller volume. For instance, a store based on a salt hydrate such as Glauber's salt — sodium sulphate decahydrate ($Na_2SO_4.10H_2O$) — will take up one-fifth of the space of a water store capable of storing the same amount of heat. Glauber's salt changes phase at around $32°C$. Other salt mixes will change phase at different temperatures. For instance, $Na_2CO_3.10H_2O$ will change at $33°C$, $Na_2HPO_4.12H_2O$ at $35°C$, $Na_2S_2O_3.5H_2O$ at $48°C$ and $NaCH_3COO.3H_2O$ at $58°C$.[11] Such salts suffer from nucleation and supercooling but stores have been developed which contain additives to overcome these problems. A range of phase-change stores is commercially available in the U.K. for storing thermal energy at different input temperatures. Over the past few years a number of houses have been built that incorporate these stores (Glauber's salt is the most commonly used) in active space-heating systems.

Converting Solar Radiation Directly into Electricity

We have already seen how solar thermal energy can be concentrated and converted into electricity via a turbogenerator in large-scale solar thermal power plants. A number of such plants have been built in many parts of the world – the U.S.A., Australia, France, Italy, Spain, Kuwait and so on. In the main, they constitute central generating units and are of large capacity, in the range 100 kW to 2 MW.

The alternative method of converting solar radiation into electricity involves the use of solar (photovoltaic) cells and is appropriate not only for large-scale power production but also for small-scale units.

A solar cell[12] consists of a thin slice of semiconducting material containing a semiconductor junction. There are electrical contacts on both faces of the slice. When the incident light (note that it is light and not heat that is important here) is absorbed on the semiconducting material, additional electrical carriers are generated. These diffuse through the semiconducting material and some of them cross the boundary between the absorber layer and the base layer. Those that cross the boundary produce the photogenerated electrical current which is collected by the contacts at the top and back of the cell and from there delivered to the load. The absorber and base layers form a semiconductor junction at their boundary and a solar cell behaves in a way similar to that of an ordinary semiconductor diode with additional electrical currents. A voltage applied between the top and back contacts will produce currents, as in any semiconductor diode, and a voltage in the forward sense will produce diode currents in the opposite direction to the photogenerated currents. A forward voltage which produces forward diode currents exactly equal and opposite to the photogenerated currents, giving zero net current, is the 'open-circuit voltage' and is the maximum voltage that the solar cell can produce. The maximum current that the cell can deliver is equal to the photogenerated current and is produced when the opposing diode currents are zero, that is, when the voltage across the cell is zero. This occurs when the cell is short-circuited.

The electrical power that can be delivered is the product of the voltage generated and the current (Ohm's law). To produce the maximum power it is necessary to generate the maximum short-circuit current and open-circuit voltage and operate the cell at its maximum power point. The efficiency with which sunlight can be converted into usable electrical power depends on the efficiency of all the stages outlined, although the practical difficulties in manufacturing cells can also reduce the efficiency of some types of cell significantly. The efficiency of sunlight absorption and conversion to additional electrons and holes depends on the material used.

The photogenerated current will be a maximum if the energy gap of the absorber is sufficiently small for nearly all the wavelengths of the sunlight to be converted to additional electrons and holes, and if the absorber is thick enough for nearly all the light to be absorbed. However, a material with a small energy gap will give large diode currents and hence the open-circuit voltage will be reduced. The electrical power depends equally on current and electricity so there is an optimum energy gap for the absorber material which will maximise the production of current and voltage. The optimum energy gap is about 1.5 eV for bright, direct sunlight.

When the absorbed sunlight has been converted to electrical carriers they diffuse through the absorber layer. The percentage that contributes to the photocurrent depends on the properties of the layer but it decreases with increasing layer thickness. A compromise must be made between the thick absorber layer needed to absorb as much sunlight as possible and the thin layer needed for efficient collection of the photogenerated layer.

The basic compromise in choice of material for a solar cell between a small energy gap to give a large current and a large energy gap to give a large voltage means that even an ideal cell made from perfect materials has a conversion efficiency of only 40 per cent. The other compromises, and the fact that real materials are not perfect, limit the practical efficiency to about 25 per cent. Mass-produced silicon cells usually have efficiencies in the range 10-18 per cent. In bright sunlight, with a light intensity of about 0.1 W/cm^2 a solar cell would generate a current of 25–30 mA from each square centimetre of cell, and a voltage of about 0.5 V.

Solar cells are usually combined in modules that are mechanically strong to allow easy handling and to resist wind-loads, provide protection against the environment and give convenient output voltage. If a number of identical cells are connected in series the total voltage output is the sum of the cell voltages while the total current is equal to the current from each cell. Thus if 24 cells each generating 1 A at 0.5 V are connected in series, the total output would be 1 A at 12 V. If each were connected in parallel the output would be the sum of all the currents at the voltage of an individual cell and 24 cells each generating 1 A at 0.5 V. Where the cells are identical the current or voltage is determined by the least efficient cell in the module. An array is a collection of modules connected in series and/or in parallel to provide the power required for a particular application.

Almost all commercially available cells nowadays are based on single-crystal silicon. Silicon cells are a by-product of the research done on silicon during its development for the electronics industry. Because the electronics industry needs silicon at a higher purity and crystal perfection than does the photovoltaics industry, a number of methods for producing cheaper silicon are being studied.

The conventional process for making silicon solar cells involves the production of trichlorosilane (a compound consisting of silicon, chlorine and hydrogen, $SiHCl_3$) from metallurgical-grade silicon, chemical deposition of semiconductor-grade polycrystalline silicon from the silane, Czochralski growth of single-crystal ingots, production of wafers by slicing this, formation of the junction and the back contact, provision of an antireflective coating, formation of the front contact, interconnection and encapsulation. Alternative methods of producing crystalline silicon cells involve replacing the Czochralski growth of crystal ingots by a casting method and production of ribbons of silicon.

Thin-film solar cells, using materials only a few thousandths of a millimetre thick, are being developed which should enable large areas of cells to be produced cheaply. At the moment such cells have efficiencies of 5–10 per cent. Cadmium sulphide/copper sulphate thin-film cells are promising. Amorphous silicon can also be produced in thin films. Other materials worked on include gallium arsenide/aluminium gallium arsenide, indium tin oxide/indium phosphide, cadmium sulphide/copper indium selenide, and so on.

Solar cells were developed originally as power sources for space satellites and costs then were high, but during the last decade improvements in manufacturing techniques and increased volume of production have reduced manufacturing costs. Additional cost reduction is likely and photovoltaic systems could well become a major energy source in the future. Since they require light rather than heat to operate, their application could be widespread throughout the world.[13]

A considerable amount of development work on photovoltaic cells and systems has been funded by national governments, particularly in the U.S.A., Japan, and Europe.

The price of a module, the biggest single cost element in a photovoltaic system, ranged from \$35 to \$40 a peak watt (Wp) in 1975.[14] (Peak power is the power produced at the nominal working voltage under standard conditions of irradiance ($1000 \ W/m^2$) and temperature ($25°C$).) The average irradiance over the year is much less than this and operational temperatures are normally higher than $25°C$. Thus the average power available from 1 peak kilowatt of solar cells with suitable storage ranges from 100 to 300 W or from 2.4 to 7.2 kWh/day, less any losses in the system (which can amount to 25 per cent). In 1982 modules with better reliability and life than those sold in 1975 were selling from \$7 to \$10/Wp. The costs of such items as the array structure, power conditioning, energy storage, marketing, transportation and installation, can increase the total system price by 2–5 times that of the module. The aim of the nationally funded programmes is to bring down the prices, first, to compete with diesel and gasoline generators used in remote regions, then to become cost-effective as distributed grid-connected generators in industrialised countries and ultimately to supply power on a large-scale for central stations. The U.S. Department of Energy's goals are as follows: in 1982, module prices of U.S. \$2.80/Wp and system prices of \$6.13/Wp for remote stand-alone systems; in 1986, \$0.70/Wp for modules and \$1.6–2.6/Wp for systems for grid-connected dwellings and intermediate load centres; in 1990, \$0.15–0.40/Wp for modules and \$1.10–1.80/Wp for central power stations.

The Commission of the European Communities has had a strong part to play in co-ordinating the European work on photovoltaic cells and systems and in funding programmes of research and development throughout the Community. Particularly useful in establishing what is the real potential for photovoltaic systems in Europe has been their series of pilot projects. There is at least one of these in each Member State and, between them, they cover a number of applications for photovoltaic systems and range in size from 30 to 300 kW. Applications include: power supply for isolated buildings and villages in Denmark, Greece, France, and the Netherlands; power management and control for the airport at Nice, France; hydrogen production and water pumping for industry in Belgium; electricity for a dairy farm on an island near Cork, Ireland; provision of an agricultural cold store in Italy; sea-water desalination in Italy; fresh-water pumping in Italy; power supply to the grid at Marchwood power station in the U.K.; power supply to FM and TV repeaters in France.

The commercial market for photovoltaic systems depends on their cost-effectiveness. The photovoltaics industry is at present marketing a wide range of photovoltaic systems for small stand-alone applications of up to 500 W. In some of these applications photovoltaic systems are economic even at today's prices. They include:

marine and aerial navigation beacons; fog signals; telecommunications relay stations; TV in remote areas; radio telephones for railways and highways; remote data monitoring, including weather stations; cathodic protection of pipelines; security systems; highway traffic-control warnings; battery-charge maintenance for emergency equipment, pleasure boats, holiday homes; pocket calculators, watches; small water-pumping systems; refrigerators; and lighting units.

The market for stand-alone systems in the size range 500 W to 150 kW is mainly occupied at present by gasoline and diesel generators. Batteries or some other form of storage are generally needed with photovoltaic systems in this area. This has an effect on the economics and also on the applications where such systems can be used. Appropriate applications include: power supply to off-grid houses and villages; power for large pumps (where battery storage may not be necessary); power for radio and television transmitters and microwave repeater stations; power for remote radar stations and other military applications.

The potential for grid-connected residential photovoltaic systems in Europe has not been studied in depth. In the U.S.A. such systems are considered by many to offer an early potential market and one that will grow rapidly. A major point in favour of residential photovoltaic systems is that the load is close to the generation source and as a result there are few transmission and distribution losses. From the user's point of view the reliability of his system is improved. From the utilities' point of view allowance must be made for the fact that much of the peak capacity would have to be provided by conventional generating plant.

The service, commercial, institutional, and industrial sectors include many businesses with electrical loads in the range 25 kW to 5 MW. Many of the Commission of the European Communities' pilot projects come into this category. The loads are intermediate between those of residences, which are usually less than 25 kW, and those of conventional central power stations with capacities between 100 MW and over 1000 MW.

The rate at which photovoltaic systems are integrated into the grid will depend on several factors. National policies may require the utilities to provide a certain proportion of generating capacity from indigenous sources to reduce dependence on imported fuel. The utilities may find that a photovoltaic plant can be built up at a rate to suit demand whereas a major thermal installation would take much longer to bring on stream. Photovoltaic systems may require less skilled operating and maintenance staff. On the other hand, limitations on land use may restrict the introduction of the large arrays associated with central generating plant.

Solar-powered Satellites

One way of providing centralised photovoltaic power generation without the use of exceptionally large areas of land is the solar power satellite concept originated by Peter Glaser of Arthur D. Little and studies by the U.S. Department of Energy and NASA.[15] The solar power satellite system, placed in geostationary orbit, would beam power in one or more receiving antennae in desired locations on earth. The satellite could consist of, say, a 10.4 km by 5.2 km array of solar cells. The electrical power would be fed to microwave generators. The phased-array transmitting antennae would be a flat structure about 1 km in diameter. The receiving antenna on earth would be about 10 km in diameter and could be located offshore as well as on land. Land use per unit of power generation would be about a quarter of that required by coal-fired steam plants, including the land required for supplying the coal. The key to the satellite is a space-transportation system capable of placing large payloads in geostationary orbit at low cost. The main argument for solar energy conversion in space is the nearly constant availability of solar radiation in orbit.

Solar Energy in U.K. National Energy Policy

United Kingdom government departments have for some years had small solar energy research, development and demonstration programmes. Although interdepartmental discussions have taken place on specific projects there has been in the past no overall government policy on the subject. This may change in the future: the recent formation of an interdepartmental energy efficiency office may have an influence on the government's activities on renewables.

The main aim of the Department of Energy's solar energy programme has been to examine the potential of solar systems for supplying energy for the U.K. in the foreseeable future and to develop systems of proven economic performance and reliability which will lend themselves to widespread adoption by the consumer. About £13 million was spent by U.K. government departments (mostly the Department of Energy) on renewables research, development and demonstration in 1983; around £1.3 million of this was on solar energy projects. Early on, the Department decided that there were economic difficulties over the use of photovoltaic conversion as a large-scale means of central electricity generation for the U.K. and so it has left support of this sector to the Department of Industry. The Department of Energy has concentrated on the areas of solar water-heating and active and passive space-heating. It is not interested in export of solar technology and products from the U.K. overseas. In 1982, the Department reported on the results of the programme so far and concluded that the potential for solar energy to make a large economic contribution to national energy supplies was in general small compared to that for nuclear and fossil fuel-based systems. It announced that it would continue to fund work on passive solar design for a period since it believed that this sector is of national interest. Active solar thermal systems, in the main, it now regarded as being sufficiently developed for industry to take up, letting the natural market forces determine whether or not the sector will grow in the future.

The Department of the Environment's interest has been on public-sector housing and planning/building regulations, and funding activities to related work in these areas. The Department of Industry is concerned with the activities of the companies in the industry *per se* and has funded developments in both the photovoltaic and thermal sectors, through its various industry support schemes.

In the U.K. householders pay taxes to the local authority based on a valuation of the house. The gross annual value of the house is the rent at which it might reasonably be expected to be let from year to year. The assessor's valuation of a house will take into account improvements if he is of the opinion that they are structural and have a bearing on the letting value. The presence, or installation, of a solar heating system in a rateable dwelling is considered in the same way as any other feature in assessing the gross value for rating purposes. The effect of a solar water-heating system in terms of gross value is dependent on the extent and efficiency of the system and is a matter for the local District Valuer and Valuation Officer to estimate on the evidence and facts of the individual case. At the moment, Section 21 of the General Rate Act 1974 means that improvements in dwellings in England and Wales that do not raise the annual value by more than £30 (and central heating to any value) will be ignored until new valuation lists come into effect. In Scotland the installation of a solar system causes an immediate increase in rates, which hardly seems appropriate in the light of rational energy policy.

Despite the lack of support from government the U.K. has managed to develop an expertise in solar technology. Many universities have departments carrying out work in this field, probably the best known being the Solar Energy Unit at University College, Cardiff, which, under Professor B. J. Brinkworth, has been working in this area since 1960.

The U.K. society concerned with solar technology is the U.K. Section of the International Solar Energy Society in London. The U.K. solar industry really began in 1973 and consists of about 70 business organisations. Most of these are small entrepreneurial companies although in recent years a few public companies have entered the market place and there are also a small number of small but experienced engineering/architectural practices working in this field. The largest number of concerns are in the solar thermal area — solar water and space-heating (both active and passive) — but a strong photovoltaics sector has been developed.

Most of the companies and practices are members of the Solar Trade Association, which aims to raise standards in the industry, to provide a greater understanding of the advantages of solar energy use for the public, government and news media, and to support industry in its export endeavours. Since 1979 members on the water-heating side have been obliged to adhere to a code of practice covering advertising, selling, installing, service and repair. The Association runs conciliation and arbitration services.

There are about 55-60 commercial organisations in the solar thermal sector employing roughly 200-300 people. Sales turnover in 1983 was about £12-15 million, around 10 per cent of which was due to overseas sales; imports amounted to around £0.3 million. About 20 of the concerns manufacture solar collectors and other components for solar domestic hot water, swimming-pool water and active space-heating. There are 20 or so installers of such systems.

Interest in passive solar design of new buildings has grown over the past few years. In 1983 there were around 500 passive solar houses in existence in the U.K. Retrofit of passive components in existing houses is still not common.

The relatively low take up of solar thermal technology in the U.K. to date has been due largely to general lack of awareness of what solar systems can do and of the benefits they can bring to the individual customer. Government's public comments through the Department of Energy have been unenthusiastic since they have generally considered the savings that can be achieved nationally in the relatively short term, rather than the investment and employment opportunities presented. The industry is small and does not have funds for education and publicity. The economic situation has caused U.K. sales to decline slightly over the last couple of years and the solar industry has tended to look away from the U.K. to the world market. The U.K. photovoltaics industry's sales are mostly outside the U.K. and are currently of the order of £2–3 million a year, representing about 5 per cent of the world market.

The Future

It seems possible for perhaps some 70 per cent of U.K. dwellings to have solar domestic hot-water systems. These would give national savings of as much as 3.6 mtce a year, which is about 1.6 per cent total current energy consumption or 5.6 per cent of the energy used in dwellings. Solar heating of the domestic hot water used in commercial and industrial buildings and of industrial process water, where appropriate, could become widespread. If all U.K. swimming pools had solar heating, the annual amount of energy collected by these systems would be equivalent to around 0.2 mtce. If passive solar measures were used in new building design and retrofitted into some 40 per cent of existing houses, the savings achieved might be 0.5–1.5 mtce a year in addition to the potential savings from non-domestic buildings.

To achieve installation of solar domestic hot-water systems in 70 per cent of U.K. dwellings by the year 2000 would represent an industry growth rate of 50 per cent per year.[16] Widespread use of solar thermal technology in the U.K. would, therefore, not only bring savings of conventional energy but create jobs in installation: with a growth rate of 50 per cent per year getting on for 200,000 jobs might be created by the year 2000. Increased sales of solar water-heating systems for non-domestic buildings and industrial process water and of manufacture and installation of passive solar components will bring additional jobs, as will increased work overseas.

Advantages and Disadvantages

The sun provides the world with sufficient energy in a given period to more than meet its requirements. The availability of this solar radiation is not dependent on political situations. The harnessing of this energy source is carried out by benign processes which are not environmentally harmful and which are, in the main, technologically simple. At a given location at a particular moment, however, there may not be enough solar energy to meet all man's needs. Therefore, not only adequate harnessing techniques but also good and cheap methods of storing solar energy from when it is available to when it is needed have had to be developed. This development work is, in the main, costly, and for some solar technologies the developed products will not be economic compared with conventional energy forms until the latter have risen in price. In countries such as France where there are no indigenous sources of fossil fuels the national government has been prepared to subsidise to a considerable degree the development of solar-energy utilisation and to help it gain a place in the market in order to hasten the day when solar processes can compete on equal terms with the conventional fuels. In other countries, such as the U.K., fossil fuels are available in adequate quantities for the short term and the national government has been hesitant to spend money to develop benign alternatives for the long-term future. Therefore, in such countries, solar products have had to compete on an equal footing with conventional fuels from the beginning. However, much expertise has been gained worldwide in companies, universities, polytechnics, and government laboratories in the last decade or so. If this is sustained then solar processes will play a real part in meeting energy needs in the future, as conventional fuels become less easily available. If governments and individuals insist on allowing short-term goals to influence their use of funds in the next few years, then the past development work may become lost and widespread use of solar technology will not be so likely before the end of the century.

References

1. Brinkworth, B. J., *Solar Energy for Man*, Compton Press, Salisbury, 1972
2. Butti, K. and Perlin, J., *A Golden Thread: 2500 Years of Solar Architecture and Technology*, Marion Boyars, London, 1981
3. Commission of the European Communities, Directorate General for Science, Research and Development, *European Passive Solar Handbook*, (in preparation)
4. Solar Trade Association Ltd, *Solar Water Heating: What's in it for me?*, Solar Trade Association, London, August 1982
5. Tabor, H., 'Solar Ponds – Progress and Potential', in *Proceedings of Conference, International Solar Energy Society, UK Section, Midlands Branch, September 1982*, UK-ISES, London, 1982, pp. 39–43
6. Ali Kettani, M., 'Review of Solar Desalination', *Sunworld*, Vol. 3, No. 3, 1979, pp. 76–85
7. Szokolay, S. V., *Solar Energy and Building*, 2nd edn, Architectural Press, London, 1977
8. Brattle, L. V., 'Cooking Fuels in Third World Countries', in Stambolis, C. (ed.), *Energy Sources for Developing Countries*, UNESCO-sponsored study, Heliotechnic Press, London, 1982, pp. 156–77
9. Tomkins, R. and Wereko-Brobby, C. Y., 'Solar Refrigeration for Developing Countries: the Prospects', in *Solar Energy for Developing Countries: Refrigeration and Water Pumping, Proceedings of UK-ISES Conference, January 1982*, UK-ISES, London, 1982, pp. 14–25
10. Gretz, J., 'Solar Thermal Power Generation. The Example of the European 1 MW(e) Power Plant "Eurelios" ', in *Solar World Forum: Solar Technology in the Eighties, Proceedings of ISES Conference, Brighton, August 1981*, Vol. 4, Pergamon, Oxford, pp. 2705–15
11. Furbo, S., 'Heat Storage Units using Salt Hydrates', *Sunworld*, Vol. 6, No. 5, 1982, pp. 134–9
12. Hill, R., 'The Properties and Production of Solar Cells', *Sun at Work in Britain*, No. 15, 1982, pp. 1–8

13. Starr, M. R., 'Photovoltaic Prospects in Europe', *Sun at Work in Britain*, No. 15, 1982, pp. 9–20
14. Treble, F. C., 'Photovoltaics – from Daylight to Electricity in One Step', in *Proceedings of Conference, International Solar Energy Society, UK Section, Midlands Branch, September 1982*, UK-ISES, London, 1982, pp. 31–7
15. Glaser, P. E., 'Solar Power Satellites', *Sunworld*, Vol. 4, No. 1, 1980, pp. 26–7
16. Turrent, D., Baker, N., Steemers, T. C. and Palz, W., 'Solar Thermal Energy in Europe: an Assessment Study', Vol. 3 of *Solar Energy R & D in the European Community, Series A*, Reidel, Dordrecht, 1983

Further Reading

BS 5918: 1980 Code of Practice for Solar Heating Systems for Domestic Hot Water, British Standards Institution, London

Lebens, R. (ed.), *Passive Solar Architecture in Europe: the Results of the 'First European Passive Solar Competition: 1980'*, Commission of the European Communities, Architectural Press, London, 1981

Palz, W. (ed.), *European Solar Radiation Atlas*, Commission of the European Communities, Groesschen-Verlag, Dortmund, 1979

Jaeger, F. (ed.), *Solar Energy Applications in Houses: Performance and Economics in Europe*, Commission of the European Communities, Pergamon, Oxford, 1981

McCartney, K., *Practical Solar Heating*, Prism Press, Dorchester, Dorset, 1978

McVeigh, C. and Schumacher, D., *Going Solar: a Practical Guide to Solar Water Heating*, Schumacher Projects, Church House, Godstone, Surrey, 1981

Solar Radiation Data for the United Kingdom, 1951–75, publication Met. 0.912, Meteorological Office, Bracknell, Berks.

Wozniak, S. J., *Solar Heating Systems for the UK: Design, Installation, and Economic Aspects*, Department of the Environment, Building Research Establishment Report, HMSO, London, 1979

7

Biomass

'What most people do not realise is the magnitude of present photosynthesis. It produces an amount of stored energy in the form of biomass which is about ten times the world's annual use of energy'

David Hall

Biomass is the generic term given to dry plant materials and organic wastes. It is also used to describe vegetation and wastes that are subsequently used for conversion into fuel. All biomass products are originally derived from photosynthesis and are potential sources of energy. For centuries biomass has been used by man in the form of wood and charcoal. The fossil fuels oil, coal, gas and peat were also once biomass, transformed over hundreds of thousands of years by natural geological forces and micro-organisms.

Since 1973, when oil prices began to rise and future shortages were anticipated, considerable interest has been shown in the possibility of using various present-day biomass sources to produce oil substitutes and chemical feedstocks. Many sources possess an average energy content of 4 kcal/g, about half the energy content of bituminous coal. The energy content of some plants is even higher than coal and as much as oil. These plants can be specially cultivated and, by means of different processes, converted into storable gaseous, liquid or solid fuels and feedstocks. Today biomass from different sources contributes 15 per cent to world fuel supplies, the equivalent of some 20 million barrels of oil per day or approximately ten times as much as is currently extracted from the North Sea.[1]

Photosynthesis

Green plants collect and store energy through the process of photosynthesis. It is the world's oldest form of energy conversion and has sustained all forms of life on this planet for over 2 billion years. Chlorophyll in the plant harnesses the energy in sunlight to build carbohydrates from carbon dioxide and water. The green plants

157

additionally use sunlight to produce other substances such as acids, hydrocarbons, fats and proteins in varying proportions depending on the species.

In photosynthesis a green plant synthesises its complex organic constituents from carbon dioxide, the source of carbon, and water and inorganic salts from the soil. The chlorophyll in the plant acts as a photosensitiser and absorbs visible light to make it available for photosynthesis. In the process the carbon dioxide is absorbed by the chlorophyll. In addition water is decomposed and the hydrogen released combines with the carbon dioxide to form carbohydrates. This reaction may be summarised as follows:

$$nCO_2 + nH_2O \xrightarrow[\text{chlorophyll}]{\text{solar radiation}} (CH_2O)_n + nO_2$$

In effect, therefore, two different reactions take place. The first is the water-splitting reaction which evolves oxygen as a by-product and is therefore essential to all life. The second is the fixation of carbon dioxide from the atmosphere to organic compounds which forms all our food and fuel resources. In energy terms the importance of photosynthesis is that the sun's radiant energy is converted into chemical energy in the carbohydrates etc., which, save for that consumed in the process of plant growth itself, becomes *stored* energy. Different plant species produce different forms of carbohydrates and hence different types of storable energy. Sugars represented by $C_{12}H_{22}O_{11}$ can be stored for relatively short periods. Cellulose and the more complex starches $(C_6H_{10}O_5)_n$ are capable of much longer storage periods.

Energy-conversion Efficiencies

The suitability of a particular plant species as a biomass energy source depends primarily on the efficiency of its conversion of solar energy into stored chemical energy. Unfortunately, compared with most other energy-conversion methods photosynthesis efficiencies are generally very low on first appearance. Only a very small proportion of all the solar energy that strikes the earth's surface is converted to energy stored in plant material. Of the 3×10^{24} J of total solar radiation that reaches the earth's surface annually only the equivalent of 3×10^{21} J is stored as biomass. Half of the total available solar energy falling on plants is lost since plants can only use radiation in the visible part of the solar spectrum, between wavelengths 0.4 and 0.7 microns, known as the photosynthetically active radiation (PAR) region. This is further decreased by losses suffered through reflection, respiration requirements and the physical requirements of 8 quanta of light being needed to fix the CO_2. Thus only about 5.5–6.7 per cent of the original light is considered practically convertible into chemical energy.[1,2] Moreover, in average year-round agricultural conditions the efficiency drops still further to about 0.3 per cent on land and even lower efficiencies in the oceans.[3] In certain crops, efficiencies are higher, up to about 1.5 per cent in typical good conditions in temperate climates and 2.5 per cent in tropical zones. Table 7.1 shows the photosynthetic efficiency for selected agricultural crops in different regions. It is this efficiency that deter-

Table 7.1 *Productivity and energy conversion in agricultural crops on an annual basis. Yields of total dry matter in tonnes/hectare/year and photosynthetic efficiency expressed as energy yield as a function of total radiation over the year*

Crop	Country	Yield	Photosynthetic efficiency (%)
Temperate			
Rye grass	U.K.	23	1.3
Kale	U.K.	21	1.1
Sugar beet	U.S.A., Washington	32	1.1
	U.K.	23	1.1
Wheat (spring)	U.S.A., Washington	30 (total)	1.1
	U.S.A., Washington	12 (grain)	0.4
	U.K.	5 (grain)	0.2
Maize	Japan	26	1.1
	U.K.	17	0.9
	U.K.	5 (grain)	0.2
	U.S.A., Kentucky	22	0.8
	Canada, Ottawa	19	0.7
	U.S.A., Iowa	16	0.5
Potato	Netherlands	22	1.0
	U.K.	11	0.5
Sorghum	U.S.A., Illinois	16	0.6
Barley	U.K.	7 (grain)	0.3
Rice	Japan	7 (grain)	0.3
Sub-tropical			
Sorghum	U.S.A., California	47	1.2
Sugar beet	U.S.A., California	42	1.2
Alfalfa	U.S.A., California	33	1.0
Bermuda grass	U.S.A., Georgia	27	0.8
Maize	U.S.A., California	26	0.8
	Egypt	29	0.6
Potato	U.S.A., California	22	0.6
Rice	U.S.A., California	22	0.6
	Australia, NSW	14 (grain)	0.4
Wheat	Mexico	18	0.5
	U.S.A., California	7 (grain)	0.2
Tropical			
Napier grass	El Salvador	85	2.4
	Puerto Rico	85	2.2
Sugar cane	Hawaii	64	1.8
Oil palm	Malaysia	40	1.4
Cassava	Malaysia	38	1.1
	Tanzania	31	0.8
Sugar beet	Hawaii (two crops)	31	0.9
Maize	Peru	26	0.8
	Thailand	16	0.5
Rice	Peru	22	0.7
	Australia, NT	11 (grain)	0.2
Rice and sorghum (multiple cropping)	Philippines	23 (grain)	0.7
Sorghum	Philippines	7 (grain)	0.2

Data from Cooper who assumes photosynthetically active radiation is 45 per cent of total radiation.
Source: Solar Energy − a UK Assessment, ISES-UK, London, 1976.

mines the energy content of dry-weight yields and hence the potential energy contribution of individual crops according to location.

In order to increase energy yields, considerable efforts have been made to raise photosynthetic efficiencies. Under good conditions algae achieve 3–5 per cent efficiencies and certain 'energy crops' have attained short-term conversion efficiencies of up to 4.5 per cent. For example, in the tropical area of Northern Territory, Australia, bullrush millet achieved an efficiency of 4.3 per cent; in the sub-tropical climate of California, U.S.A., Sudan grass has reached 3 per cent; and in temperate areas of the U.K., sugar beet has achieved 4.3 per cent.[4]

Sources of Biomass

There are three main sources of raw materials from which biomass can be obtained: the first is by the harvesting of natural vegetation, the second is by the cultivation of a special 'energy' crop grown for its energy or combined energy content, and the third is through the use of organic waste products obtained from industrial, domestic or agricultural sources.

1. Natural Vegetation

Biomass from natural vegetation has been used mainly in the form of *wood* from trees. Wood is 50 per cent cellulose, 25 per cent hemicellulose and 25 per cent lignin. Dried wood has an average energy content of 1.5×10^7 kJ per metric tonne. When forests are cut down for paper-making or timber, over 50 per cent of tree growth is usually wasted. This waste nevertheless contains cellulose for possible conversion into fuels such as methane or charcoal, or for direct combustion after drying. If trees are not burned directly to provide heat energy, *charcoal* can be produced by heating the wood in the absence of air ('destructive distillation'). Charcoal, a traditional fuel in many rural communities, can have about double the energy content per unit weight of wood. It also burns more slowly. The main disadvantage is that a significant amount of the energy content of wood is lost in the conversion process. It requires between 3 and 12 tonnes of wood (depending on the type of wood and the kiln) to produce 1 tonne of charcoal which has an energy content of only 2 tonnes of wood. In some processes, such as electricity generation or combustion in industrial boilers, direct use of wood is preferable, especially if the industrial application is situated near to the fuel-wood supply source. If this is not the case, conversion to charcoal is economically attractive because of the very high cost of fuel for transport that would otherwise be involved. Charcoal is also more suitable where very high temperatures are required, such as for cement and steel production, and has a lower sulphur and ash content more appropriate to certain specialised applications. On the other hand, when converting natural vegetation to charcoal, care must be taken that the very large fuel requirements of heavy industries do not exceed the scope of reforestation programmes.

About 90 per cent of the world's energy stored in biomass is in the form of trees.

This is mainly used in Third World countries (see chapter 11). In some regions, it accounts for as much as 95 per cent of total energy consumption. The advantage of natural vegetation as a fuel, especially for remote communities, is that it is frequently free and locally accessible despite the fact that annual yields are, at best, only 50 per cent of those of specially grown 'energy crops'. With 'energy crops', density can be maximised and planting kept uniform to facilitate mechanical harvesting. However, natural vegetation contains a variety of species to stabilise the ecosystem which makes large-scale mechanical harvesting difficult. A more urgent problem in many countries, particularly in the Third World, is that cutting and burning of wood is uncontrolled and not accompanied by soil enrichment and replanting programmes. The result is serious ecological and environmental deprivation through soil erosion, flooding and desertification. In 1970, it was estimated that more than 100,000 hectares of land in Libya, Algeria, Morocco and Tunisia alone were lost to the desert annually through human activities. In the 1980s the world loses, on average, an area of forest the size of England and Wales annually through tree-felling alone. The fuel-wood crisis has become the second energy crisis.

2. Energy Plantations and Energy Crops

Many countries have investigated the possibility of planting crops for use as biomass. According to Kemp, one of the earliest pioneers of the energy plantation concept,[5] the 200-mile Paulista Railroad built in 1903 to serve São Paulo in Brazil was the first recognised example. Six-hundred eucalyptus plantations supplied wood to fuel the steam locomotives which continued to be operated by this power source up until 1958 when the engines were replaced by diesels.

Several authorities have estimated the land area needed for energy-crop cultivation in order to provide for a country's energy requirements. Calculations for the U.K. in 1976 estimated that at a 1 per cent conversion photosynthesis efficiency, one-tenth of total energy requirements could be met by about 9 per cent of the land area,[5] although the energy expended in collecting and subsequent processing of the biomass was not included. By comparison, all Ireland's energy needs could be met by 11 per cent of its land.[6] In Australia in 1977, it was calculated that, allowing for energy inputs in harvesting and processing, 3 million hectares of land would produce one-third of total energy requirements.[7] In the U.S.A., it was calculated that a national total of some 430 million dry tonnes of biomass was available each year from agricultural residues and this could theoretically supply 31 per cent of the nation's electricity, 20 per cent of the natural gas or 8 per cent of its oil requirements. In practice, only about one-fifth of this might be collectable.[8] Further studies estimate that biomass fuels could supply some 9–20 per cent of the total 1980 U.S. fuel consumption by the year 2000, with wood, forage crops and municipal waste making the main contributions.[9]

In addition to these theoretical calculations, the many practical problems encountered in growing, harvesting, processing and distributing large quantities of biomass must be considered.[10] First, the agricultural conditions should favour maximum productivity. The light intensity, temperature, type of plant, nutrients and water supply, and freedom from disease are crucial determinants of plant growth. Ample irrigation and highly nutritious soil conditions (notably the non-renewable fertiliser phosphate) are therefore prerequisites for high-yielding crops, even in sunny, warm conditions. Second, it is obviously important that the energy used in producing the biofuel is less than the energy obtained in the output, unless this energy form is not available from any other source. In very large plantations covering thousands of acres, with large and expensive mechanical-handling equipment, energy-intensive processing technologies to produce the fuel, and long-distance transport, the 'energy ratio' can easily become unattractive.

Many short-rotation fast-growing tree plantations in fact achieve yields of well over 20 tonnes per hectare per year of harvested dry material. However, these require a water supply of at least 500 l/m^2/year. Tree crops can be difficult to grow in arid regions although, once established, they help to encourage rain and can dramatically affect the climate. An example is Les Landes on the French Atlantic coast where thousands of hectares of pine forests were planted on the dry wind-swept sand dunes during the nineteenth century. The region now enjoys a plentiful rainfall and is one of France's main timber-producing areas.

Energy crops and biofuels can best be cultivated at competitive costs by choosing the appropriate plant species, planting density and harvest schedule for each plantation site for its specific market requirements. Fuels can be used on site to power industrial machinery, or converted to electricity or other energy forms for use elsewhere. Usually methanol is produced from wood through gasification to produce synthesis gas which is then transformed into methanol by catalytic conversion. However, where electricity generation rather than oil substitution is the main requirement, well-designed wood-burning equipment and modern generation techniques can produce combustion efficiencies almost as high as those for oil-burning facilities.[2] In Brazil, for example, tree energy plantations are being developed to provide methanol, charcoal (mainly for the steel industry) and for burning directly to produce electricity from steam. This appears to be a short-term economic, rather than a long-term ecological solution.

Some Examples of Energy Crops. Since the productivity of different tree and plant species varies considerably under different conditions, much research has been undertaken to discover the most suitable species for intense 'energy farm' cultivation and to find ways of increasing conversion efficiencies.

Eucalyptus and pine forests were an early biomass source. By 1980 Australia was producing significant quantities of methanol from eucalyptus as a substitute for rising oil imports, while Sweden, Finland and other Scandinavian countries also planned to utilise their pine forests for fuel.

Sugar cane is a particularly versatile and interesting energy source. As already mentioned, it has a comparatively high rate of energy conversion. Approximately half its carbohydrate yield is in a form that can be easily processed into sugar or converted to ethanol by direct fermentation. It yields about 70 litres of ethanol (ethyl alcohol) per tonne and ethanol can also be made from molasses, a by-product of sugar production. Bagasse, the fibrous woody residue after the juice has been extracted, can be used to fuel the sugar-processing plant or to power the alcohol distillery. At present most sugar is consumed as food, but there is an increasing emphasis on its fuel value especially in countries where there is a surplus and where fuels have to be imported. Sugar may also be converted into esters through chemical reaction with natural oils, such as coconut oil or tallow, to produce biodegradable detergents. As esters are derived from non-toxic natural products, they may be used in soaps, cosmetics, shampoos and animal feeds in place of some of the petroleum by-products. By 1980 Japan was marketing sugar-based detergents, and other petrol-importing countries were considering replacing petroleum chemical imports by domestically produced sugar esters. Other chemical derivatives of sugar are used in emulsives, lubricants, plasticisers, resins, adhesives and paints. As with similar bio-mass sources, new derivatives are extending the possible range of applications to replace the increasingly expensive oil by-products. Whether sugar is used as food, animal feed or fuel will depend on national requirements in each country, and whether surplus sugar crops are exported or converted to fuel will be a matter of economic or political policy.

Sugar beet and other crops are used to produce alcohol through a more expensive and elaborate process although this frequently results in a net energy deficit. The main benefit is to obtain a liquid fuel where no alternative equivalent is available. World production of cane and beet sugar molasses was 33.5 million tonnes in 1978/79 representing a surplus of some 5 million tonnes, much of it originating in countries with inadequate fuel reserves. This surplus is therefore a useful potential ethanol source and oil substitute.

Sweet sorghum is now being cultivated as an energy crop although it was previously considered only for its food value. It has a short growing season of 3–4 months and can be grown on sugar-cane land between plantings, thus extending the alcohol-production season and reducing the capital charges on the fermenting and distilling plant. Sweet sorghum can yield up to 4000 litres of ethanol per hectare assuming two crops a year, which is exceptionally high. It also has the added advantage, together with certain other plants such as maize and sugar cane, that it fixes carbon dioxide via a C_4 photosynthesis pathway and unlike P_3-type plants such as wheat, can grow at higher light and temperature regimes using less water.

Alder, spruce, willow, sycamore, and *poplar plantations* are also favoured for energy farms since they are quick-growing and can be densely planted. The wood may be cut at regular intervals since new shoots sprout continuously from their stumps.

Soya beans, sunflower and palms can provide oil, which may be used directly in diesel engines, or blended with diesel oil. Groundnut and sesame are other possibilities. Oil produced from plants such as sunflower (see figure 7.1) can be refined in much the same way as crude oil.

Experiments with *grain* in Nebraska, which benefits from occasional huge grain surpluses, showed that a bushel of mildewed grain could be fermented by micro-organisms to produce just over 2 gallons of alcohol. When the alcohol is mixed with gasoline in a 1 to 5 blend to make gasohol it may be used in a normal car engine without adverse effect. If the car engine is modified the efficiency can be greatly

Figure 7.1 *Sunflowers*

improved. Tests on cars running on gasohol alone demonstrate an improved fuel consumption in the order of 5 per cent with a higher octane rating (120 against 98 from 4-star petrol) and less pollution. The cooler burning temperatures of alcohol allow engines to run efficiently at higher compression ratios.

The *wild rubber plant, Euphorbia Lathyris (caper spurge)* and *Euphorbia Tirucalli* both possess a latex with a very high hydrocarbon content similar to crude oil. It has been estimated by the Nobel prizewinner Melvin Calvin that these plants could provide over 20 barrels of oil equivalent per acre at a reasonable cost after only 7 months of cultivation.[11]

Copaiba Multijuga is a species of copaiba found mainly in Brazil which forms oil in its leaves and is capable of producing up to 7 litres of oil in 2–3 hours when a hole is bored in the trunk. Although diesel trucks have been run on oil taken directly from the tree this source is not yet (1983) commercially competitive, but offers possibilities for the future.

Aquatic plants. Experiments have suggested that large-scale aquatic energy plantations may also be developed, since photosynthesis takes place whenever there is light together with sufficient nutrients to sustain plant life. However, ocean areas are likely to be comparatively low in productivity unless natural flows or artificial devices, such as ocean thermal energy or wave energy systems, bring the nutrients to the sunlit surface.

The *water hyacinth (Eichhornia Crasspipes)* is a fast-growing weed found in freshwater rivers, streams and drainage ditches in many parts of the world, especially in tropical countries. It was first developed by NASA as an energy crop for the produc-

tion of methane and to purify water contaminated by toxic waste. It has the
capacity to double its volume in 8-10 days in warm nutrient-enriched water and to
produce yields of over 70 tonnes per hectare per year on a dry-weight basis. In
addition to its phenomenal growth rate it has the ability to absorb metals and other
toxic substances such as lead and mercury from industrial waste water, sewage or
other contaminated water. A hectare of Eichhornia Crasspipes grown on sewerage
nutrient may yield 0.9-1.8 tonnes of dry matter each day, optimum growth rates
being obtained between latitudes 32°N and 32°S.[12]

Cassava (manioc) is a most promising biomass source since it may be cultivated
on the poorer marginal lands and has fewer fertiliser requirements. It is a highly
efficient converter of solar energy into starch which gives a high yield of ethanol.
As it requires labour-intensive cultivation with small commercial energy input, it is
particularly suitable for developing countries and thus research is now directed
towards increasing yields and also resistance to disease.

Finally a concept propounded by the late Richard St Barbe Baker and still in its
infancy is that of *agroforestry*. Certain tree species, for example, the honey locust,
can supply both food and fuels while at the same time enriching the soil. Such three-
dimensional crops avoid the food *versus* animal feed or fuel dilemma and are of
great value in reforestation programmes. Studies are currently under way to discover
the most appropriate agroforestry crops for specific types of environment and in
1980 the International Council for Research in Agroforestry in Nairobi published
a manual on the subject.

3. Organic Wastes

Organic wastes suitable for conversion into biofuels include: industrial waste such as
paper and pulp; urban garbage; domestic sewage; and agricultural and forest wastes
(stalks, dung, leaves, branches, undergrowth and so forth). Urban and industrial
wastes have the advantage that they are more easily collected and otherwise create
an increasing disposal problem. Further details may be found in chapter 10. Domestic refuse and agricultural wastes (frequently by-products of food production) are
other valuable potential energy sources. Much forest waste from the timber and
paper industries used to be disposed of by burning on site because of the high cost
of collection, transport and handling. Now, wood chippings from sawmills are
frequently used by the paper industries to power their plants. The integrated
harvesting of crops for both food and fuel is also under development in many areas,
the fuel components being supplied as the by-product (formerly waste) of the crop.
Examples include cereals such as maize, where residues may be harvested for their
energy content (rather than burned off) and then fermented to produce ethanol. A
number of crops have cellulose residues which in an era of cheap energy and costly
labour have proved more economical to destroy than to collect and process. High
fuel costs are now leading to a re-assessment of the role of such wastes both as
sources of energy and fertilisers. In the U.K., the Energy Technology Support Unit
(ETSU) estimated in 1980 that 20 mtoe per year or the equivalent of 15 per cent of

the oil and gas used could be obtained from biomass in the form of wood, straw and other waste residues by the end of the century.

Biomass Conversion Processes

There are several well-known processes by which energy can be derived from organic materials. Apart from direct combustion (to produce heat, or electricity from steam) the main processes are the aqueous processes (anaerobic digestion, alcoholic fermentation and hydrogenation) and the dry chemical processes (pyrolysis and hydrogasification). Figure 7.2 shows potential biomass conversion methods, fuels and chemical by-products. It can be seen that the major fuels that can be produced, including methane, methanol, ethane, char, ethanol and oils, will give direct heat and hence steam and electricity, as well as transport fuels. Ammonia and resins are also produced, as are a variety of chemical feedstocks which have immense value in replacing petrochemicals. The direct production of oils from certain plants, and animal power may also be regarded as biomass energy sources.

In the *anaerobic digestion process* organic matter is fermented in the absence of oxygen to produce methanol. Most biomass, except for wood, may be anaerobically digested. The process has been used for many years to power the machines in sewage-treatment plants (over 5000 in the U.K.). It is also used in the fermentation of simple sugars to form ethanol. Methane itself, as has already been seen in chapter 3, is a high-grade fuel suitable for cooking, heating and certain industrial applications, having an energy content of 40 MJ/m^3 — a potentially valuable energy source in areas with biomass but no natural gas reserves. In an early industrial example in 1975 the U.K. company Biochemicals Ltd implemented a method for full-scale anaerobic treatment of medium and high-energy wastes in the brewing industry. In this particular anaerobic digestion process the bacteria converted the organic polluting material from distilleries to gas since distillery effluents contain a large proportion of biodegradable matter which is broken down to form methane and carbon dioxide. Waste heat from the factory was used to maintain the temperature in the digester at a stable 35-38°C. The methane was then recycled to the factory as fuel, considerably reducing overall energy costs.[13]

One of the best-known applications of the anaerobic digestion process is the extraction of methane from human and animal wastes. Several countries such as India, Korea, Indonesia, Taiwan and China have developed biogas plants to provide methane and fertiliser in remote rural areas (see figure 7.3). Since rising oil prices have deprived many Third World countries of the kerosene fuel on which they previously relied, biogas production is increasing (see chapter 11). In this process the waste material is usually mixed with water to form a slurry which is maintained at a constant temperature within the range of 15-60°C, while a series of bacteria decompose the organic matter and synthesise methane. Biogas normally consists of 56-70 per cent methane, 43-29 per cent carbon dioxide and 1 per cent hydrogen sulphide. The main problem with biogas digesters is that seasonal temperature shifts can halt or slow down the digestion process. This is particularly difficult for countries such

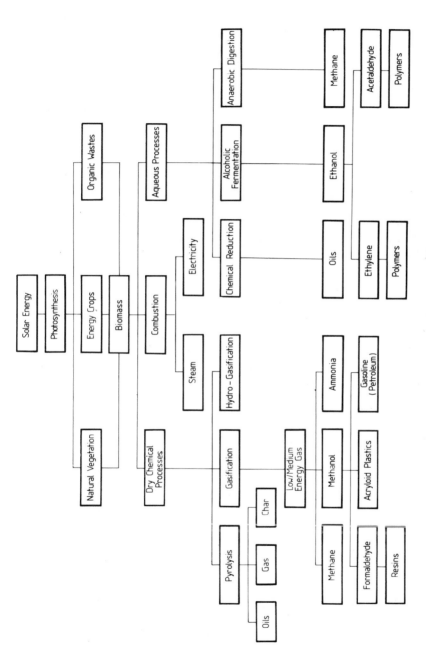

Figure 7.2 *Biomass sources: energy-conversion processes and products*

as China and Korea where winters can be cold and methane production falls at a
time when fuel requirements increase. By increasing the temperature of the slurry
the amount of biogas is increased and the retention time decreased. In the large-
scale digesters such as those in North America or the Kibbutz Zikam in Israel,
integrated systems are used where the slurry is artificially heated, the gas used to
raise steam to power internal combustion engines which drive electricity generators
and the waste heat or a portion of the steam recovered for heating the digester
contents. Such integrated units, if supply is well matched to demand, can be 80 per
cent efficient according to a recent report by Stewart Crocker. Fluctuation in
acidity or rate of feed can also affect production. A third problem is that of finding
the most cost-effective scale for biogas plants. Previously, it was thought that to
build the larger community plants was more economical but studies in India have
shown that this may not necessarily be the case. Local cooperation is essential.
Despite the fact that sometimes equipment is expensive, methane digesters have
been developed using both domestic sewage and animal dung. The domestic version
suffers from the drawback that the average amount of sewage produced by a typical
(British) family will yield only about 0.1 m^3 of methane per day, while it might
require an average of 0.9 m^3 for cooking alone. Even in Third World countries
family-sized plants frequently have insufficient waste to satisfy their fuel needs
unless the family possesses livestock. This form of methane production is a more
valuable resource on farms where animal waste is abundant and the fuel can be used
for cooking, lighting, heating, pumping or for electricity generation. Also the
nutrient-rich sludge residue is an excellent fertiliser, possessing a higher nitrogen
content than the original dung; it is also used as animal and fish feed. In the U.K.
for example, in Edenbridge, Kent, in 1979 a prototype commercial methane
digester went into production using the dung of 320 Friesian càttle. The methane
powered a gas generator which produced up to 29 kW of electricity[14] to supply the
farm while the residue was fed back on to the land as fertiliser.

Alcoholic Fermentation

Another biomass conversion method is alcoholic (aerobic) fermentation, in which
materials containing simple sugars and starches can be fermented and distilled into
ethanol and ethane gas. For ethanol, the ligno-cellulose content of plants is first
hydrolysed into sugars, lignin and resin, using enzymes or chemicals. The glucose
sugar is then fermented by yeasts and distilled into ethanol, as is shown in figure
7.4. Ethanol (a liquid fuel) has the added advantage that it can replace oil for certain
uses such as fuel for transport. In fact, even during the Second World War, ethanol
derived from plants such as potato and sugar beet was blended with gasoline and
used widely in Europe for industrial and other purposes. The main disadvantage of
ethanol production is that it can require more energy for the distillation process
than is available in the final product if the distillation is inefficient. Ethanol
(C_2H_5OH) has an energy content of 22 MJ per litre and methanol (CH_3OH) rather
less. However, 109 J are required to produce 100 J worth of ethanol, although a
further 14 J may be derived from the by-products. In some cases, this may be com-

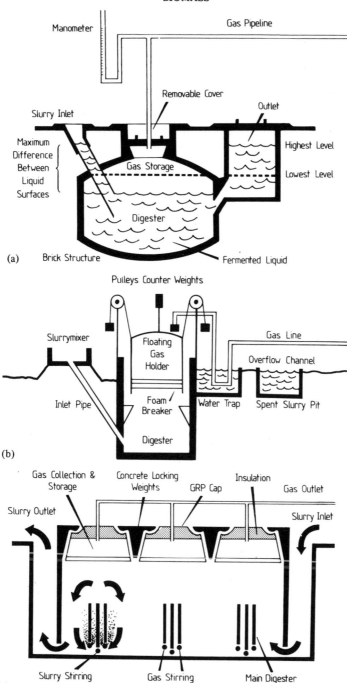

Figure 7.3 *Three types of methane digester: (a) typical Chinese;*
(b) typical Indian (Gobar); (c) commercial size, Edenbridge, England

pensated by ensuring that the biomass is derived from waste and therefore energy used in producing the initial biomass source can be discounted. If, as is likely, water-absorbent materials can be introduced to replace part of the distillation process there could even be an overall energy gain of about 5:1.[15] Absorbers that take water from the ethanol concentrate may be recycled and solar dried or produced from biomass materials as primary feed to the fermentation plant, further reducing the energy requirements. Already by 1980, several motor vehicle manufacturers, such as Volkswagen, Fiat and Chrysler were marketing cars to be run off methanol and 95 per cent hydrated ethanol.

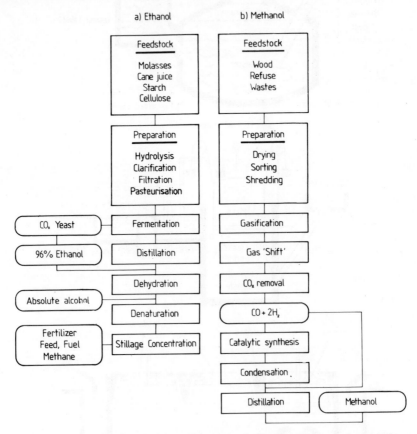

Figure 7.4 *Flow diagrams for the formation of ethanol (with by-products) and methanol from biomass feedstocks* (source: J. Coombs, *Solar Conversion through Biology*, Paper No. BA142/CIF, Tate and Lyle Ltd)

Hydrogenation

This is the chemical reduction of biomass with carbon monoxide and steam to produce heavy oil. However, this method requires very great pressure (more than 100 times atmospheric pressures) and, even when using urban refuse, forest and agricultural residues, it is not cheap to operate.

Pyrolysis

In this process the biomass is heated to temperatures between 500 and 900°C in the absence of oxygen. Products include a mixture of gases, methanol, light oil and flaky char, the proportion of each being a function of generating conditions. Pyrolysis is a particularly useful process with woody biomass which cannot yet be digested anaerobically. True pyrolysis is endothermic and requires an external heat source. However, some hybrid systems employ combustion to produce heat. By 1980, over 12 different pyrolysis systems had been developed. These included: the Garrett 'Flash Pyrolysis' process which produces oil with a fairly low energy content; the Monsanto 'Langard' gas pyrolysis process, which produces steam with a 54 per cent overall efficiency; and the Union Carbide 'Purex' system involving high temperatures and producing low-energy gas.[16]

Hydrogasification

This is the process by which a carbon source is treated with hydrogen to produce high-energy hydrocarbon gas. Methanol may also be produced by *gasification* followed by catalytic synthesis, as is shown in figure 7.4. At present, the primary source of methanol is natural gas, but several small-scale trials are being carried out to investigate the potential of biomass, as well as of coal gasification.

Biomass Research and Development Programmes – Some Examples

Basic research either to increase the effectiveness of biofuels or to discover new ways of using photosynthesis to produce fuels directly is under way in several countries.

Australia

Extensive studies have been made of the potential biomass energy contribution of special crops including eucalyptus, cassava, hibiscus, Napier grass and sugar cane. Of these, it was found as early as 1976 that cassava production and processing costs compared favourably with the price of conventional fuel alcohol.[7] By the early-1980s several ethanol-production plants were under way as well as methanol production from forest sources.

Brazil

At the time of writing Brazil has by far the largest and most ambitious biomass programme in the world, with biomass supplying about 25 per cent of the country's total energy requirements, mainly in the form of electricity, charcoal or alcohol derived from sugar cane. In 1981/82 the Brazilian government was spending 1.3 billion dollars on subsidising the National Alcohol Programme,[1] launched in 1975 as part of the National Fuels Programme which was itself a direct result of the 1973 oil crisis. Even in 1982 Brazil was spending two-thirds of its total foreign income

(about 11 billion dollars) on imported oil, mainly for transport; hence the importance of developing alternative indigenous transport fuels. By August 1977, 141
distilleries had been built and by 1979 Brazil was producing 1.6 billion litres of
ethanol annually. This was progressively mixed with gasoline (petrol) for public use
until by 1980 a blend of 20 per cent ethanol, 80 per cent gasoline (gasohol) had
been achieved nationally without engine modification. By 1983 the programme had
been expanded to produce 5 billion litres of ethanol and it was hoped to increase
this to 11 billion litres by 1987. By 1980 the motor industry had, in fact, produced
over 250,900 cars modified to run off hydrated alcohol (about 95 per cent ethanol
by volume). These now account for 25 per cent of all new cars sold in Brazil and
are cheaper to purchase and run than ordinary cars owing to government subsidies.
Gasoline was not sold on Sundays and the alternative alcohol fuel was some 10–20
per cent cheaper – an added incentive to manufacturers and motorists. The
Government's argument in favour of such heavy subsidies was that not only did
they increase Brazil's energy autonomy, but that each barrel of imported oil
eventually cost the country 100 dollars rather than the official market price of 34
dollars (1982 prices) when financing charges and opportunity costs were taken
into account. Brazil is also experimenting with the extraction of oil from peanuts,
palm oil and other hydrocarbon-producing plants for the purpose of blending the
products with diesel.

The Brazilian National Alcohol Programme envisaged the planting of some 3
million hectares of energy crops (0.35 per cent of Brazil's total land area). Originally it was hoped that by using its surplus of land and agricultural resources it would
become almost self-sufficient in energy by 1985, producing an estimated 10×10^9
litres of ethanol as a substitute for oil imports. This would guarantee future fuel
supplies and also stimulate new industries and create employment. However it now
seems likely that the original targets may not be met before the late 1980s.

Canada

In 1977 Canada launched a 6-year $30 million ENFOR research programme which
suggested that by 1990 8–10 per cent of her total primary energy could be provided
by forest biomass.[17]

China

China had more success than India with small-scale methane digesters, and claims to
have 8 million in operation with about one million being installed annually. The
systematic production of biogas in rural areas (mainly in the province of Szechuan)
only began in 1970 for the purpose of improving rural sanitation, producing fertiliser
and methane fuel. Animal and human as well as crop residues are used to fuel two
main types of digester. The more common family-sized unit consists of a small pit
of 8–10 m³ and produces gas for cooking and lighting. The second type is built by a
production team of between 100 and 200 members, has a capacity of some 100 m³,
and is used to power machines, pump water, and generate electricity.[18]

E.E.C.

Most E.E.C. countries have developed biomass energy research programmes since 1973 although some countries began earlier. The E.E.C. depends on imported oil for over 50 per cent of its energy requirements, so biomass sources derived from wastes or food surpluses and biomass fuels that replace petroleum products are particularly attractive. Several estimates have been made as to the practical biomass contribution. The Groupe de Bellerive in France suggested that, without serious land-use conflict, 11-14 per cent of the E.E.C.'s total hydrocarbon requirements could be met by biomass specifically grown for its energy content.[19] In 1982 another study suggested that the E.E.C. could produce 7 per cent of its projected energy demand for 1985 (about 85 mtoe per year) without major upheavals in current agricultural or forestry practice.[20] A more radical programme which would convert some of the agricultural surpluses into fuel could contribute up to 20 per cent. In 1975 the E.E.C. launched a major research programme into biofuel development. In 1982 38 projects were already underway, in addition to others undertaken by member countries. Further details of individual projects are given in a 1983 assessment *Evaluation des Programmes de Démonstration de la Communanté dans le Secteur de l'Energie*, published by the E.E.C. Commission.

France

Since 1975 France has undertaken one of the most ambitious biofuels programmes in Europe in an attempt to increase her production fivefold from an estimated 2 mtoe to 10 mtoe between 1980 and 2000. The programme, which in 1981/82 accounted for over $200 million per year or 40 per cent of the French renewable-energy programme, includes many different forms of biomass conversion from methane and charcoal production to the manufacture of straw burners and trucks designed to run off charcoal. 1.5 to 2 million hectares of France's 30 million hectares now under cultivation are required to produce 25 per cent of the country's needs in vehicle fuels. The French government hopes that petrol substitutes known as 'carburols' will eventually substitute up to 50 per cent of petrol consumption.

India

This country has pioneered small-scale anaerobic digestion processes mainly using cow dung rather than agricultural or domestic wastes. As early as 1971 Ram Bux Singh, a leading proponent, estimated that a 5-cow plant would pay for itself in just 4 years.[21] The dung is mixed with water, passed into an underground tank and maintained at a temperature of about 35°C, sometimes with supplemental heating by solar heat or methane. Although the anaerobically digested dung from one cow is estimated as sufficient for an Indian villager's cooking needs, progress in introducing these biogas (or gobar gas) units has been extremely slow, partly because of local custom, such as villagers preferring to use dried dung directly as a fuel in the traditional way. Also units are comparatively expensive to purchase since the storage tank in which the gas accumulates is capped with a steel cylinder. By 1980 over 100,000 biogas plants were thought to have been installed in rural communities although by no means all of these were functioning. They included some larger units of 80–100 m³ capacity.

Ireland

A short-rotation forestry scheme growing willows and alder on peat bogs has been introduced.

New Zealand

Among other forms of biomass, ethanol production using sugar beet feedstocks is being developed.

The Philippines

By 1980 25 power station sites had been located with the aim of providing fuel from nearby tree plantations. The fast-growing Ipil-Ipil tree (which can grow 6 m high with a diameter of 75 cm within 12 months) is thought to be as economic as imported oil for this purpose, with its attendant long-term economic advantages.

Sweden

Sweden currently derives about 8 per cent of her total energy requirements from biomass, using pine forests, forestry waste products and short-rotation crops such as very high-yielding willows which have been propagated especially for the purpose. The use of biomass sources for energy is likely to increase considerably in all the Scandinavian countries owing to the availability of large areas of land, the low population densities and the availability of capital to invest in new projects. Up to 3.4 million hectares could be suitable for 'energy forest' projects.[22]

U.K.

In contrast with most of the countries so far mentioned, the U.K. has had virtually no biofuels programme although since 1977 there have been a number of studies estimating both yields and conversion possibilities of potential biomass sources. In 1979 a U.K. Department of Energy study suggested that by the year 2000[23] up to 8 per cent of total energy requirements could be provided by biomass including urban, industrial and forestry wastes. This would mainly consist of 6 mtce per year recovered from the 24 mtce per year wastes at present unused. The E.E.C. study already quoted[20] estimates a possible annual U.K. biofuel contribution of 17 mtce, while the *maximum* technical potential from organic wastes, energy crops, natural vegetation and forests was estimated by ETSU at 50–60 mtce annually. It seems likely that the greatest contribution will lie in the processing of organic waste. Pilot trials funded by the Energy Conservation Demonstration Projects scheme in collaboration with the Department of Industry are under way. Other U.K. research is concerned with: the anaerobic digestion of wet biomass to produce gas; thermochemical processing to produce gaseous and liquid fuels; and short-rotation forestry. Unfortunately, for such projects to make any substantial impact the Department of Energy would need to allocate considerably greater funds, as have other governments of E.E.C. member countries. By 1981/82 only £0.5 million had been spent of the total £2.4 million allocated to biofuels research schemes. The main investments so far have come from industry.

U.S.A.

By 1983 the U.S.A. was deriving about 3 per cent of its total energy requirements (the equivalent of one million barrels of oil a day) from different forms of biomass, mainly from wood and crop residues. Wood, in the form of forests, represents 50 per cent more in energy terms than total U.S. conventional oil reserves and contains the same amount of energy as known gas reserves.[2] Several states and federal bodies had developed programmes for producing biomass fuels by the early-1980s. Ethanol, mainly derived from maize, was being produced for blending with gasoline in a 1 to 9 fuel mix. This gasohol was on sale at over 11,000 filling stations by early-1982. The cost was only a few cents more to the customer, and studies showed that even a 25 per cent blend of ethanol when used in unmodified engines had no detrimental effects on component parts; moreover the increased octane ratings improved the economics. In the 2 years from 1980–1982 the cost of U.S. gasoline more than doubled from $12 to $25 per litre ex-refinery while the cost of ethanol production rose by only 15 per cent from $28 to $32 per litre. Taking the octane enhancement into consideration, therefore, grain alcohol is already almost competitive with gasoline in the U.S.A.

Apart from state tax incentives, tax credits, low interest loans and purchase guarantees, by 1982 the U.S. government was also heavily subsidising a crash programme to further develop alcohol-production schemes with the aim of producing 7.6 billion litres of alcohol a year by 1990.[9] The U.S. Department of Energy also provides free information and literature in an attempt to encourage farmers and local communities to produce their own fuel.

Methanol production is also being investigated in the hope of reducing costs and using inedible residues rather than potential food crops. By 1983 plants were producing methane from sawmill and logging wastes, such as at Winchester, New Hampshire. At the time it was considered that smaller plants more suited to decentralised communities would be uneconomic although the size/efficiency/cost/energy ratios are under review.

Silviculture biomass plantations are another aspect of U.S. biomass research and development in an attempt to increase the per hectare energy contribution of various short-rotation tree crops.

Zimbabwe

In 1980 the government opened an alcohol distillery to produce ethanol from sugar cane grown in the Triangle area of the South East lowlands. By 1982 this ethanol, blended in a 15 to 85 per cent mix, was already saving $10–12 million per year in foreign currency. It was planned to increase the ethanol proportion to 25 per cent with a new refinery and distillery. Energy yields of sugar cane in this area are extremely high at about 120 tonnes per hectare per year, and the stillage from the fermentation process is used as a fertiliser on the plantations to increase the yields further by an estimated 6 per cent. Zimbabwe's current ethanol output from cane is about 40 million litres annually, but the output is expected to increase rapidly.[2]

Future Possibilities for Biomass Development

One promising area lies in 'growing' fuel for farm machinery. Many studies suggest that if a farmer devoted about 10 per cent of his land area to growing oil-bearing plants such as sunflowers or peanuts he could run all the diesel-powered machines on the farm. Although these oils are usually blended with gasoline, experiments are under way to discover whether it is more cost-effective to use these oils directly or first to refine them, and what the respective energy balances are.

Another approach has been to experiment with some of the unnumbered species of bacteria, fungi, algae and protozoa for their potential for converting biomass to usable energy. In the U.S.A., for example, it has been found that algae may be cultivated in sunlit tanks and fed on wastes for conversion to fuel in a digester. Where insufficient sunlight is available, waste heat from industrial plants could be used. Figures suggest an energy production of 37 kW/ha from algal ponds on the basis of only a 2–3 per cent conversion efficiency. In California, yields in excess of 100 tonnes/ha/year (dry weight) have been obtained. The blue-green algae are the most rich in protein for conversion into energy with some 60–70 per cent of extractable protein, while the green algae have 50–60 per cent. In an Australian experiment an alga, *Botryococcus braunii*, has been shown to yield 70 per cent of its extract as a hydrocarbon liquid resembling crude oil and further developments are under way. The solar energy stored in the different types of algae and bacteria could also produce different types of raw materials and pharmaceuticals, replacing those currently being produced by the petrochemicals industry. It may even be possible to extract minerals from the sea using algae in photobiological reactors.

Finally, instead of producing H_2 from H_2O, it is possible to produce H_2O_2 (a fuel) which blue-green algae will secrete under certain conditions. There are substantial difficulties still to be overcome, as there are with attempts to emulate synthetically the ability of plants to build up CO_2 into carbohydrates. Some experts hope for a breakthrough in both these areas before the end of the century.[3]

Disadvantages and Advantages of Biomass as an Energy Source

Despite enthusiastic research and development since 1973 there are the familiar limitations concerned with large-scale development of new energy forms and as usual, choices are involved. Although an estimated 200 billion tonnes of photosynthetic product is generated each year, much of this is too highly dispersed (forest canopy, uncollectable agricultural waste, inaccessible marine vegetation or algae) to be a useful energy source, since the energy spent in harvesting, drying and processing would exceed the energy gain. The wet equatorial regions which have the greatest biomass potential are often far from centres of civilisation or industry and are the most ecologically sensitive. The temperate lands, where energy use is highest, tend to have fewer biomass possibilities. In the U.S.A. and the U.K., while providing a useful alternative fuel, biomass even if developed from all possible sources would be unlikely to contribute more than about 20 per cent of total present energy use. In most tropical countries it could provide more diversified fuels than at present. Yet many of the developing countries with potential biomass resources do not have access to the necessary capital to develop them.

Scarce or unsuitable land can limit biomass growth while good land is at a premium for agriculture and food production. Sometimes the soil does not contain sufficient nutrients or water is unavailable to utilise marginal lands, while the energy costs of fertilising and irrigating these might dramatically reduce the net energy gain. In countries that experience extreme temperature fluctuations, the digestion processes of certain types of biofuel processes that need to be maintained at constant temperatures would be halted, and an additional fuel source would be required.

Where animal wastes are used as biofuels other than through methane digestion with a return of the digested sludge to the land, there is the danger of robbing the soil of vital nutrients. Agricultural residues are often needed to feed livestock, retard soil erosion or enrich the soil; if burnt they cannot do these jobs. Also the availability of biomass sources is highly seasonal and may not always coincide with labour availability or fuel requirements. The addition of fuel-storage capacity might well add to the costs. There is the danger that the cultivation of successive energy crops might decrease the land's productivity or resistance to pests, disease and climatic vicissitudes, and the danger of ecological imbalances produced by monocultures. There is also the possibility that the development of biomass resources, unless undertaken by governments or public bodies, might accentuate the gap between the landowners and the rural poor; or that multi-national companies might invest in 'energy farms' only for short-term profit rather than concern for long-term economic and ecological stability. For small-scale rural schemes to be effective local co-operation is essential. Frequently tradition, age-old custom, apathy or ignorance act as obstacles, as do existing agricultural and forestry practices in developing and developed countries alike. Also experience in biomass techniques is frequently lacking, and the risks and the uncertainties of ultimate yields tend to discourage initial investment. Contrary to popular belief, the production of biofuels, like fossil fuels, can produce various forms of pollution. For example, for every litre of ethanol derived from sugar cane there are 8–10 litres of stillage. In the South-west of Brazil this was initially dispersed into rivers causing widespread pollution, until it was later found valuable as a fertiliser when fed to the land, or as animal feed when dried.

A final disadvantage is the low photosynthetic conversion efficiency of most green plants and the further loss incurred in converting these biomass sources to fuels. At present many biomass conversion processes result in a net energy deficit, and so are acceptable only where alternative fuels are unavailable.

Against all these factors must be weighed the very considerable advantages that

could be achieved by a careful development of renewable biomass resources. Biomass fuels are extremely varied, have many useful by-products, and can replace at
comparatively low cost many of the uses of fossil fuels. Particularly attractive is the
fact that, under certain conditions, they can provide liquid hydrocarbon fuels to
replace petroleum, a possibility not open to other forms of renewable energy. They
are environmentally more benign than many fossil fuels, such as coal and oil, and
help to maintain the carbon dioxide balance in the atmosphere. Plant matter contains less than 1 per cent sulphur as compared with the higher sulphur content of
fossil fuels, the treatment of which adds greatly to capital costs. The energy produced is safe, generally non-polluting, and may easily be stored and safely transported. Biofuels can be produced and used locally with systems designed to meet
local requirements. There is a wide range of conversion technologies to produce a
diverse range of products. Because biomass may be grown almost everywhere, access
to its energy is not so susceptible to international or political pressures. It helps to
increase a country's energy autonomy and decrease balance of payments deficits.
Despite the pressure for agricultural land much potential cropland exists in areas
where food production cannot be sustained. In some cases agricultural lands can be
used in an off-season to grow energy crops. Side-by-side cropping and year-round
harvesting, if carried out on a scientific basis, would increase the ecological diversity
and minimise soil depletion and vulnerability. The new and important agroforestry
developments, especially when applied to reforestation schemes, will help to avoid
the food/fuel dilemma. Also the increased planting of trees and crops helps to
counteract the CO_2 problem caused by burning fossil fuel. The increasing use of
domestic, agricultural and urban wastes to produce fuel is one of the most promising
economic applications of biomass technology. In this respect methanol production
may eventually have an advantage over the production of ethanol where edible
crops are used as feedstocks.

   Although there are many proven techniques which merely require implementation to produce biofuels, some aspects of biomass technology are still in their
infancy compared, for example, with the agricultural where experience and necessity have led to ever-increasing yields and more efficient production. With increased
interest and investment, however, both plant yields and improved conversion
technologies in the production of biofuels will enhance their economics and contribution. This in turn will increase local employment and new industries. Of the
five main routes by which biomass is converted into fuel (direct combustion, direct
production of oil from plants, fermentation, biomethanation and pyrolysis) combustion is the most significant, although even here efficiencies may be significantly
increased. It is also likely that biofuels will be found increasingly useful when
employed in conjunction with other energy sources.

   The development of biomass for energy is essentially a conservation strategy
since the products are designed to replace the fossil fuels on which we have come
to rely. The selection of particular biomass crops and energy systems depends
on local conditions and is partly determined by the type of fuel required. However,
ultimately the success of any biomass strategy will depend on the energy ratio and
also the long-term state of the soil from which the crops derive. To quote Denis

Hayes 'There is a popular tendency to think of renewable resources as infinite resources, but this is a confusion of size with duration. If care is taken biological crops can be cultivated in perpetuity. But with short-sighted management, energy crops (like all other biological systems) can simply collapse.'[24]

References

1. Hall, D. O., Photosynthesis for Energy, Lecture, August 1983
2. Hall, D. O., 'Biomass for Energy: Fuels Now and in the Future', *Royal Society of Arts Journal*, No. 5312, Vol. CXXX, July 1982, Table 1, p. 459
3. Hall, D. O., 'Solar Energy Use Through Biology – Past, Present and Future', *Solar Energy*, Vol. 22, No. 4, 1979
4. *Solar Energy – a U.K. Assessment*, UK-ISES, London, 1976
5. Szego, G. C. and Kemp, C. C., 'Energy Forests and Fuel Plantations', *Chemtech*, May 1973, pp. 275–84; 'The Energy Plantation', *Proc. 79th Meeting Amer. Inst. Chem. Engineers, May 1975*
6. Lalor, E., *Solar Energy for Ireland*, Report to National Science Council, Dublin, February 1975, pp. 40–7
7. Boardman, N. K. and Saddler, H. D. W., 'Biological Conversion of Solar Energy, Its Potential Contribution to our Energy Requirements', *Sun at Work in Britain*, No. 4, December 1976
8. Alich, J. A. *et al.*, *An Evaluation of the Use of Agricultural Residues as an Energy Feedstock – a Ten Site Survey*, Stanford Research Institute, E (04-3) 115, Department of Energy, Washington D.C., 1977
9. Tyner, W. E., 'Biomass Energy Potential in the United States of America', *Mazingira*, Vol. 5, No. 1, 1981, pp. 44–53
10. Coombs, J., Hall, D. O. and Chartier, P., *Plants as Solar Collectors, Optimising Productivity for Energy*, Reidel, Dordrecht, 1983
11. Calvin, M., 'Petroleum Plantations for Fuel and Materials', *Bio Science*, 29 September 1979
12. Wolverton, B. C., Barlow, R. M. and McDonald, R. C., *Application of Vascular Aquatic Plants for Pollution Removal, Energy and Food Production in a Biological System, Bay St Louis, Mississippi*, NASA, Houston, Texas, 1975
13. *Brewing and Distilling News*, January 1980, p. 5
14. British Anaerobic Biogas Association, U.K., December 1979
15. Brown, L. R., *Food or Fuel, New Competition for the World's Cropland*, Worldwatch Paper 35, Worldwatch Institute, Washington D.C., 1980
16. Floueko, G. C. and McGauhey, P. H., 'Waste Materials', in Hollander, J. J. (ed.), *Annual Review of Energy*, Vol. 1, Annual Reviews Inc., Palo Alto, California, 1976
17. 'Bioenergy: Canada's Trees Prodded for Potential', *Canadian Renewable Energy News, May 1980*
18. Van Buren, A., *A Chinese Biogas Manual*, Intermediate Technology Development Group, 1979
19. Tucker, A., *The Guardian*, 31 January 1980, p. 13
20. Hall, D. O., 'Energy for Biomass in Europe: the 1982 Situation', in Strub, A., Chartier, P. and Schleser, G. (eds), *Energy from Biomass, 2nd E.C. Conference*, Applied Science Publishers, London, 1983
21. Ram Bux Singh, *Biogas Plant, Ajitmal, Etawah, India*, Gobar Research Station, 1971

22. Berkovitch, I., 'Sweden's Energy Forest Prospects', *Coal and Energy Quarterly*, No. 32, Spring 1982, pp. 27–35, NCB, London
23. *U.K. Department of Energy Paper No. 39*, HMSO, London, 1979
24. Hayes, D., *The Solar Energy Option*, Worldwatch Paper 19, Worldwatch Institute, Washington D.C., 1978, p. 24

Further Reading

Slesser, M. and Lewis, C., *Biological Energy Resources*, Spon, London, 1979
Solar Energy – a U.K. Assessment, revised edn, UK-ISES, London, 1976, chapter 9
Hall, D. O., Barnard, G. and Moss, P. A., *Biomass for Energy in the Developing Countries – Current Role, Potential, Problems, Prospects*, Pergamon, Oxford, 1982
Brown, L. R., *Food or Fuel, New Competition for the World's Cropland*, Worldwatch Paper 35, Worldwatch Institute, Washington D.C., 1980
Strub, A., Chartier, P. and Schleser, G., *Energy from Biomass, Second E.C. Conference*, Applied Science Publishers, London, 1983
Sarkanen, K. V., Tillman, D. A. and John, E. C., *Progress in Biomass Conversion*, Academic Press, New York, 1982
Meynell, P. J., *Methane – Planning a Digester,* 2nd edn, Prism Press, Dorchester, Dorset, 1976

8

Renewable Sea Energy Sources

A. OCEAN THERMAL ENERGY CONVERSION (OTEC)

By harnessing the small temperature differences between the warm surface of the ocean and the cooler deep water, energy can be generated using a heat engine. Although technically possible, at the present time (1984) no commercial system is yet operational. Views also differ on the projected generation costs.

The OTEC Plant in its current form uses a closed Rankine cycle with working fluids such as ammonia, hydrocarbons or halocarbons, although by 1984 the U.S.A. was modestly funding research into an *Open Cycle Plant* which uses seawater as the working fluid. With the *Closed Cycle* the fluid is evaporated in a boiler located in the warm surface water, expanded through a turbine over a pressure drop of some 3.5 bar and condensed in a heat exchanger by the deep cold water before being pumped back into the boiler to begin the next cycle. The efficiency of the process is very low, at between 2 and 2.3 per cent.[1] It is also true that a significant proportion (some 20 per cent) of the energy extracted must be diverted into pumping cold water from the depths. Another disadvantage is that to produce a useful amount of electricity the turbine systems must be larger than the conventional steam turbines and operate where the temperature differences between the surface and deeper water are at least 15–20°C. This naturally restricts their use. On the other hand the technology is straightforward, some relatively low-cost materials can be used and, where thermal temperatures of the ocean are sufficient to operate a system, the energy provided is free, inexhaustible and non-polluting. Such facts tend to out-weigh the objection of the low energy-conversion rate, especially since the high capital investments are balanced by low operating costs.

The basic principles of an OTEC plant are illustrated in figure 8.1. The most commonly envisaged plant is a large offshore floating or partially submerged structure, although plants may more economically be land-based or rest on an ocean shelf. The electrical energy provided can be brought ashore via underwater cables or be used offshore, for example, to produce chemicals.

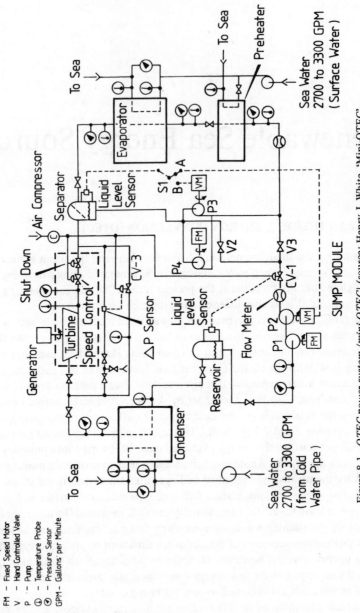

Figure 8.1 *OTEC power system (mini-OTEC)* (source: Henry J. White, 'Mini-OTEC', *International Journal of Ambient Energy*, Volume 1, No. 2, Construction Press, 1980, p. 76)

OTEC Developments

As early as 1881, a French Scientist, Arsène d'Arsonval, in the *Revue Scientifique* outlined plans for a heat engine working on the Carnot cycle, using the temperature difference of water as a source of power. Later, in the early twentieth century another eminent French engineer, George Claude, troubled by bleak energy prospects, experimented on ocean thermal power, first in the laboratory and later at sea. In 1926 Claude reported his first findings to the French Academy of Sciences — that in the tropics he and Paul Boucherot, a colleague, had discovered that the temperature of the water 1000 m below the surface was about 4°C while at the surface it was 27-29°C. 'This fact,' he commented, 'no doubt, will be the starting point of a grandiose solution to the problem of the use of solar heat.'

After building a short-lived experimental 22 kW OTEC plant at Matanzes, Cuba, which operated on an open Rankine cycle but with an overall efficiency of less than 1 per cent, Claude constructed another in 1935 on a ship which he took to Brazil hoping to use the electricity generated to make ice to sell in Rio. Owing to the turbulence of the wind and waves he kept losing the connecting pipe. Disillusioned, and with true Gallic *élan*, he blew up the plant and ship to the accompaniment of the *Marseillaise*. As with many other renewable energies, interest in the concept of ocean thermal energy conversion faded during the era of cheap fuels.

In the 1960s Dr Clarence Zener revived Claude's ideas in the U.S.A. He calculated that the energy potential stored in the oceans was 'more than all the coal and oil on earth.' Moreover, coming from the sun it was renewable. While working with Westinghouse Inc. he attempted to design a plant consisting of boiler, engine and condenser fitted vertically and floating about 70 m below the surface, so that the surface waves and storms could not reach it. Zener together with Dr Abraham Lavi worked on the concept of a closed Rankine cycle to increase the conversion efficiency. The driving-fluid was to be vaporised by heat from the warmer surface water and condensed by cold water drawn up from a depth of about 600 m (approximately 2000 ft). In the past the problem had been to find a suitable vapour cycle but now this new concept was to form the basis of future large-scale theoretical investigations.[2]

William Heronemus, Professor of Civil Engineering at the University of Massachusetts from 1967, together with the Anderson father and son team, helped to initiate a U.S. programme in ocean engineering, and to develop a Mark I OTEC plant. This was to be anchored in the Gulf Stream off Miami to generate 400 kW of electricity. The proposed site was 25 km east of the University of Miami and in 3600 m of water where the Gulf Stream provides a 17°C temperature difference. As a result of several preliminary studies of this type the U.S. National Science Foundation selected ocean thermal energy conversion as one of six areas of solar energy in which it would finance research.

Another early mini-OTEC installation was built at Keahole Point off the coast of Hawaii in 1977. Although its rated output of 50 kW in effect contributed only 18 kW net, the economics and practicability of larger OTEC power stations had been explored, the inter-relationship of components tested and much valuable data gained. Both the evaporator and condenser consisted of expensive titanium plate-type heat exchangers in this case. Interestingly the data also indicated that micro-fouling in deep cold water was substantially less than in warmer waters. Future U.S. Department of Energy plans are for a 40 MW plant at Kahe Point, Hawaii, and a 48 MW plant for the island of Guam, both design schemes being land-based.

The Europeans and the Japanese have been prominent in the field of OTEC research. The Japanese national 'Project Sunshine' for solar energy has within it a limited programme to develop a large floating OTEC plant, while industry has concentrated on shorter-term technical and commercial developments. Industry's immediate target is the construction of land-based plants with capacities of up to 10 MW.[3] The Tokyo Electrical Power Services Company, Shimizu Construction and Toshiba have successfully operated a pilot plant on the Pacific island of Nauru since the end of 1982. Although there are only a limited number of sites for OTEC, mostly in the tropics, these early commercial developments are helping engineers gain valuable experience of operating conditions. The Japanese government contributed 40 per cent of the cost of the Nauru plant. By 1984 the main research in OTEC systems was being carried out by Japan, the U.S.A., France, India, Sweden and the Netherlands (Tokunoshima, Jamaica, Tahiti and Bali being some of the favoured sites).

Some Unresolved OTEC Issues

Economic Feasibility

Obviously the cost of electricity from OTEC varies with the location of plants, the ocean temperature differences at a given site and the method of transporting power to a point where it may be used. The cost of the plant itself depends largely on the cost of the condensers, the largest and most expensive components.

In 1966, it was calculated that a plant based on the Claude process could be built for the same capital cost as that of a commercial power plant using fossil fuel. Zener estimated that it could produce electricity at about half the cost of conventional plants. Nevertheless at that time the figures did not compare favourably with the anticipated low cost of nuclear fuels and this reason together with the low efficiency factor (as opposed to 30–35 per cent efficiency for conventional power plants) partly explains why OTEC projects failed to attract funds. Later when it was discovered that the cost of nuclear power was rising rapidly there was a wave of renewed interest in OTEC systems.

In 1974 Heronemus had hoped that if the total cost per installed kW of his Mark I type plant could be less than $1100 by 1980, the Florida market would be won over from ideas of fossil or nuclear fuels. He argued that each $100 in capital costs below this $1100 would pay for another 200 miles of transmission distance and, if backed by appropriate individual and national commitment, OTEC could supply markets as far afield as Chicago. Unfortunately, as with nuclear power, neither the projects nor the costs have lived up to these early expectations and there is still wide-ranging uncertainty about costs relating to the size of plant, its location and the exact nature of components. One of the major technical and economic problems is bringing the energy ashore, since the waters needed for OTEC are at least 100 km offshore in most areas. Obviously since transmission is a costly factor where the electricity cannot be used at sea, economic advantage depends on building OTEC plants as near as possible to the places where the electricity is to be used. Examples of suitable sites are certain offshore islands in the tropics where the sea-bed drops away rapidly so that plants could be placed within easy reach of the mainland and electricity transmitted by cable. The main technical problem that remains is the cold-water pipe which is undergoing drastic design modifications. The Japanese are currently experimenting with its replacement by twin underwater tunnels of 3 m diameter, which draw the deep cold water to the on-shore Narau plant.

Environmental Issues

Several studies have been carried out in various areas to provide data on OTEC plants before full-scale tests commenced. In one example Zener and Lavi found that corrosion could be halted by using aluminium with a self-protecting oxide while chlorine gas (used in a small concentration) could prevent fouling by microbial growth without endangering marine life. It has even been suggested that the artificial up-welling of nutrient-rich deep water caused by OTEC systems will prove beneficial to fish breeding since the remains of decayed marine life, instead of settling on the ocean bed, will be drawn up towards the surface and remain within the cycle of marine biological production. However, most environmental studies have concentrated mainly on the effects of the environment on the power plant rather than the effects of the plant on its surroundings. Very little is known, for example, of the impact of large-scale heat extraction by numerous OTEC plants on marine life or on the overall temperature of the surrounding seas.

Possible Applications

In the U.S.A., studies have been carried out to find energy-intensive products that could be made on a floating industrial complex. By 1984 the two projects with the greatest economic potential were a sea chemicals complex (sodium sulphate, potash, magnesium, sodium hydroxide and chlorine) and an organic chemicals and plastics complex (for vinyl chloride, polyethylene oxide, polyethylene and ammonia). Another possible location is Jamaica, which possesses 50 per cent of the bauxite reserves of the Western Hemisphere. Current practice is to ship this ore to the U.S.A. to be refined into alumina and then to transfer it to other sites with cheap power for final reduction. Zener has suggested that a more economical process would be to refine the bauxite into alumina and effect the electrolytic reduction into metallic aluminium on the island itself, using the cheap and abundant electric power generated by OTEC plants. On the other hand, even a 100 MW station without associated industrial plant would require a semi-submerged steel or concrete vessel at least the size of a large North Sea oil platform. To be competitive with fossil fuels or nuclear power, OTEC plants of 250–400 MW would have to be built and the sheer size may pose hitherto unknown construction and maintenance problems. However, unlike many other renewable energy sources OTEC systems can provide base-load electrical power which is an added advantage. Besides improved fishing, an important by-product could be potable water (a 5 MW open cycle plant would produce 90,000 litres/day).

Other Potential Thermal Sea Energy Sources

Salinity Gradients

Although only recently recognised, the amount of energy potentially available from salinity gradients (the difference between the salt content of different bodies of water) is very great. The salinity energy density of a fresh-water river with respect to the sea can be equivalent to the energy density of a 240-m high dam; that of ordinary sea water with respect to a coastal pond of concentrated brine may be equivalent to a dam 3.5 km high. Fresh water or sea water flowing into brine at a rate of 1 m^3/s releases energy of 30 MW. However, common salt solutions release very little energy of dilution as heat. Instead, it is converted into increasing disorder, or entropy.

The approaches suggested for harnessing power from salinity gradients include: (i) vapour exchange between two solutions (preferably at an elevated temperature), called inverse vapour compression; (ii) simple osmotic change against a hydrostatic pressure, called pressure-retarded osmosis; and (iii) the dialytic battery, which can be thought of as reverse electrolysis. Only the first of these does not require membranes, with their attendant expense, relatively short life, polarisation problems, and high pumping costs. The technique of inverse vapour compression was therefore that adopted by Isaacs, Olsson and Wick in their pioneering tests in 1979/80.

As an indicator of the energy potential of salinity gradients, Wick has suggested that the subterranean salt domes normally drilled for oil, such as along the coasts of the Gulf of Mexico, could yield even more energy from their salt. Although the same principles as OTEC are involved, and salinity is generally regarded as a renewable-energy source, the costs of further development are extremely high,[4] and the most practical and cost-effective extraction processes have not yet been determined.

Ocean Currents

These have also been considered as potential sources of sea energy. Currents have varying velocities according to location, some of the highest velocities occurring in the equatorial regions and along western ocean boundaries. As currents have relatively low energy densities, it is necessary to intensify their velocities to drive turbines or paddle wheels, and a number of devices such as the Venturi shroud or the dynamic dam have been considered. Although the general principle is that of installing some static structure to concentrate energy in a specific part of a large energy flow, there are still many technical problems to be overcome before ocean currents can make any significant energy contribution.

Other Potential Sources

These include marine nuclear resources, submarine geothermal sources and the latent heat of icebergs, although not all of these are by any means renewable. Both fissionable elements, such as uranium and thorium, and fusionable elements, such as lithium, and hydrogen in the form of deuterium, are present in varying but abundant quantities in the sea. The harnessing of energy from submarine geothermal springs and submarine volcanoes are other possibilities which may be achieved at some future date. The utilisation of temperature gradients from the melting of icebergs under pressure, to be used for drinking water in arid regions, is yet another potential energy source.[5]

B. TIDAL POWER

For centuries machines that obtain power from the tides have been in use in Europe and elsewhere. The *Domesday Book* mentions a tide mill at Dover which was thought to have been built between 1067 and 1082. People have long known that it is possible to trap water when the tide is high and to release it when the tide falls, thus providing an energy source.

Tides are caused primarily by the gravitational pull of the moon on the seas, to a lesser extent by the pull of the sun and also by the rotational force of the earth itself. A combination of all three results in the Tide Generating Force (TGF) which is strongest when the gravitational pull of the sun and the moon are in the same direction (the high spring tides) and least when the sun and moon are pulling at right angles to one another (the low neap tides). The spring tides can be up to three times as high as the neap tides. The moon circles the earth every 24 hours and 50 minutes and in this period the tide rises and falls twice — once through the gravitational pull of the moon on the seas nearest to it, and once when the moon is on the far side of the earth as a result of centrifugal forces caused by the earth's rotation.

The tides can be used to generate power only when they produce a head of water high enough to drive a turbine or mill. Normally, the tidal range has to be over 1.5 m. In the open seas the tidal range is less than this, averaging 1 m or less. In certain suitably shaped estuaries, however, it can be much higher, and ranges of up to 17 m have been recorded. Amplification of the tides is caused by such local effects as funnelling, shelving and resonance. The energy available from tidal ranges is a function of the surface area of the tidal basin and the square of the tidal range. The tidal range increases up the estuary whereas the basin area decreases so the optimum location of a tidal barrage is where the sum of these two conflicting tendencies is the greatest.

The energy from the tides can be harnessed in two main ways: either by using the tidal currents directly to drive slow-speed water wheels or turbine–generators; or by constructing a tidal barrage and driving the generator as the barrage is filling up or emptying.

The Direct Use of Tidal Currents

Tide Mills

Tidal power has been harnessed by tide mills for over 1000 years. Tide mills situated in estuaries with large tidal ranges have provided power usually for the same purposes as the traditional watermill. The first reference to tidal power comes from the Muslim geographer, Al-Maqdisi Shams Al-Din, who prior to his death in A.D. 1000 wrote about the incoming tide irrigating the orchards of Basra, carrying ships to the villages twice a day and working the mills situated near the mouth of the river and its tributaries.

At least 11 tide mills had been built in Britain by 1200. By 1600 there was a total of 89 tide mills in England alone and, as population and energy demand increased, this tidal energy was harnessed wherever suitable sites could be found. Even in the nineteenth century, 27 new tide mills were built and some mills even operated until the Second World War. Altogether there is evidence of 170 or more in operation along the southern coasts of Britain.[6]

Other European countries, notably France, the Netherlands, Belgium, Portugal and Italy, experienced similar developments in tidal power. During the last century France had about 80 tide mills, with 12–14 situated in the Rance estuary where the world's first major tidal-power barrage is in operation. As with the larger tidal schemes, the inflowing tide was trapped by gates across the mouth of the enclosed basin, which closed immediately the tide changed direction so that the enclosed water could be stored and released to drive the undershot wheels of the mills.

A traditional tide mill was normally able to work only 4–5 h in every 12 which partly explains why they came to be abandoned in the era of cheap and continuous energy supplies. The old tide mills were mini-barrages across suitable inlets. They should not be confused with machines similar to underwater 'wind turbines' which are designed to harness tidal stream power.

Tidal Stream Power

Since the power densities in tidal streams can be relatively high, following the late-1970s the subject attracted some attention. In the U.K., Dr Peter Musgrove of Reading University and Peter Fraenkel of the Intermediate Technology Development Group developed a water rotor with a vertical axis (similar to Musgrove's wind turbine — see page 234), the blades rotating as a result of the water motion. This vertical axis underwater turbine was originally developed with a 3-m diameter blade for use in tidal streams, rivers and fast-flowing canals for irrigation pumping applications, and possibly for the construction of reliable systems for small-scale electricity generation.[7] Efficiencies of up to two-thirds of the maximum theoretically possible 59.3 per cent were achieved with model tests in 1979. Useful power densities can be obtained in currents of 1 m/s (2 knots) or faster: a current of 2.1 m/s (4 knots) corresponds to a power density of 4.4 kW/m^2.

The advantages of using underwater turbines extracting tidal stream power for small-scale electricity generation are firstly the greater choice of location and secondly the considerably reduced capital expenditure needed before power can be generated. The disadvantages of such turbines are the low density of power (a tidal stream machine of 9 m diameter releases about 150–200 kW while a tidal barrage turbine of 9 m diameter gives about 45,000 kW); and difficulties of installation, access for maintenance and risk of accident caused by ships, etc.

Tidal Power Barrages

There are three main ways of harnessing tidal power using barrages. The simplest is a one-way generation system which uses the high tides to fill a basin or estuary behind a tidal barrage by opening sluice gates. The gates are closed when the tide turns, and the water is released through a row of low head turbines during low tide. Such schemes can operate for 3–5 h during the 12-h tidal cycle. A second way is to use reversible turbines which can generate power during both the ebb and flood tides. These can be used as long as water levels on either side of the barrage are unequal, that is, for four periods per day. Thirdly, two barrages can be constructed to create two basins at different levels. By pumping water from one to the other, flow rates can be evened out and continuous energy provided, although at lower efficiencies than the other two methods.

Existing Schemes

The best-known tidal barrage is at the La Rance estuary near St Malo in Brittany, France, which commenced operation in 1961. This is a 750-m long dam containing 24 reversible Kaplan-type bulb turbines, with a capacity of 240 MW. The dam encloses a basin of about 20 km^2. A much smaller barrage exists at Kislayaguba, near Murmansk in the U.S.S.R. Built in 1968, it has a peak output of 800 kW. Numerous small tidal pumping stations have been reported in China, including a 3 MW tidal barrage.[8]

Proposed Schemes

Many other tidal schemes have been proposed in various countries ranging from the U.S.S.R. to Canada, the Channel Islands, Australia and the U.K., but at present under 40 viable locations have been identified. Notable examples include the following.

The Bay of Fundy, north-east Canada, has the world's highest tidal ranges of up to 17 m in its upper regions, the average being around 11.5 m. A prototype turbine has been installed in an existing barrage at Annapolis Royal, a small inlet on the coast of the Bay and another 18 MW scheme is under construction.

Kimberley, Western Australia, is another possible site for tidal barrages with an installed capacity approximately double that of La Rance.

The Severn Estuary, provides the most promising site in the U.K. with a possible annual energy output of 12–20 TWh, although studies have shown that Morecambe Bay, the Solway Firth, the Wash and the Humber also have large potential resources of tidal power.

There have been many suggestions for the development of tidal power in the Severn Estuary which experiences very favourable tidal conditions.

Interest in a Severn Barrage was aroused as long ago as 1925, when the government appointed the Brabazon Committee to inquire into its feasibility. Subsequently in 1966, Professor E. M. Wilson of Salford University, having calculated the power available from six different sites on the Severn, put forward the single basin 'Wilson Scheme'. In 1971 Dr T. Shaw of Bristol University proposed a two-basin scheme.

A further Severn Barrage study was carried out by a new Severn Barrage Committee set up in 1978 under Sir Hermann Bondi to advise government on the desirability of a tidal scheme in the Estuary. Its report,[9] issued in 1981, once more concluded that it was technically feasible to enclose the estuary by a barrage to be used for the generation of electricity. Three schemes (figure 8.2) were put forward by the Committee. The report summarised the choice of schemes as follows:

'(a) Outer Barrage — a single basin ebb generation scheme based on a barrage running from just east of Minehead to Aberthaw. This would cost about £8900 million, including an allowance for contingency, and produce about 20 Terawatt hours a year (about 10 per cent of the 1980 electricity demand in England and Wales) from an installed capacity of about 12,000 Megawatts. This scheme would effectively exploit the energy potential of the estuary.

(b) Inner Barrage — a single basin ebb generation scheme with a barrage running from Brean Down to the vicinity of Lavernock Point on the Welsh coast. This would cost about £5600 million, including an allowance for contingency, and produce about 13 Terawatt hours a year (6 per cent of the 1980 electricity demand) from an installed capacity of 7200 Megawatts. This scheme would be the most cost-effective.

(c)   Staged Scheme — as (b) but with a second basin bounded by a dam branching off from the Inner Barrage near Brean Down and running to just east of Minehead, enclosing Bridgwater Bay. This two-stage scheme would generate almost as much energy as the Outer Barrage and would have a similar economic performance. If this second basin were to be operated in flood generation mode, electricity would be produced in four blocks extending over about 20 hours each day. A decision on this second stage could be deferred for many years without affecting the development of the first stage.'

Of these three, scheme (b), the Inner Barrage scheme, was the most attractive on economic grounds. In the Committee's view an Inner Barrage would be broadly within the range of future costs of generation (including capital and operational components) from conventional coal and nuclear plant. The faster that fossil-fuel prices rise the greater is the value of tidal power. However, according to this report, if the proportion of nuclear power in the total electricity-generating system were to increase rapidly, the value of tidal power would reduce. Unfortunately, insufficient emphasis was placed in the report on the non-economic advantages of having an indigenous renewable-energy source.

Environmental Issues

Earlier assessments of the environmental effect of a barrage on the tidal range had been contradictory but the Severn Barrage Committee's 1981 report predicted with

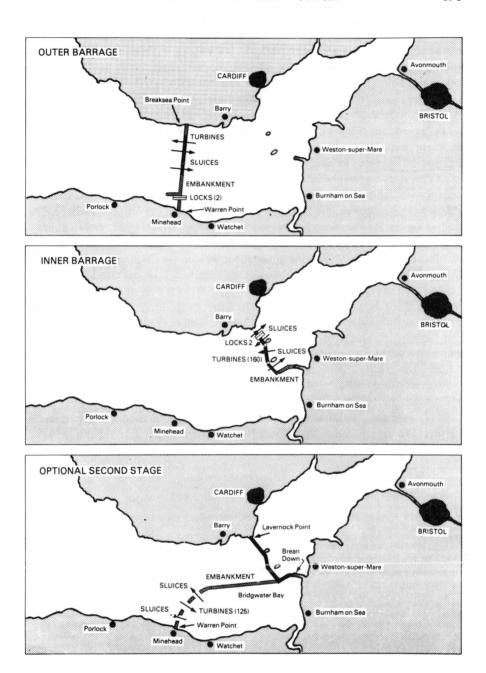

Figure 8.2 *Three barrage schemes proposed for the Severn in 1981*
(source: *Tidal Power from the Severn Estuary*, Energy Paper No. 46,
Department of Energy, HMSO, London, 1981)

some confidence that the effect of an inner barrage would be to reduce the tidal range immediately to seaward of the barrage by about 11 per cent. The barrage would also alter the pattern of water flows and water levels above the barrage through the daily tidal cycle. These effects have been predicted by modelling and the report claims that its predictions have an accuracy of ± 0.2 m, although 'possible errors arising from the way the models represent the various components and method of operation of the barrage have not yet been estimated'.[9]

Because of the complexity and uncertainty of the subject, the precise extent of sedimentation, deposition and erosion with a tidal barrage across the Severn are unknown. However, the Severn Barrage Committee felt sufficiently confident to recommend that a further £20 million feasibility study be undertaken in order to resolve these issues and prepare proposals leading to full implementation. After considering these proposals, the U.K. government decided in 1983 to commence a £10 million appraisal of the Severn Barrage together with private industry. As a direct result of the U.K. Energy Act 1983, which allowed the generation of electricity by private organisations, in 1984 the government was jointly funding a study of the Severn Tidal Power Group (STPG) on the technical and financial viability of a barrage built and operated by the private sector. There is also an investigation of a much smaller tidal barrage downstream from the Severn Bridge which would involve a substantially smaller capital cost (£885 million compared with £6 billion) and provide a second bridge and road link across the estuary. Although the generated capacity would be less (1050 MW as opposed to 7200 MW proposed by the Bondi scheme) the smaller scheme has many advantages since it would combine a minimum construction distance (4 miles as opposed to 11) with maximum tidal range and a much shorter payback period.

Many environmental issues surround the development of any large-scale tidal barrage, and there is still sharply divided opinion on their possible impact. Most likely changes would stem from the effect that a barrage would have on the tidal range and hence on water velocity, salinity, dispersion of pollutants, sewage, water-flow patterns, sediment movements and so on.

Some of the main considerations to be examined before work on a Severn Tidal Barrage could begin would include:

(1) General regional development patterns. Any development on the scale of the proposed barrage would affect a wide region and large numbers of people living and working near the estuary.
(2) Possible road links and the associated engineering costs.
(3) Ports. The benefits to ports above the barrage from the partial stabilisation of tidal regions would need to be assessed against costs associated with the installation of locks capable of handling the largest anticipated vessels, delays at the locks and a small drop in water levels at high tide.
(4) Source of construction materials. This would need careful consideration. The noise, dust and vibration caused by large-scale rock quarrying and transporting considerable masses of materials would need to be assessed.
(5) The effects on landscape and amenity. Landscape quality could be reduced by the change from a natural open estuary to an enclosed basin. Quarrying and roadworks might also adversely affect the landscape.

(6) The effects on holiday resorts and recreation facilities in the location of the barrage require consideration. Generally raised water levels at Western-super-Mare could be of benefit to holiday-makers.

(7) Ecology. As mentioned, changes in tidal height, salinity, pollutant dispersion and other factors could cause changes in the general ecology of the estuary and adjacent salt marshes. There is also the possible problem of reduced siltation and less movement of mud banks which could affect the sea fisheries, plankton, and wild life in general.

(8) Fresh-water fisheries. Ways of maintaining migratory routes through a barrage for fish such as salmon would have to be sought and costed.

(9) Nature conservation. Ornithologists have expressed the fear that the reduced tidal range would adversely affect the mud flats which are the winter habitat of large populations of wading birds and wildfowl. Other important sites both inside and outside the barrage, primarily salt marshes and coastal wetlands, could also be affected by changes in sedimentation and land drainage.

(10) Pollution. At present the industrial and sewage effluent discharged into the estuary is mixed and dispersed very rapidly. If current flow and water mixing were reduced behind a barrier, local pollution could increase. Moreover, if suspended sediment settled more rapidly in the calmer waters, light penetration could increase with the possibility of troublesome algae developing. A barrage might therefore require new controls over discharges to the estuary.

(11) Agriculture. Much of the agricultural land surrounding the Severn Estuary is below high-tide levels. The alteration in tidal regime above a barrage with higher average water levels could make new land drainage work necessary and might increase pressures for the reclamation of estuarial salt.

(12) Floods and sea defences. A barrage would obviously reduce the risk of sea flooding in the area it enclosed but new measures might be needed to guard against river flooding. Sea defences would need modification if tidal ranges were altered outside the barrage or if erosion in some areas was increased.

Apart from the environmental debate, the greatest obstacle to the implementation of a Severn Barrage Scheme is the commitment to a huge project that would generate a significant proportion of the country's electricity only after several years, the supply source being highly centralised. The Severn Barrage Committee recommended a programme that would require a period of 15 years from the start of a 4-year acceptability study to the time when electricity was actually produced.

The main advantage of tidal schemes is that, given a suitable site, the energy is inexhaustible and reliable and the technology exists and has been proven. The U.K. is fortunate in having this potential resource which is by no means universal. Also, unlike many of the alternative energy sources, tidal power can supply large amounts of high-grade energy. Since a tidal barrage could have a life span of up to 140 years, it offers a long-term security of energy supply at costs that are in the same range as the fossil-fuel alternatives and could be considerably more advantageous by the time the Barrage has been completed.

C. WAVE POWER

Waves are caused by the wind blowing over the surface of the sea. They can be up to 30 m high in a severe storm. Wave energy is the greatest where the winds are strongest and this occurs in between latitudes $40°$ and $60°$ N and S of the Equator. Winds may be local and generate waves directly, or they may be up to several thousand miles away and generate 'swell seas' which eventually result in local wave action. Waves from the 'swell sea' may arrive from directions different from waves generated by local winds.

Measuring Wave Energy Availability

The waves most suited as an energy source are 'deep-water waves' (where the depth of the water in which they move is greater than half a wavelength). A deep-water wave travels across the sea but, in contrast with inshore breakers where the mass of the water moves bodily with the wave, the water that forms the troughs and crests of the deep-water wave does not move in the direction of wave propagation but simply oscillates up and down as the wave progresses.

The amount of energy in a wave[10] is difficult to calculate precisely because the sea state differs widely at different times and places. Theoretically, the total energy per metre of wave front for an idealised, sinusoidal, small-amplitude, monochromatic wave with a straight wave front is

$$\frac{g \rho L H^2}{8}$$

and the power (P) is given by

$$P = \frac{g \rho L H^2}{16T} \text{ per metre}$$

where g is the gravitational acceleration (9.81 m/s^2), ρ is the density of the water, L is the interval between successive waves, H is the wave height between peak and trough and T is the number of times the water surface moves upwards through its mean level per second. In reality, the power (in kW/m of wave) is much less than its theoretical level and approximates to

$$P \approx 0.55 \, H^2 T$$

Actual inshore measurements at South Uist have shown that average power levels of some 40–50 kW/m can be expected in water of about 50 m depth.

Although data on wave energy are still relatively scarce, sufficient information has been collected to give an approximate idea of wave energy available around the U.K. Figure 8.3 shows the main locations.

Wave Power Systems

Since wave power is a diffuse form of energy, efficiency losses are high and wave machines demand large and costly structures. Nevertheless, in the U.K. there has been a substantial amount of research and development on wave power, especially since the 1973 oil crisis. In 1974 a report entitled *Energy Conservation* by the

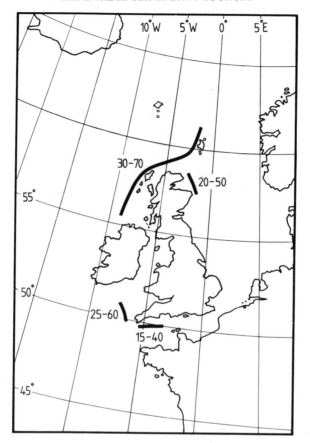

Figure 8.3 *Possible locations for wave energy devices* (source: HMSO, London)

Central Policy Review Staff first highlighted wave energy as a promising renewable-energy source. In 1976 the government allocated £1 million to study designs for wave energy converters. These included the Cockerell Raft, the Salter Duck, the Oscillating Water Column, the Lancaster Flexible Bag, and the Russell Rectifier. In 1978, Mr Alex Eadie, Parliamentary Under Secretary of State for Energy, announced an increase in the U.K. research budget for wave energy to £2.5 million. This was increased to over £3 million for 1980/81 but, having spent a total £15 m over 7 years, government support of wave energy projects was substantially withdrawn as a result of the 1982 ACORD report and before many of the R & D projects had been completed. The decision was severely criticised on the grounds that, while the U.K. was leading in the field of research, the U.K. government had accepted the wrong economic criterion (5p/kWh at 1982 prices) and an unfairly short timescale compared with other major energy technologies.

Following a recent classification,[11] wave power systems can be grouped under four main headings: passive reservoir systems, terminators, attenuators, and point absorbers. Some examples of each are described below and are illustrated in figure 8.4.

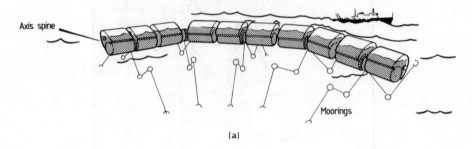

Duck Section containing gyroscopes,
processing in response to wave motion to drive hydraulic pumps

Axis spine

Moorings

(a)

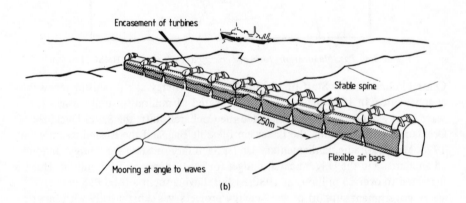

Encasement of turbines

Stable spine

250m

Flexible air bags

Mooring at angle to waves

(b)

Figure 8.4 (above and opposite) *Some examples of wave energy devices:*
(a) the Salter Duck (Edinburgh); (b) the Sea Clam (Lanchester);
(c) Belfast Buoy; (d) Bristol Cylinder

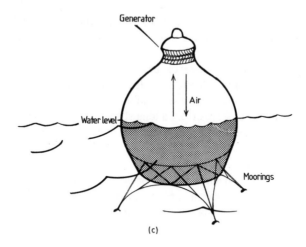

(c)

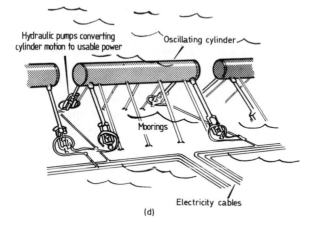

(d)

1. Passive Reservoir Systems

These are large immovable structures in which the wave crests enter a reservoir, drop down through turbines, and exit into the sea at the level of the wave troughs. Such systems are, however, extremely large and costly, in relation to wave energy conversion efficiencies. Two systems investigated were the *Russell Rectifier* and the *Mauritius Scheme*, the latter consisting of a sloping ramp over which the waves break into a reservoir from which the water returns to the sea via an ultra low head water-turbine.[12]

Diffraction power is another reservoir-based possibility for wave energy which was investigated by the Central Institute for Industrial Research in Oslo in the late-1970s. The scheme envisaged fixing a series of concrete blocks offshore in one of the bays of the Oslo Fjord. Waves from a long stretch of sea would be channelled through these blocks and concentrated on to a stretch of 400 m of coast. The result would be to amplify their size to 15–30 m high, thereby increasing their force. On reaching the coast the waves would be channelled into a reservoir where power could be generated using conventional hydroelectric methods. Preliminary estimates suggested that 10 km of Norwegian coast could generate 800–2000 GWh of electricity per year, although it remains to be seen how much of this would be collectable.

2. Terminators

These are free-floating, partially or fully submerged devices aligned horizontally and broadside to the wave direction in which vertical oscillations are created under the impact of the oncoming waves. Their length is similar to that of the wave or longer. The oscillatory motion is converted into power since, as the waves rise and fall, a water column is moved inside the device which in turn drives a turbine. The main advantage of such devices is that they can theoretically absorb 100 per cent of the incident wave energy and when fully submerged avoid damage through storm conditions (although secure mooring may be difficult to maintain).

A well-known example of a wave-power terminator is the *Salter Duck* developed by Stephen Salter at Edinburgh University,[13] in which gyroscopes are used to drive power-producing pumps on a series of cams or 'ducks' rotating around a large floating cylindrical spine. Another terminator system is the *Bristol Cylinder* where the oscillatory motion of the device is converted to usable power by means of hydraulic pumps in the mooring legs.

Other systems include the *National Engineering Laboratory (NEL) Breakwater Device*, which is mounted on the sea-bed and also uses the oscillating water column to drive a turbo-generator. An 11-strong consortium in Glasgow plans to build the U.K.'s first wave-operated power station and may later adapt this design for developing countries. The *Sea Clam*, developed at Lanchester Polytechnic, consists of a series of flexible airbags made of tough rubber mounted along a rigid hollow spine of reinforced concrete. The passing waves alternately compress the air in the bags forcing it through the turbine to the spine, and then in the troughs to allow the air to return through the turbine to the bag. The system is called the 'clam' because the original design included steel flaps or 'shells' to protect the bags from the full force of the waves, but in the course of development it proved possible to dispense with these and so considerably reduce the construction costs.

Another system suggested by the Lockheed Corporation in the U.S.A. is known as the *Dam–Atoll* and consists of a dome-shaped structure floating just beneath the ocean surface.[14] The prototype, under test in 1980, was 6.5 m in diameter. The device incorporates characteristics of both dam and an atoll – hence the name. Waves, normally arriving about every 10 s, enter an opening at the top of the unit just near the ocean's surface. A set of guide vanes at the opening causes the entering water to spiral into a vortex, held inside a deep central core. This swirling column of water in the central core turns a turbine, the unit's only moving part, which in the full-scale version can provide a continuous electrical power of 1–2 MW.

3. Attenuators

These are structures placed parallel to the direction of the waves and are at right angles to the wave crests. The waves run over them progressively yielding up their energy. Theoretically, attenuators can absorb all the wave energy, but in practice efficiencies only half those achieved in terminator devices can be achieved in attenuator structures of the same volume. Examples of attenuators include the *Vickers Attenuator*, which is a submerged stationary duct system. Originally conceived as a resonantly oscillating water column on the sea-bed powered by the static head of the waves passing above it, it was successively modified to bring the device nearer to the surface.

Other systems are the *Lancaster Flexible Bag* in which air is forced through bag compartments to drive a turbine as the waves travel along the device; and the *Japanese Wave Power Machine*. This latter has been taken up as a joint IEA project with the U.K., Ireland, the U.S.A., Canada and Japan. A special boat-shaped buoy research vessel, the Kaimei, was fitted with a number of oscillating water column chambers driving air turbines. It was designed by Yoshio Masuda of the Japanese Marine Science and Technology Centre. Kaimei was the first wave-powered device to transmit electricity to land (December 1979) and was also the largest sea-powered electric generator.[15] The 80-m ship cost $50 million and delivered 2 MW of electricity to the Japanese grid.

4. Point Absorbers

These devices are small free-floating buoys with no structural connection between each other. One such was designed by Dr Alan Wells while at Queen's University, Belfast. As waves cause the water level inside the buoy to rise and fall, the air movement causes the turbine to rotate and drive a directly coupled generator. Under a licence agreement signed in 1982 development of five prototype 100 kW generators, designed by the Belfast team under Professor Adrian Long, was assigned to the Fuji Electric company of Japan.

Other U.K. universities including Salford and Sussex have also been experimenting with various combinations of floating and submerged buoys or cylinders. Similarly, a research team in Norway has been examining the power absorption from a system of identical oscillating buoys. A buoy of 16 m diameter subjected to a maximum oscillation height of 8 m was considered to be capable of generating 4 GWh/year off the western Norwegian coast.[16] Another wave power plant consisting of 3-m diameter concrete buoys grouped in units with four rows of 80 buoys was being studied in Sweden. An estimated 24 MW could be generated from 21 units (6720 buoys) moored between 2 and 8 km from the shore. Features of the system included the ability of the buoys to be submerged during a storm or in the presence of drifting ice, and the use of linear electric generators in each buoy. Work is also being done in Norway, Bristol and elsewhere to use projecting walls on wave-energy devices to create an extra resonant mode to improve performance so that a smaller device may be used to capture the same energy.

Future Potential of Wave Energy

With the exception of the Kaimei, the energy contributions of all the schemes outlined above lie in the future and in the words of Dr C. Grove-Palmer, former wave energy manager of ETSU, the second law of wave energy is that 'The device you last thought of is the most attractive.'[17] Yet there are considerable advantages. As with wind energy, the greatest wave energy availability occurs in mid-winter which coincides with the highest energy-demand period. This makes it a very appropriate power source for distribution through a grid system. It may also be used on a smaller scale in the decentralised applications already described. Wave power is also non-polluting and continuously available along most windward coastlines, although the energy potential varies from place to place. Other advantages are that the energy is perpetual. Despite the size of most wave energy devices, once the capital equipment has been installed, total operating costs should be relatively low and unaffected by the inevitable rise of fossil-fuel prices. The technologies are likely to require few highly specialised operating staff, although according to the U.K.'s Department of Energy Paper 39 (published 1979) the considerable manpower required for the construction of wave energy converters might provide employment for people skilled in ship-building, platform construction and power-plant manufacturing. It has also been suggested that by installing a line of wave energy machines between 3 and 10 km offshore, an area of calmer waters would be created between the machines and the coast line. This could encourage fish-breeding and also provide safer fishing grounds.

The disadvantages of wave energy systems are the high initial capital costs and the costs of transmitting the electricity into the grid. Ideal wave conditions for these devices are not available in most developing countries except for parts of the West African coast, Indonesia, Papua New Guinea and Mauritius. Technical problems are also being encountered in establishing a robust and reliable structure capable of trouble-free operation in even the worst coastal storm conditions. Again, experience with North Sea oil rigs has shown up the problems of seaweed obstruction, barnacle encrustation and corrosion. Possible hazards to navigation or interference with marine life are other unknown factors, as is the visual acceptability of such devices.

If these disadvantages are overcome, wave power could be an important and inexhaustible energy option for the future.

References

A. OTEC

1. Lestor, R., 'Energy from the Oceans', in *Energy Crisis in the Eighties*, Crown Agents, London, July 1980, p. 26; Georghiou, L. and Ford, G., *South Magazine*, November 1982 and *Physics Technology*, Vol. 13, 1982, p. 15
2. Behrman, D., *Solar Energy – the Awakening Science*, Routledge and Kegan Paul, London, 1979, pp. 206-11
3. Ford, G. and Georghiou, L., 'Japan Stakes its Industrial Future in the Sea', *New Scientist*, 3 June 1982, p. 649
4. *Soft Energy Notes*, Vol. 3, No. 4, August/September 1980
5. Isaacs, J. D. and Schmitt, W. R., 'Ocean Energy: Forms and Prospects', *Science*, Vol. 207, 18 January 1980

B. TIDAL POWER

6. Minchinginton, W. and Meigs, P., 'Power from the Sea', *History Today*, Vol. 30, March 1980
7. Musgrove, P. and Fraenkel, P., *Natural Energy and Living*, No. 8, 1980, pp. 15-18
8. Flood, M., *Solar Prospects: The Potential for Renewable Energy*, Wildwood House, London, 1983, p. 132
9. *Tidal Power from the Severn Estuary*, Energy Paper No. 46, HMSO, London, 1981

C. WAVE POWER

10. *Wave Energy*, Energy Paper No. 42, HMSO, London, 1979
11. Count, B. M., Fry, R. and Haskell, J. H., 'Wave Power: the story so far', *C.E.G.B. Research*, No. 15, November 1983, p. 14
12. Bott, N., 'Blue Oil from the Sea', in *Must the World run out of Energy?*, Crown Agents, London, 1978, pp. 23-8
13. Salter, S. H., Jeffrey, D. C. and Taylor, J. R. M., 'The Architecture of Working Duck Wave Power Generators', *Naval Architect*, January 1976, pp. 21-4
14. *Financial Times*, 11 May 1979
15. *Pacific Friend*, June 1980
16. Platts, M. J., 'Large scale electricity generation from sea waves', *Seminar, Wind and Wave Energy*, Heliotechnic Educational, London, 1981
17. Grove-Palmer, C. O. J., 'Wave Energy in the United Kingdom, A review of the programme June 1975-March 1982', *Proceedings of the Second International Symposium on Wave Energy Utilization, Trondheim, Norway, June 1982*

Further Reading

A. OTEC

McCormick, M. E., *Ocean Thermal Energy Conversion*, Wiley, 1981
Marine Resources Project, Ocean Thermal Energy Prospects and Opportunities,
  MRP, University of Manchester, 1981
Knight, H. G., Nyhart, J. D. and Stein, R. E. (eds), *Ocean Thermal Energy Conver-
  sion*, Lexington Books, Lexington, Massachusetts, 1979
Ford, G., Niblett, C. and Walker, L., 'Ocean Thermal Energy Conversion', *Institute
  of Electrical Engineers Proceedings*, Vol. 130, Part A, 1983, pp. 93–100

B. TIDAL POWER

Severn Barrage Committee, *Tidal Power from the Severn Estuary*, Energy Paper
  No. 46, Volume I, HMSO, London, 1981
Shaw, T. L. (ed.), *An Environmental Appraisal of Tidal Power Stations: with
  particular reference to the Severn Barrage*, Pitmans Advanced Publishing Pro-
  gram, Fearon Pitman Inc., Marshfield, U.S.A., 1980
Charlier, R., *Tidal Energy*, Van Nostrand Reinhold, Wokingham, 1982

C. WAVE POWER

Shaw, R., *Wave Energy: A Design Challenge*, Ellis Horwood, Chichester, 1982
Ross, D., *Energy from the Waves*, 2nd edn, Pergamon, Oxford, 1981
Proceedings of the Second International Symposium on Wave Energy Utilization,
  Trondheim, Norway, 1982

GENERAL

Workshop on Alternative Energy Strategies, *Energy: Global Prospects 1985–2000*,
  McGraw-Hill, New York, 1977
Flood, M., *Solar Prospects: The Potential for Renewable Energy*, Wildwood House,
  London, 1983
Dawson, J. K., 'The Renewable Sources of Energy in the United Kingdom', *Proceed-
  ings of the Institution of Civil Engineers*, Part I, Vol. 74, February 1983

9

Renewable Land Energy Sources

A. GEOTHERMAL ENERGY

Geothermal energy is technically an 'alternative' energy source rather than a renewable one, although for research and funding purposes, it is frequently classed as a renewable-energy form. It is, however, the only one that is not directly or indirectly derived from the sun. Geothermal heat is present throughout the earth's crust in the form of hot dry rocks and is also available in certain regions as hot water or steam from underground reservoirs. The heat is obtained largely from the *magma* — a mixture of high-temperature gases and molten rock which lies underneath the earth's crust, approximately 30–45 km below the surface of the oceans and land masses. Most natural geothermal activity occurs in places where the tectonic plates which comprise the earth's crust meet and the magma approaches the surface through fissures as volcanic activity or hot-water springs. Global plate boundaries are shown in figure 9.1. which also indicates the main areas of known geothermal potential. Geologists currently believe that the continental crust itself possesses an internal heat or energy source due to radioactive decay of potassium, thorium and uranium. It is also possible that appreciable quantities of heat come from the heat released by exothermic chemical reactions within the crust, from the friction generated in faults by the sliding action of huge rock masses caused by gravitational and tectonic processes, and from the latent heat released by the crystallisation or solidification of molten rocks on cooling.

The amount of geothermal heat available is generally proportional to the square of the depth below the earth's surface. In geologically stable areas, thermal gradients average between 25°C and 35°C per km depth, that is, for each 30–40 m depth the temperature rises by 1°C, although in areas of high tectonic activity they can rise to 80°C/km or beyond.

High gradients may also be found where rock masses have been thrust upwards from high-temperature depths in regions of recent mountain building such as the zone stretching from Italy through Greece, Turkey and northern India to southwest China. Also, as has been mentioned, where a rift occurs between tectonic

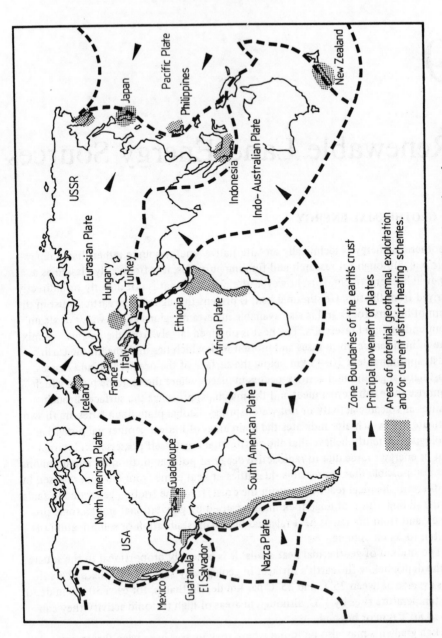

Figure 9.1 *Global plate boundaries and areas of exploitable geothermal energy (adapted from World Energy Atlas and Aquater Geothermal Resources Map)*

plates molten rock may rise to the surface as in the African Rift Valley area of Kenya, Uganda, Zaire and Ethiopia or in the Iceland mid-Atlantic ridge. These plate boundary zones cover perhaps 10 per cent of the earth's surface and the higher thermal gradients obviously offer the most attractive prospects for geothermal development when there is a market nearby.

Like oil, geothermal energy in the form of hot water or steam is usually extracted by means of drilling which with current costs can extend to about 2-3 km in depth, although in areas where the rocks are permeable water can circulate naturally to even greater depths. Ground water rarely circulates naturally below a depth of 4 km, however, so that the temperature of geothermal hot water and steam is generally below 300°C. (Since it is under pressure geothermal water can exceed the normal boiling point of 100°C.) Recent experience of mining and drilling for oil has shown that the earth's heat may be tapped even in regions remote from plate boundaries using current drilling technology. Most oil fields, for example, are also potential sources of geothermal energy. This is because saline water usually underlies the oil pool and often this is at a considerable temperature because of the depth of working. An example is the 'geopressurised' water beneath the Gulf Coast of Texas and Louisiana, which is seriously being considered as an energy source to follow the extraction of the oil and gas in the region. Dissolved methane gas has also been discovered in this and other areas with geothermal potential: an added economic incentive to developers.

Geothermal heat from low-enthalpy aquifers is usually obtained by the operation of twin pipes (doublets) equipped with pumps. The first pipe is for extraction. When the hot water reaches the surface it passes through a heat exchanger. After the heat has been extracted the cooled water is replaced by being pumped back into the ground through a second pipe to prevent the land above the underground reservoir from subsiding and to maintain the pressure of the reservoir. This process is known as re-injection.

The re-injection of the cooled water obviously tends to cool down the aquifer from which it was drawn and thus to shorten the life of the well. Only where the rate of heat extraction is limited to match the rate at which the water is naturally heated up again, known as the 'equilibrium capacity' of the well system, can geothermal energy be regarded as a renewable-energy source. Although the design of extraction systems and knowledge of the behaviour of geothermal fields is improving, most geothermal wells do cool down gradually, and may be regarded as having a life expectation of anything between 20 and 80 years.

The quantities of gases emitted from a large geothermal station in a steam field and of chemical elements brought to the surface in a hot-water field can be very considerable, entailing safety, environmental and disposal problems.[1] Table 9.1, which is based on an analysis of various existing fields, shows both the range of quantities of gaseous emissions for a 100 MW generating plant in a steam field, and the range of quantities of certain chemical elements brought to the surface by a 100 MW generating plant in a hot-water field. It can be seen that up to 800,000 tonnes per year of certain chemicals may be brought up in the hot water.

Geothermal steam may be accompanied by up to 5 per cent gas, mainly carbon

Table 9.1 *Gaseous emissions and elements brought to the surface by 100 MW geothermal generating plants in steam and hot-water fields*

Steam		Hot water	
Chemical constituent	*Output (tonnes/year)*	*Chemical constituent*	*Amount brought to surface (tonnes/year)*
Carbon dioxide	10,000–400,000	Sodium	20,000–200,000
Hydrogen sulphide	500–5,000	Potassium	2,000–100,000
Ammonia	100–1,500	Lithium	200–1,000
Mercury	0.01–0.05	Calcium	100–10,000
Radon	10–1,000*	Chloride	50,000–500,000
		Fluoride	30–300
		Sulphate	100–1,000
		Carbonate	0–5,000
		Boron	200–2,000**
		Silica	10,000–30,000
		Arsenic	50–500

*These quantities in Curies per year.
**Occasionally as high as 20,000.
Source: A. J. Ellis, 'Geothermal Energy Utilization and the Environment', *Mazingira*, Vol. 5, No. 1, 1981.

dioxide. Small amounts of other gases may be present, the most noxious being hydrogen sulphide which is both poisonous and has an objectionable smell. The small amounts of radon and mercury that are emitted are quickly dispersed in the atmosphere and are not thought to create a health hazard.

In hot-water fields, the composition of chemicals in the water can vary widely between fields. Most common are high concentrations of chloride/sodium/potassium in solution, but as table 9.1 shows, appreciable quantities of other chemicals and some trace metals may also be present. Particularly troublesome are waters with very high concentrations of salt as these usually also contain large amounts of heavy metals, such as manganese, iron, zinc, copper, lead and arsenic, which if discharged into local rivers or lakes can cause serious pollution.

One method of dealing with unwanted gaseous emissions such as hydrogen sulphide is to segregate them and then treat them chemically. For liquid emissions, methods include use of evaporation ponds and desalination (which have the advantage that the dissolved salts such as potassium may be recovered), and conventional water-treatment processes for filtering heavy metals. The most favoured method, however, is deep re-injection of the cooled waters into the supply aquifer. This prevents any environmental damage, but it too has its disadvantages, notably the gradual corrosion and the deposition of silica within pipes and channels, and, as mentioned, the fact that the continual re-injection of cold water can cool down the aquifer. To eliminate corrosion, specially treated steel or fibre-glass pipes are often used together with heat exchangers treated with anticorrosives. Another approach is to render the water acid or alkaline.

Geothermal heat extraction can sometimes create physical subsidence of the surrounding land if very large quantities of water are withdrawn and not re-injected. This has already happened at Wairakei in New Zealand. The high-pressure ejection of steam and water can be noisy and efficient silencers have to be installed to prevent damage to hearing, while the construction of cooling towers, pipeline networks and electric pylons increase costs and the visible environmental impact.

Useful Geothermal Energy from Aquifers: Main Categories

Extractable geothermal heat from aquifers can be divided into four main categories.

(1) High-grade Energy – Dry Steam Sources

In certain locations, geothermal energy emerges in the form of dry steam, at temperatures from 150°C to more than 350°C at the point of extraction. This can be used to provide electricity with a conventional steam turbine. Several conditions are necessary for the existence of a high-grade steam field. First, there has to be a body of magma within about 10 km of the earth's surface. Second, this intrusion must be overlain by a layer of permeable rock such as sandstone or limestone, which is in hydraulic contact with the surface (by way of faults or outcrops) so that water can percolate for heating into steam. Third, an overlying layer of impermeable rock must exist to prevent the steam escaping on a significant scale through natural convection. Fissures in this rock layer can give rise to geysers and hot springs as is shown in figure 9.2. Unfortunately there are very few places where all these con-

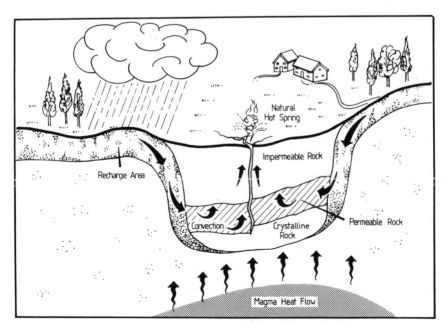

Figure 9.2 *Cross-section of a typical geothermal aquifer*

ditions occur naturally to give rise to a dry steam field. The main ones are located
in Italy, Japan and the U.S.A. The earliest example of an installation designed to
generate electricity from geothermal steam was at Larderello in Northern Italy,
where production commenced in 1904. By 1912 there was a commercial plant
generating 250 kW. Its capacity was extended to 406 MW and by 1979 there were
altogether 17 power stations in Italy generating electricity from geothermal steam.
The other main dry steam installations are at The Geysers near San Francisco in the
U.S.A. (the world's largest at a rated 900 MW) and at Matsukawa in Japan. There
are other smaller dry steam-generating plants such as Kawah Kamojang in Indonesia
(some 30 MW in 1984) with a planned expansion of 110 MW within the next four
years as part of Indonesia's proposed 1000 MW expansion by 1995.

(2) Medium High-grade Energy – the West Steam Type

This second type provides geothermal energy in the form of wet steam and water at
temperatures of between 150 and 250°C. Those sites combining high temperatures
with good permeability are the most worth while to exploit. Wet steam fields are
estimated to be 20 times more common than dry steam fields and can be found in
countries such as Guadeloupe, Iceland, Mexico and New Zealand. Electricity generation is less efficient because of the lower energy content. Also there are
operational problems due to turbine blade corrosion caused by water droplets and
scale formation caused by dissolved salts and minerals. These are reduced in dry
steam conditions. Notwithstanding the problems, by 1983 over 3100 MW(e) of geothermal electrical capacity from both wet and dry steam fields had been installed
worldwide, with the U.S.A., Italy and the Philippines supplying almost three-
quarters of this. A minimum of 17,644 MW(e) is anticipated by the year 2000.
Table 9.2 illustrates the situation.

Electricity from wet steam geothermal fields is not usually generated in normal
steam turbines unless the input temperature of the steam is at least 150°C, and
preferably over 200°C. At lower temperatures, it is cheaper to adopt a binary
technology whereby the heat in the steam or hot water is transferred to a liquid
with a lower boiling point which then drives the turbine. Such a process has also
a high efficiency which makes the economics more attractive. The U.S. company
Turbines International developed a bladeless turbine designed specifically for
electricity production from geothermal water and steam fields using temperatures
as low as 100°C. It remains to be seen whether this will significantly increase the
possibilities of electricity production in the medium/low-temperature fields. Currently power stations using high-temperature geothermal sources also tend to have
a low thermal efficiency and much of the available heat is rejected as hot waste
water. The economic potential of such sites could be enhanced by the introduction
of combined heat and power schemes (see chapter 10) using the low-grade 'waste'
heat for district heating. An early example of such a development is at Svartsengi,
in south-west Iceland.

Table 9.2 *World geothermal electrical-generating capacity,
1983 and 2000 (estimated)*

	1983		2000	
	MW(e)	*% of world total*	*MW(e)*	*% of world total*
USA	1284	40	5824	33
Mexico	205	6	4000	23
Japan	228	7	3668 +	21
Philippines	593	19	1225 +	7
Italy	457	14	800	5
El Salvador	95	3	535	3
New Zealand	202	6	382 +	2
Costa Rica	0	0	380 +	2
U.S.S.R.	11	–	310 +	1
Turkey	0.5	–	150	1
Nicaragua	0	0	100	1
Indonesia	32	–	92 +	1
Iceland	41	–	68 +	–
Ethiopia	0	0	50	–
Kenya	30	–	30 +	–
France	0	0	–	–
China	8	–	–	–
Others	4	–		
TOTAL	3190	100	17644	100

Sources: U.N. Conference on New and Renewable Sources of Energy, Nairobi, 1981;
Di Pippo, R., *Geothermal Resources Council Bulletin,* January 1984.

(3) Low-grade Energy – Hot Water Sources

Naturally occurring hot water between 50 and 100°C is common in all the areas within tectonic plates where there are water-bearing rocks. The geysers and hot springs frequently manifested in these areas have been used by man for thousands of years both for baths and for medicinal purposes. The Roman civilisation used natural hot springs wherever possible when building the elaborate baths for which its architecture is renowned. In fact most of Europe's currently fashionable spa towns, such as Bath in England, Aix-les-Bains in France and Baden Baden in Germany, have grown up on the sites of natural hot-water sources formerly discovered by the Romans. Today, however, the vast majority of the natural hot water used for bathing is found in Japan. Installed low-temperature geothermal capacity in 1980 for various countries is given in table 9.3. It should be noted that table 9.3 is expressed in thermal megawatts, a measure of the amount of heat available, whereas table 9.2 is in electrical megawatts, a measure of the amount of electricity output. Each megawatt of electricity output will represent a heat-energy input several times greater.

Table 9.3 *World installed low-temperature geothermal capacity in 1980,*
in thermal megawatts

	For bathing		For other purposes		Total	
	MW(t)	%	MW(t)	%	MW(t)	%
Japan	4394	82	81	3	4475	56
Hungary	547	10	619	23	1166	15
Iceland	209	4	932	35	1141	14
U.S.S.R.	0	0	555	21	555	7
Italy	192	4	73	3	265	3
China	7	0	144	5	151	2
U.S.A.	4	0	111	4	115	1
France	0	0	56	2	56	1
Czechoslovakia	8	0	35	1	43	1
Romania	0	0	30	1	36	0
Austria	3	0	2	0	5	0
TOTAL	5364	100	2644	100	8008	100

Source: U.N. Conference on New and Renewable Sources of Energy, Nairobi, 1981.

(4) Very Low-grade Energy

The lowest-grade thermal temperatures are between 20 and 60°C and are generally
to be found in reservoirs near the surface of the ground. They are therefore less
costly to exploit and are particularly suitable for such purposes as fish farming and
agriculture. The possible industrial applications are also immense and their availability is very widespread.

Useful Geothermal Energy from Hot Dry Rocks

The natural geothermal resources considered above are limited by the occurrence of
a water source. However, besides these geothermal aquifers another source of geothermal heat is also to be found in impervious hot rock deposits such as granite.
These hot dry rocks can be found anywhere under the earth's surface. Drilling and
heat extraction could be economic only down to about 6 km, the problem being
to ensure that the technology of the extraction process is capable of generating
the amount of energy extracted. These developments are still in the research stage
and are a long way from being a proven technology. However, with improved
drilling techniques and hybrid applications, heat extraction from greater depths
may prove possible.

In order to obtain energy from hot dry rocks it is first necessary to drill a well
through the outer sedimentary cover of the earth into the hot impermeable base
rock. Then a 'fracture' is made in an area of rock at the base of the well to allow an
injected circulating fluid (such as water) to flow through the system and be heated

by the rocks. Finally, the water is extracted through a second well and the heat is converted into electricity at the surface through a system of heat exchangers, pumps and turbo-alternators. The heated water may also be used as direct thermal application. The process is shown diagrammatically in figure 9.3. One of the most

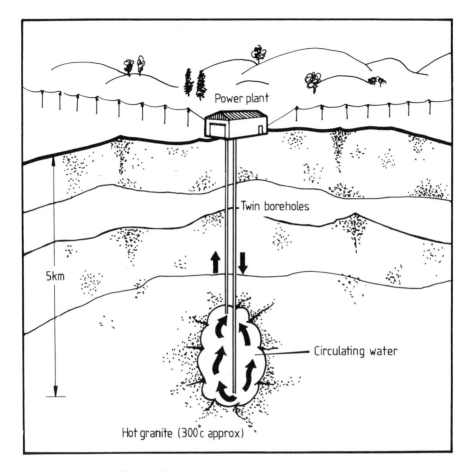

Figure 9.3   *Typical hot dry rock fracturing system*

ambitious hot dry rock geothermal energy research programmes has been carried out by the Los Alamos scientific laboratory in the U.S.A. under the auspices of the U.S. Department of Energy. In 1972 an experimental site was selected at Fenton Hill in New Mexico and a 785-m well was drilled. Hydraulic fractures were created and another outlet-well connected. In 1978 a series of drawdown tests was conducted with very encouraging results. By late 1979 some 10 million kWh of thermal energy had been extracted from the site at power levels of 3–5 MW(t). As a result of these successes, drilling at greater depths began in 1979 with the objective of developing a hot dry rock (HDR) system of commercial proportions. This system

will comprise two wells at a depth of about 4 km, where the rock temperature is
275°C, and should be capable of producing 20–50 MW of thermal power for at
least 10 years.[2] Such a scheme could make a very considerable contribution to the
energy needs of a sizeable city.

Since the heat from hot dry rocks is extracted through the medium of water, the
uses to which it may be put are similar to those for hot water or steam extracted
from aquifers. In both cases the crucial variable is the temperature at which the
water (or steam) reaches the surface. Table 9.4 gives an indication of the uses
appropriate to the different temperature ranges.

Table 9.4 *Potential uses of geothermal fluids of different temperatures*

200–400°C	Electricity generation combined with other applications
150–200°C	Industrial process heat, industrial drying, refrigeration, electricity generation (binary cycle)
100–150°C	Space-heating, industrial drying, evaporation processes, salt recovery, food canning, binary cycle electricity
50–100°C	Greenhouses, space-heating, domestic hot water, industrial drying, refrigeration
20–50°C	Bathing, fish hatching, animal rearing, soil warming, fermentation processes, crop drying

Source: A. J. Ellis, Geothermal Energy Utilization and the Environment, *Mazingira*, Vol. 5,
No. 1, 1981.

The U.K. Geothermal Programme

In the U.K., officially sponsored geothermal energy research began in 1976 and was
initially concentrated on aquifer, as opposed to hot dry rock, resources. It was felt
that since successful exploitation of the latter lay at least 10 years in the future, it
was expedient to wait until other countries had developed a technology which the
U.K. could emulate.[3] Also, as the U.K. is in a tectonically stable area, the geothermal gradient is not high (about 30°C per km) and heavy drilling costs make deep
drilling for higher temperatures uneconomic at current comparative fuel costs.

Aquifer research in the U.K. was previously confined to gathering data about
potentially suitable geothermal sites until 1976 when the Department of Energy
allocated £3.65 million for technical assessments and associated field studies.
The first geothermal aquifer in the U.K. at Southampton was successfully drilled in
1982 to produce hot water at 70°C. This hot water, which forms part of a larger
Wessex Basin reservoir, will hopefully be used to heat the new civic complex in the
city. Humberside lies over a similar hot water aquifer and an exploratory borehole
is being drilled to tap hot water at temperatures of 63–77°C, believed to be lying
2–3 km below the region.[4] Such a development would cost approximately £1.3
million and could be used as a production well for supplying district-heating in
Cleethorpes if the heat flow proved great enough. Other potential sites exist in the

Bath and Bristol area, Lincolnshire, Northern Ireland, the Central Valley of Scotland and in the Wessex Basin. If any of these sites possess accumulated hot water in sufficient quantities it may be used to supply heat to small district-heating schemes, but in general, as there has been no volcanic action for some 55 million years, the average thermal gradients are comparatively low. While the hot spring at Bath reaches a temperature of 49°C this is thought to be the result of local patterns in the flow of water underground. When the Gas Council drilled in the sandstone and salts of the Cheshire/Shropshire basin it found a temperature of only 59°C at a depth of 3 km — a gradient of less than 20°C/km. The hot rock deposits in Cornwall and parts of Scotland may be able to supply higher-grade heat, and work is being conducted on an experimental basis. The Camborne School of Mines in Cornwall has developed a technique for carrying out carefully controlled explosive shots at the foot of a well. The resulting fracture is then extended by hydraulic pressure in order to achieve better fracturing and more economic extraction of energy from the rocks. This technique has proved so successful — reducing the resistance to flow in the fracture by more than a factor of 10 without leakage — that the U.K. Department of Energy together with the EEC agreed to fund a project to create a fracture system with 2 million m^2 of fracture surface.

In June 1983 the Energy Secretary, Mr Peter Walker, endorsed an £11 million Cornish 'Hot Rocks' project by the Camborne School of Mines to investigate further the existing site at Rosemanower quarry near Falmouth. £1.5 million is to be allocated on related background studies by the Institute of Geological Sciences. The results from the major experiments carried out over the past 3 years at a depth of about 2000 m have apparently been encouraging but further research is required to prove the practicability of the techniques applied. The next phase of research will involve drilling a deeper hole to make a connection from below to the large reservoir of fractured hot rock already created around the existing pair of boreholes. The Department of Energy estimates[5] that the exploitable hot dry rock resources of the U.K. could be used to generate at least 2500 terawatt hours of electricity, representing about 10 years' current electricity consumption.

Advantages and Disadvantages

Since geothermal energy exists at every point under the earth's surface, its theoretical potential is enormous and has been calculated at about 4×10^{13} GWh or some 450,000 times the current world energy consumption. However, such estimates rely heavily on guesswork and in any case only a small fraction of this energy can ever be utilised since most lies between 25 and 45 km in depth which is well out of the range of current drilling techniques. It is thought that wells of over 6 km in depth could be feasible by the year 2000, but such developments are still technically a long way from proven applications. In assessing the practical potential of this resource in any given location there are as usual conflicting advantages and disadvantages to be weighed. Some of these are considered below.

Costs. Since wells can have widely varying success rates, there is a considerable investment risk unless geothermal energy in usable quantities has already been discovered as a by-product of other forms of geological exploration such as drilling for oil. Consequently the enormous capital outlay and high risks deter private development unless backed by heavy subsidies and nearby markets assuring adequate financial returns.

Life expectation. The high capital costs of geothermal wells demand that the maximum use be made of them to produce an adequate return on the investment. If this results in heat being extracted at a rate above the equilibrium capacity, the well will cool down, more or less gradually depending on the amount of heat extracted, and have a finite effective life. Insofar as this takes place, geothermal energy cannot be regarded as a renewable-energy source.

High-technology Option. Apart from the capital investment already mentioned, the advantages of geothermal resources in providing indigenous energy can be offset outside seismic zones by the sophisticated technology required. This will favour its development mainly in the richer industrialised countries which can afford such a large-scale power-generating system. Figure 9.4 gives some idea of the level of technology involved. Environmental considerations such as noise, pollution, and the availability of water as the heat-transfer medium may also restrict geothermal developments in certain areas.

Figure 9.4 *Typical geothermal installation in a dry steam field (Lardarello, Italy) showing hot-water pipes and electricity-generating plant* (source: ENEL)

Other obstacles to development include a shortage of trained technicians and experience, difficulties in obtaining information, and the difficulties of technology transfer, especially where appropriate geothermal resources exist in developing countries. Another problem is the proprietary nature of information about geothermal energy obtained by private companies through oil drilling and the lack of appropriate institutions to utilise and market these resources in developed countries. One of the main limitations to the use of geothermal energy is its geographical inflexibility. The high-grade sources may be used to generate electricity and where a national grid exists the energy can then be transported in this form. The much more common low-grade sources, however, require the heat produced to be used at or close to the point of extraction. This means that low-grade sources that are far from centres of human habitation and economic activity are of little use. The need to site geothermal wells close to populated areas also aggravates any environmental and pollution problems that may be involved.

On the more positive side geothermal wells could become an increasingly sound investment where high geothermal gradients are found near to the site of potential local applications. Over 50 countries have already identified geothermal energy potential and further exploration is being conducted by various means. Apart from electricity generation there already exist a number of economic and useful thermal applications and the expansion of schemes especially for medium and low-grade applications is almost certain.

Whether electricity generation from geothermal energy is an attractive option will depend on the price of alternative sources such as fossil fuels, and — because of the high capital costs but subsequent nil fuel costs — on current and prospective interest rates. The answer will be different for each proposed project but the World Bank had already suggested in 1980[6] that electricity from geothermal energy was broadly competitive with, and in some cases distinctly cheaper than, the relevant alternatives.

B. WATER POWER

Water power has been used extensively for thousands of years. Its main applications are the age-old water wheel, hydraulic-powered pumping devices and the more modern hydroelectric generating turbine.

Water wheels, like windmills, were originally used to turn mill stones for grinding corn and subsequent uses included the operation of bellows in iron forging, pumping out water from mines and also filling canal systems. In the eighteenth century the pioneers of the Industrial Revolution used water wheels to power woollen and paper mills and for wood and metal-working.

Hydropower is provided by the pressure or velocity of water flowing from a higher level to a lower level so that the potential energy (the energy of position measured above a fixed level) may be converted into kinetic energy (the energy of motion). This motive power may be harnessed to provide mechanical energy by means of water wheels or converted into electricity by a turbine. There also exist

pumped storage schemes, such as at Dinorwig in Wales where 'off-peak' electricity
is further used to pump surplus water to a higher reservoir to allow it to flow down
to generate more electricity at times of high demand. To calculate the theoretical
power available in a given stream or river, the following equation is used:

$$kW = \frac{H \times Q}{102}$$

where H = head measurement in metres and Q = flow in litres/second. In practice
typical efficiencies of water wheels are between 40 and 75 per cent and of water
turbines between 60 and 90 per cent or more depending on the size and type of
equipment used.

Hydroelectric power was first developed in the U.S.A. in 1882 and gradually
spread throughout the world in regions where there was a suitable river source and
an adjacent market for the electricity generated. In Switzerland, for example, at the
end of the 1940s there were some 7000 small hydroelectric power stations serving
isolated communities and small local industries. From 1950 to 1973 the importance
of these small-scale hydroelectric schemes, as of other small renewable energy
applications, declined in most industrial countries, as a result of the increasing
availability and diverse uses of cheap oil. However, ability to transmit electricity
over long distances fostered development of major hydroelectric power schemes.
During the 1970s there was a growth of 3.6 per cent per year in world hydropower
generation and by 1983 hydroelectricity supplied 24 per cent of the world's electricity or about 7 per cent of world primary-energy consumption. After the fossil
fuels – oil, coal and gas – it had become the most important energy resource, contributing about twice the world's total nuclear capacity.

Both water wheels and turbines deliver their power as torque on a shaft with
pulleys, chains or gear boxes connected to this shaft to provide mechanical or
electrical power. Water wheels generally operate at 2–12 revolutions per minute
(rpm), and although they can be 'geared up' to produce electricity this is costly and
inefficient. Water wheels are therefore best suited for generating motive power, while
the usually smaller, enclosed, higher rpm turbines are used for electricity generation.

Water Wheels and Pumps

Water wheels are generally large-diameter, slow-turning devices driven by the weight
of the water. Traditionally they are made of wood or iron and they will operate
where there are substantial fluctuations in the flow rate, although these will affect
the number of revolutions per minute.

During the nineteenth century there were over 20,000 water wheels operating in
England while Cumbria and Cornwall alone have more than 5000 sites of former
wheels. Some of the wheels in Yorkshire, Lancashire and Derbyshire generated
enough power to operate small factories, yet by the 1970s it was estimated that
there were only about 500 water wheels and small turbines functioning in the whole

of the U.K. This energy source fell into disuse with the advent of centralised electricity supplies. With the post-1973 fuel price increases, however, private and national interest in small-scale water power systems and the future potential of hydropower were revived in the U.K. and elsewhere.[7]

There are basically two categories of water wheel, the undershot wheel and the overshot wheel, with many individual design and size variations.[8]

Undershot wheels are the earliest and most basic in design. They can operate on fairly low heads of 1–3 m with flows from one-tenth of a cubic metre per second upwards. The water undershoots the wheel as the name implies. Louis XIV used wheels of this type 14 m in diameter, to operate the fountain pumps at Versailles. Overall efficiencies of the earlier wheels tended to be low — those serving Versailles had an overall efficiency of only 10 per cent — until the beginning of the nineteenth century, when the shape of the paddles was changed and the design modified. The Poncelet undershot wheel with its curved blades, buckets or vanes, was designed to reduce the energy loss through shock as the water hit the flat paddles, and remains the basic and most efficient design for low-technology devices.

Breast wheels are undershot wheels, generally used for heads between 2 and 3 m, with a structure or breast built to fit close to the rim of the wheel so as to direct the water into the buckets and blades along its lower half. Because they are more expensive and complicated to construct and maintain than the other basic types of water wheel they are not widely used.

Overshot wheels were developed towards the end of the eighteenth century and have proved most efficient for heads between 3 and 10 m. Overshot wheels gradually came to replace undershot wheels especially for the higher heads of water, and efficiencies of up to 80 per cent can now be achieved. The rotation and hence power is derived partly from the impact of water entering the bucket from the top but largely from the weight of the water as it descends in the buckets. During seasons of diminished water supply power output is reduced, although efficiencies are increased since less water is spilled from the partially filled buckets. In most overshot wheels the water flows along a channel, trough or sluice and through a sluice gate which regulates supply.

Figure 9.5 contrasts basic designs for undershot and overshot wheels.

Other types of water wheel include the *impulse wheel* where the motive force is derived from the impact of a high-velocity jet of water producing kinetic energy, the *reaction wheel* which relies on pressure energy being partly changed into kinetic energy, and the *piston wheel* where water under pressure is used to drive an engine. Modern water wheels sometimes use new materials such as fibre-glass since the traditional iron wheels with their large, heavy and expensive structures are regarded as uneconomic in relation to the power generated. Also water wheels normally work at lower efficiencies than turbines and so are generally unsuitable for electricity generation. To increase the wheel's speed of rotation to that needed by an alternator to generate electricity requires the installation of very complicated and expensive initial gearing, so turbines have proved more appropriate. Some fibre-glass wheels have been installed to provide electricity but their efficiency and long-term durability have yet to be proven.

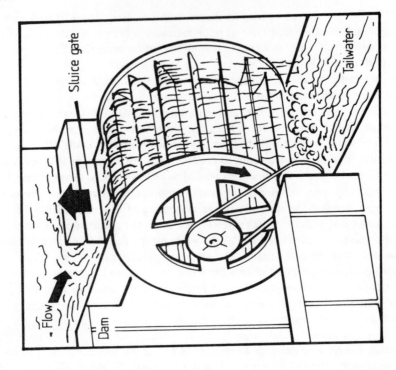

Figure 9.5   *Basic undershot and overshot wheel construction*

The hydraulic ram is a simple low-technology water-powered pump which raises water, sometimes to heights of over 100 m. This is done by converting some of the kinetic energy of flowing water into potential energy thereby enabling a proportion of the water to be raised to a level higher than that of the supply source. It was developed in France by J. Montgolfier towards the end of the eighteenth century and its use was subsequently adopted in rural areas of England and other industrial countries. Although, in common with the water wheel, its popularity declined when centralised electricity supplies were substituted, it is now being revived as an energy converter for water power, particularly in developing countries. However, it can function only if the water supply source is steady and reliable. A cross-section of an hydraulic ram is illustrated in figure 9.6. The reservoir head causes flow in pipe (a) induced by valve (b) being open, so allowing water to escape. As flow increases, valve (b) closes (it is an inward opening valve and pressure on its underside causes it to shut, moving upwards, when the pressure is greater than the valve weight) giving

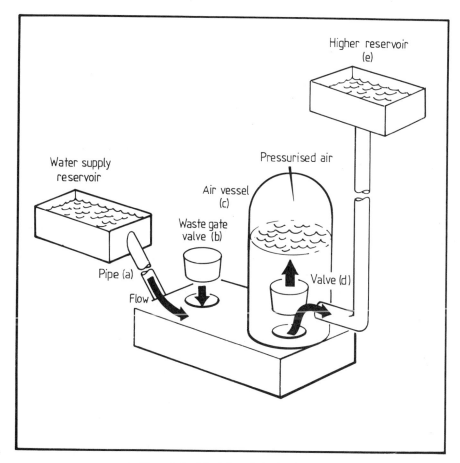

Figure 9.6 *Principles of a hydraulic ram*

rise to a rapid increase in pressure which forces open valve (d); water escapes into
the air vessel and compresses the air so that water flows away along the rising main
to the higher reservoir (e). When that pressure subsides the entire cycle is repeated.

Water Turbines[9]

There are two basic types of water turbine — impulse and reaction — with many
variations within each category depending mainly on the direction of flow through
the turbine. The suitability of a particular installation at a given location depends
mainly on the pressure or head of water available and the power output required.
Water turbines, in contrast to water wheels, are usually comparatively small constructions (sometimes being only 12 inches in diameter); however some extremely
large turbines also exist, and in general the design of a water turbine is more complicated than that of a water wheel.

Impulse Turbines

These basically consist of two parts: a nozzle and a vertically installed turbine wheel.
The head-water flows down a pipe or duct into the nozzle which causes it to be
directed under pressure on to the curved blades of the turbine wheel. The amount
of energy available in the water flow is a function of the velocity of the water and
the pressure in the pipe. The main types of impulse turbine are the Michell/Banki
Turbines and the Pelton Wheel. Michell, an English engineer, first developed the
concept of a radial or 'cross-flow' impulse turbine at the beginning of the twentieth
century and the use of Michell turbines was extended throughout Europe. A German,
Banki, working quite independently, introduced the same cross-flow concept to the
U.S.A. where most similar installations are known as Banki turbines. *Michell/Banki*
turbines are very widely used for small hydropower installations since they are less
expensive than other types, extremely versatile and relatively simple to construct,
and so may be assembled in rural workshops. This type of turbine is really an
enclosed and improved version of the water wheel and can take water directly from
streams where the flow is very variable.

In the Michell Turbine the wheel comprises two parallel discs joined at their rims
by a series of curved runners. As the water comes out of the nozzle it strikes the
runners as they pass, thus transferring energy to the turbine wheel. The water then
flies across the hollow centre of the wheel until it strikes the inside of the curved
runners on the other side, pushing the wheel round further in the same direction.
The water then drops down and exits through a 'tailflow'. About three-quarters of
the power is developed at the first strike and one-quarter during the second. The
construction and operating principles of a Michell/Banki turbine are shown in
figure 9.7.

The Pelton Wheel is a widely used type of impulse turbine which, given a
sufficiently high head (typically over 150 m), can develop high speeds (up to 1000
rpm) suitable for a.c. power generation. It is particularly suitable for installation in

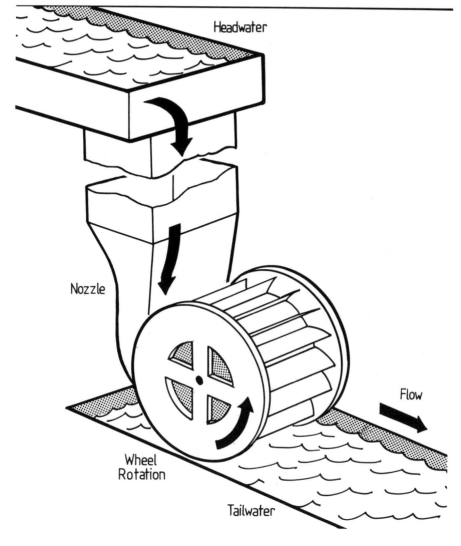

Figure 9.7 *Michell/Banki turbine*

fast-flowing mountain streams since the higher the head, the smaller the wheel and flow volume required. However some small units can operate efficiently on heads down to 20 m. Lower heads require larger wheels and greater flow rates to generate the same power. Water is delivered to the Pelton Wheel through a feed pipe or penstock which may be many hundreds of feet long, depending on the head. The wheel itself has specially shaped buckets or cups mounted on the perimeter. A high-speed jet of water delivered through a horizontal nozzle is directed into the cups at the bottom of the wheel causing it to rotate. Conversion efficiencies can be up to 93 per cent although the Pelton Wheel is initially more complicated and expensive than the Michell/Banki Turbines.

Other types of impulse turbines include the Jonval axial flow impulse turbine, the Girard axial and radial impulse turbines (which work with heads of up to 500 m producing efficiencies of around 45 per cent), and the Ossberger cross-flow turbine in which the vanes are parallel to the shaft axis and the water enters radially to the shaft. The Turgo impulse turbine consists of a Girard axial-flow runner used in conjunction with one or more nozzles of a Pelton Wheel type. It has a smaller diameter than the Pelton Wheel and runs at twice its speed. This reduces overall costs because there is less need to gear up for high-speed electricity generation.

Reaction Turbines

Unlike the vertical impulse turbines these wheels are horizontal and are encased in a housing which is completely filled with water. All reaction turbines operate on the principle of a change of pressure across the wheel or runner blades, the water passing through the wheel, thus transferring its energy and leaving with both velocity and power diminished. As reaction turbines have low efficiency in part-flow or low-head situations, they may have variable vanes or runners to maintain their efficiency level by adjusting to variable rates of flow. This naturally adds to installation costs but increases running efficiency. In some situations, depending on power requirements at a given location, the simpler version may be preferred. By flowing through the curved vanes or runner blades the water causes the wheel to rotate by the 'reaction' or pressure of the water. The main types of reaction turbine are: *outward-flow turbines* where the water flows into the centre, passing through fixed radial guide blades into moving blades adjusted to a correct angle — and then out again; *inward-flow turbines* which work on the opposite principle, with water leaving in a direction parallel to the axis; *axial-flow turbines* where the water enters at the sides and flows parallel to the axis; and *mixed-flow turbines* which are a combination of the inward-flow and the axial-flow, in which the water hits the turbine end-on in much the same way as a water wheel and turns 90° through the vanes, thus imparting energy and driving the mechanism. Most reaction turbines are specifically designed for a limited range of flow and head conditions. Although expensive and complicated compared with some other types, they can achieve efficiencies of up to 93 per cent when properly regulated to the load and flows of a given location. Some examples are given below.

Francis radial reaction turbines have a very high maximum efficiency under full-flow conditions but a much lower part-flow efficiency. Typically suited locations range from a few metres head to 100 m or higher depending on the required output.

Propeller turbines are simpler in construction and are ideally suited for low heads from 1 to 20 m. As they are easy to instal, these turbines may also be built into existing dams and weirs. (The turbines have runners very like the propellers of a ship, and one variation, known as a *tube turbine*, can be likened to a propeller and shaft installed in a pipe to provide a very simple and cheap application.) Unfortunately it is impossible to describe within the space available the many other variations on the examples described above. The main disadvantage of fixed-blade propeller turbines is their low part-flow efficiency.

Kaplan turbines are also propeller-type reaction turbines but have runner blades which may be adjusted to suit different flow conditions. However the additional mechanical complexity and expense makes them suitable only for larger power plants.

Vertical-axis river turbines were developed in the late-1970s by the Intermediate Technology Development Group (U.K.) and Reading University, mainly for water pumping in fast-moving irrigation canals. These turbines work on the same principles as small vertical-axis windmills installed under water and are also capable of generating local electricity supplies with fairly low heads of water.

Table 9.5 sets out the specific characteristics of the main types of water wheel and water turbine.

Table 9.5 *Water wheel and water turbine specifications*

Wheel or turbine type	Range of head (ft)	Wheel or runner diameter (ft)	Efficiency (%)	Ability to handle changing		Technology of construction	Construction materials
				Q flow	H head		
Water wheels							
Undershot	6–15	3H (H = head in ft)	35–45	good	fair	low	metal/wood
Poncelet	3–10	2H–4H (> 14)	60–80	good	fair	medium	metal/wood
Breast	6–15	H–3H	40–70	good	fair	low	metal/wood
Overshot	10–30	0.75H	60–85	good	none	low	metal/wood
Water turbines							
Michell (Banki)	15–150	1–3 +	60–85	good	good	medium	welded steel
Pelton Wheel	50–4000	1–20	80–94	good	fair	medium/high	steel, cast iron or bronze
Francis	100–1500	1–20	80–93	poor	poor	high	cast or
Kaplan	14–120	2–30	80–92	poor	good	high	machined
Propeller	8–200	2–30	80–92	poor	poor	high	steel

Source: R. Merrill and T. Gage (eds), *Energy Primer*, p. 109, © 1978 Portola Institute, reprinted by permission of Dell Publishing Co. Inc.

Hydroelectricity Generation

Hydroelectricity generation may be on a small, medium or large scale depending on the size of the river and the energy requirements. To simplify, the description below will focus on large and small-scale hydropower schemes.

Large-scale Hydropower

Large-scale hydroelectric schemes have steadily spread in recent years wherever there has been a suitable water source since they are usually more cost-effective than fossil-fuelled power stations. These systems typically consist of a large dam, penstocks, and banks of turbines and generators (see figure 9.8).

Figure 9.8   *Tongariro Hydroelectric Power Station, New Zealand*
(source: Sir Alexander Gibb & Partners)

The world's largest hydroelectric station at the Itaiupu dam which commenced generating in 1984 is situated on the Parana River between Paraguay and Brazil. With an installed capacity of 12,600 MW from 6 turbines, it is the focus of a highly centralised supply system and will supply about one-third of Brazil's current energy requirements.

As with other renewable-energy systems, however, the availability of large-scale hydropower sources varies considerably from region to region. Some countries such as Switzerland, Norway, Sweden, Canada and Brazil have abundant mountain lakes and rivers and rely heavily on hydroelectric power for their energy supplies. In other flat, arid regions the opportunities are very limited. Large-scale hydroschemes are also very costly and capital-intensive and are thus more readily developed in the rich industrial countries than in developing regions, even if the possibilities exist. Table 9.6 compares the annual technically usable hydropotential with that actually being harnessed or under construction in various regions by 1980. The table shows that

Table 9.6 *World hydropower potential 1980 (annual rates)*

Region	Technically usable potential (A)		Hydrocapacity operating or under construction (B)		Percentage (B) of (A)
	10^{12} kWh	(MJ)	10^{12} kWh	(MJ)	
Africa	3.14	(11.30)	0.20	(0.72)	6.4
North America	3.12	(11.23)	1.43	(5.15)	45.8
Latin America	3.78	(13.61)	0.65	(2.34)	17.2
Asia (excluding the U.S.S.R.)	5.34	(19.22)	0.55	(1.98)	10.3
U.S.S.R.	2.19	(7.88)	0.46	(1.66)	21.0
Oceania	0.39	(1.40)	0.08	(0.29)	20.5
Europe	1.43	(5.15)	0.94	(3.38)	65.7
WORLD	19.39	(69.80)	4.31	(15.52)	22.2
Africa, Asia, Latin America	12.26	(44.14)	1.40	(5.04)	11.4

Source: Adapted from World Energy Conference, 1980. Survey of Energy Resources, Table 4.7.

the industrial regions of Europe and North America are already exploiting a substantial proportion of their usable resources while in the less-developed areas the potential remains largely untapped.

In the U.K., hydroelectric power generation contributed 2.3 mtce (1.4 mtoe) or 0.7 per cent of total U.K. energy consumption in 1983. Except for north Scotland and Wales, virtually all the possible sites for large-scale hydroelectric schemes have already been used up so that only an estimated additional 2 mtce per year of hydroelectricity is expected from such sources. On the other hand, wherever there are rivers and lakes there are usually smaller tributaries and streams often not included in official resource estimates.

Despite its initial high costs, the competitiveness of hydropower is increasing owing to improvements in hydrotechnology. Cheaper dams using local materials instead of concrete are sometimes built. Better-designed concave-faced 'arch' concrete dams are also reducing construction costs, as are more efficient turbines and generators. Also, since increased water storage and a greater head can be obtained from higher dams, some recent developments have extended existing dams with consequent cost savings.

By 1982 several novel schemes to harness water power on a large scale had been proposed or were under way. One of the more ambitious is Israel's plan to utilise the 396-m drop between the Mediterranean and the Dead Sea by linking the seas with a 100-km canal to power a 240–570 MW, four-turbine power station. It is hoped that this station will eventually supply 18 per cent of Israel's power needs. Similarly Egypt is planning to build a canal from the Mediterranean to the Qattara Depression where a 30,000 km^2 lake will be created some 60 m below sea level. The proposed scheme will generate about 2400 MW of electricity.

An even less conventional idea under consideration is to harness summer melt-water from the Greenland ice cap, from which it is estimated some 120,000 MW of electricity could be generated.

Mini Hydroelectricity Schemes

Although interest in small-scale hydroelectricity schemes declined after 1950, there has been renewed interest since the early-1970s when oil prices started to escalate, especially in simply constructed mini hydro-plants with a capacity of a few kW up to 10 MW.

In some industrialised countries where sites for large constructions are running out, the possibility of using smaller rivers and streams is now under consideration. In France, for example, over 3000 potential sites were identified in 1981 for systems of up to 4.5 MW, and in Sweden 1300 potential sites for systems of between 1 and 1.5 MW were found. In the U.S.A. it was also discovered that an additional 55,000 MW of electricity could be generated by installing better turbines and generators at existing dams.

In several developing countries such as Pakistan, India and Indonesia there are numerous existing irrigation canals with low-head hydropower possibilities. In some countries there is also great potential for new small-scale water-power projects which can be constructed more rapidly, require less funding, and entail fewer transmission costs and, as always, fewer environmental problems than larger schemes. In China, for example, a policy initiated by Chairman Mao of 'converting hazards into benefits,' controlled flooding by diverting the waters of the Yellow River to irrigate farmlands and provide electric power while at the same time stimulating rural industry. These small-scale rural initiatives began in 1958 with very small water turbines of an average 35 kW output and resulted in there being over 80,600 small hydroelectric stations in 1982 with an average capacity of 100 kW. Many other countries where large-scale hydropower is being installed, such as Brazil, also have plans for developing mini hydrocapacity. Nepal is currently developing systems of between 5 and 50 kW and surveys of potential mini hydrosites have been carried out in Indonesia, Malaysia, the Philippines, Papua–New Guinea, Sri Lanka and Tanzania.

The Future Role of Water Power

Although the technology is well established, since it has been used widely for nearly 100 years, the initial costs of building a conventional hydroelectric power station are nearly always high. The main costs are in the associated engineering work, for example the construction of the dam (where necessary), and the transmission of the electrical power over long distances from the central power source. This initial outlay proves a drawback to the developing nations, especially as highly centralised electrical supplies cannot satisfy the needs of remote outlying villages. A further drawback to the expansion of hydroelectric power, especially in large-scale applications, is the limited geographic availability of sites and suitable locations. This applies especially in industrial countries such as those of Western Europe, including France, Germany and the U.K.

Large-scale hydroelectric power schemes also have potentially far-reaching environmental, social and economic hazards which must be taken into consideration. The building of a dam and upstream flooding of river valleys can reduce water flow downstream. This in turn can affect the soil structure and cultivation of the surrounding land. The natural ecological balance may be destroyed, fish breeding may suffer or a bacterial imbalance in the river may be created. The fertile delta silt which is normally carried downstream frequently settles in the reservoirs behind the dam and can cause serious engineering problems. There are also health hazards, as with Ghana's Lake Akosombo, where there has been a major increase in the disease bilharziasis (schistosomiasis) since the snails carrying the disease are now able to breed rapidly whereas the previous flow of the river kept the species in check. It is also thought that earthquakes may be a consequence of building large dams in seismic areas owing to the pressure of the water and the inevitable re-adjustments that take place.

While in developed countries large-scale hydroelectric stations are welcomed to supply high-density population areas, in developing countries they can be a threat to local agriculture and promote an unhealthy trend towards urbanisation. For example, the Kariba Dam project flooded the fields of countless 'unproductive' small farmers and no attempt was made to rehabilitate them or provide alternative agricultural development schemes. Many of the dispossessed rural poor drifted to the towns to become unemployed or, if they were lucky, to find work as unskilled labourers!

The main advantage of all water power systems is the ability to convert a renewable-energy source to electricity at high overall efficiency. A second advantage is that there is no atmospheric pollution in generating this electricity and no fossil fuels are burned. A third advantage is that so long as there is a reliable source of water it should be possible to provide a scheme having a fairly constant output. Other advantages are that water power systems, once installed, can provide a free source of indigenous energy which does not rely on foreign fuel imports or foreign exchange rates. Most systems are easy to operate and cost little to maintain. Mini hydroschemes are particularly interesting for remote rural areas where there is not a heavy demand for electricity, provided that the installation costs can be kept low. The above factors also make hydropower attractive in industrial countries where suitable sites are available, particularly in cold and temperate climates where the periods of peak flow tend to coincide with those of peak domestic demand.

C. WIND POWER

As mentioned in chapter 1, wind energy was an important power source in many countries until the advent of centralised electricity supplies. Modern technology is now seeking to improve on the traditional machines and to find ways in which the new wind generators, sometimes referred to as *aerogenerators* or *wind turbines*, can efficiently supplement electricity grid systems, provide independent power sources for remote regions, and supply basic fuel and pumping needs for Third World countries.

Winds are mainly caused by the uneven levels of solar radiation that fall upon different regions of the earth. The resulting differential air pressure causes convection currents by which cool polar air is drawn towards the Equator to replace the warm air which rises and moves towards the poles, but these major flows are affected by the rotation of the earth and the net effect is an anticlockwise air flow in the Northern Hemisphere and a clockwise flow in the Southern. In latitudes from about 35° to 50° N and S of the Equator, air masses moving faster than the earth lead to two zones of prevailing Westerlies. There are two polar masses of high pressure that outflow as prevailing Easterlies. At about 30° N and S of the Equator, the air moves at about the same speed as the earth, thus leading to a belt of low-speed winds or calm.

Winds also vary seasonally and between day and night. For example, air over land tends to heat up faster than air over water during the day and cool down faster at night, bringing onshore and offshore winds, respectively. In any given locality, winds are further influenced by the topography; by hills, valleys, woods, water, tall buildings and so on. And within these configurations smaller high and low-pressure zones give rise to the many winds, eddies, currents and gusts. Despite their local unpredictability, winds tend to have a dependable annual pattern as can be seen on the isovent map of the British Isles (figure 9.9) which offers a guide to the annual average hourly windspeed in different areas in metres per second. Whereas in certain areas of the west coast of Scotland, the Western Isles and Ireland annual average wind speeds of 7 m/s are recorded, the Department of Energy has identified 10,000 potential sites for windmills in the U.K. overall, with 3100 hilltop locations with wind speeds of 8.6-13.3 m/s and some 6900 coastal sitings with wind speeds of 5.9-7.7 m/s.[10] For small and medium output applications a 2.7 m/s windspeed is generally considered the lowest for any practical use.

There are at least 57 winds recognised by name, each having its own characteristics like the different ocean currents. Some of these winds are renowned like the *Mistral*, the Nor'easter, the *Harmattan* and the monsoons. There are the Trade Winds,

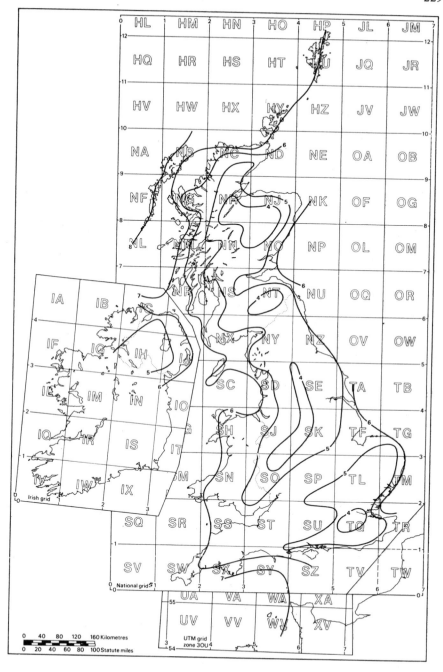

Figure 9.9 *Annual average hourly mean wind speed (metres per second) in the United Kingdom at 10 m above ground in open country. Applies at altitudes up to 250 m above sea level.* (Source: Building Research Establishment)

moderate but very steady in character, and the Westerlies, sometimes blustery, sometimes calm and wet.

The wind has been harnessed for power since time immemorial. Main applications have been to power sailing ships and to drive windmills. Modern versions of both exist as well as wind-powered electricity generation. Sailing ships were known in Egypt as long ago as 5000 B.C.; and primitive windmills are first thought to have been used for irrigation in Babylon (around 2000 B.C.) and also in China and Japan. Windmills were widely used in Persia at around 200 B.C. to grind corn. They came to Europe by the twelfth century and by the nineteenth century existed in nearly every country in large numbers. The traditional four-bladed Dutch windmill had a wooden blade with a diameter of 10 m or more and an output of up to 50 kW. For hundreds of years windmills were one of the main sources of power in Western Europe from the Greek Islands to East Anglia, from Portugal and France to the Outer Hebrides. By the early-nineteenth century about 10,000 windmills were in use in the U.K., 10,000 in the Netherlands and a proportionately high number in Denmark where they were first used on a large scale to produce electricity in the 1890s.

Gradually, with the coming of steam power in the nineteenth century, windmills fell into disuse as a prime source of mechanical power, although many are still being used today in rural areas for irrigation and other agricultural purposes. In fact, over 6 million multi-bladed wind pumps were manufactured in the U.S.A. between 1850 and 1940 and over 1 million wind pumps of different designs are still functioning worldwide with their use increasing, particularly since 1973. These are especially useful for irrigation purposes in developing countries. They are usually capable of pumping water from wells, rivers or lakes without the need of a back-up power source. A disadvantage is that they can be used only where wind and water are simultaneously available. In India, for example, the best wind conditions occur during the monsoon when water is plentiful.

Modern Wind Energy Conversion

Only a very small portion of the total wind energy available is extractable owing to geographical factors. Nevertheless, in 1981 this was estimated to be 20 million MW by the UN Conference on Renewable Energy Sources: an impressive figure equal to twice the world's energy consumption.

The extent to which winds can be converted into useful work depends on: (i) the speed of the wind; (ii) the area of wind falling within the span of the windmill rotor; (iii) the proportion of available wind energy extracted by the blades; and (iv) the overall efficiency of the mechanical parts of the windmill itself. The key relationships are:

(i) The power (P) available in the wind increases as the cube of the wind speed (V) according to the equation

$$P = \tfrac{1}{2}\rho A V^3$$

where ρ is the wind density and A is the cross-sectional area of the rotor.

(ii) A linear increase in the diameter of the rotor results in a square law increase in the area covered by the rotor.

(iii) The theoretical maximum amount of kinetic energy in the wind that can be recovered by a windmill is 59 per cent (A. Betz).

(iv) Frictional losses in the transmission, gear box and generator reduce the available energy by about 20 per cent.

These relationships indicate that there are major advantages in building large-diameter windmills in areas where wind speeds are high, and when rotor blades are well designed and components well matched.

Main Types of Wind Energy Conversion Systems

There are two major types of wind energy converters: horizontal axis and vertical axis. The rotors on a horizontal-axis wind converter revolve around a horizontal shaft, that is, the blades rotate perpendicularly to the ground, while with a vertical-axis converter it is the other way round.

The traditional windmills, such as the Dutch post mills, were horizontal lift machines where the sails moved slowly and had high solidities (the ratio of blade area to swept area). The latter enabled them to have a high starting torque and be self-starting but they ran at low efficiencies with a lower power-to-weight ratio. Modern wind turbines have been developed mainly since 1973 drawing largely on modern aircraft technology. They have low solidities with blades that are aerofoils achieving high rotational speeds and high efficiencies. While vertical-axis machines are omnidirectional, horizontal-axis machines need to be guided into the wind either by a tail vane or by using canted blades as on aircraft wings. Contemporary research and development is aimed at achieving the most efficient and cost-effective design and materials for a given application in a given type of wind regime. Good aerodynamic design is required to give high lift and good structural design to permit low construction costs, but with a structure strong enough to withstand heavy loads on the tower and blades even in severe gale conditions. Figure 9.10 illustrates basic shapes of some main types of wind energy converters.

1. The Horizontal-axis Wind Turbine

Because of the great advantages of scale mentioned above, recent technical developments have focused mainly on large wind turbines capable of generating several megawatts of electricity for the national grid. Interest began in the U.S.A. in the 1930s when Palmer C. Putnam together with Morgan Smith Engineering Inc. designed and built the world's largest wind turbine on a Vermont hill, named Grandpa's Knob. The 1.25 MW turbine was built on a tower 34 m high, had two blades 53 m long, and commenced operation in 1941. When a spar snapped in 1945, the project was abandoned but much valuable technical data had been collected on large-scale wind-turbine construction and operation, and subsequent research was directed to overcoming the material and engineering flaws.

In 1970 Professor W. E. Heronemus of the University of Massachusetts, U.S.A.,

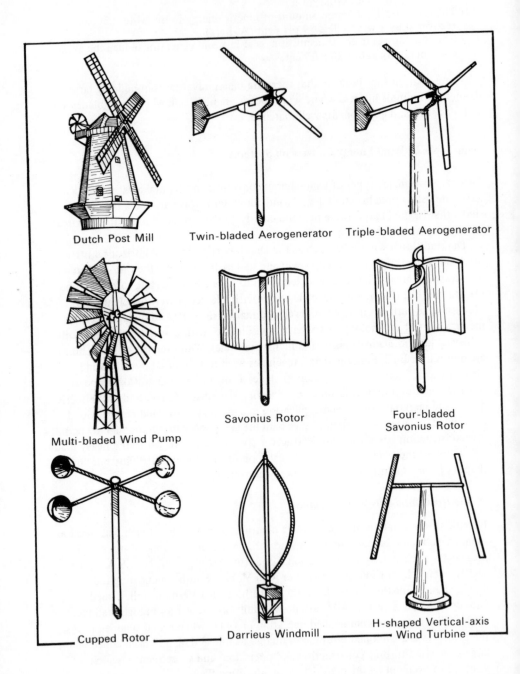

Dutch Post Mill   Twin-bladed Aerogenerator   Triple-bladed Aerogenerator

Multi-bladed Wind Pump   Savonius Rotor   Four-bladed Savonius Rotor

Cupped Rotor   Darrieus Windmill   H-shaped Vertical-axis Wind Turbine

Figure 9.10 *Some main types of wind converters*

took the lead in re-opening feasibility studies of large-scale wind energy conversion with a series of designs based largely on the Smith–Putnam windmill. One of his proposals envisaged a series of towers 600 ft (183 m) high, with 20 two-bladed wind turbines mounted on top, on each square mile of the 300,000 square-mile area of the Great Plains, so providing the equivalent of half the 1971 U.S. installed electrical-generating capacity. In 1973 he proposed an offshore chain of wind generators as an alternative to a proposed nuclear-power station at Long Island, New York. Although these schemes were not implemented at the time, interest in wind power continued to grow in the U.S.A.

After the 1973 oil crisis, official government interest in large-scale electricity-producing wind turbines was rekindled in the U.S.A. and the American Space Agency NASA introduced a research programme (entitled 'MOD') with the development of a series of increasingly large machines. A first prototype 100 kW wind turbine (MOD-0) was built in 1975; four 200 kW turbines (MOD-0A) followed in 1977; a 1.5 MW turbine was built in 1979 (MOD-1); and three 2.5 MW turbines followed in 1981 (MOD-2). By 1983 about 3000 wind turbines had been installed in the U.S.A., with an installed capacity of about 300 MW.

In the U.S.A. developments have been supported by a 1978 Public Utilities Regulatory Policies Act (PURPA) requiring utilities to buy power from small generators. These, together with state and federal tax credits, are increasing the interest of individuals and private investors in smaller machines and wind farms. As an example, Altamont Pass in California which enjoys ideal wind conditions ranging from 12 to 19 miles per hour, has become a popular development site for wind-energy farms. By 1982 US Windpower was selling wind-generated electricity produced from 140 machines to the local utility Pacific Gas and Electric; the Fayette Manufacturing Corporation had 40 machines installed and Windmaster Inc. and another small farm each had five.

Interest in large-scale energy developments has also been growing in Europe in recent years. In *Denmark* a 2 MW 'Tvind' machine (two blades of 53 m, tower 54 m high) was constructed as a community project in 1978 by the Tvindskolen College in Jutland. This was followed by the government-sponsored refurbishing of the smaller 200 kW Gedser wind turbine in 1977, and the construction of two 630 kW machines at Nibe. Private owners feed surplus into the national grid.

In *Sweden* a major wind energy programme costing up to £25 million commenced in 1977 with the commissioning of an 18 m diameter, 63 kW machine, WTS 1. By 1982 two further large machines (WTS 2 and WTS 3) had been constructed to evaluate a wide range of technologies. Both were twin-bladed 78 m diameter horizontal-axis machines, with WTS 2 rated at 2.2 MW in a 13 m/s wind and WTS 3 (a slightly larger machine) rated at 3 MW in a 14 m/s wind. WTS 4 is rated at 4 MW.

In the *Netherlands* research into wind power by 1984 had including the testing of both horizontal and vertical-axis 25-m diameter machines. Future development plans include a medium-sized wind turbine of up to 0.5 MW plus a new larger prototype with an installed capacity of 3 MW. Government participation in these investments amounts to 40 per cent of investment costs.

West Germany has also embarked on an extensive wind energy programme, including the construction of a 3 MW wind turbine in 1981 (Growian I, currently the largest in the world) and the single-bladed monopterus. Other designs are planned as well as feasibility studies on wind energy farms and small-scale developments.

The *U.K.*'s programme currently includes the detailed design of a 60-m horizontal-axis windmill (costing £6.2 million) which should be commissioned by 1986 at Burgar Hill, Orkney. This is being funded by a consortium of private companies and government agencies. A 20-m diameter, 250 kW prototype was erected on the same site in 1983 and is now feeding 180 kW into the island's grid. It was followed by a 300 kW machine and a 25-m diameter experimental vertical-axis wind turbine at Carmarthen Bay. The CEGB plans to erect a 4 MW turbine at Richborough near Ramsgate, Kent, to serve a community of about 4000 people, and is to lead an international study into offshore wind power.

2. The Vertical-axis Wind Turbine

These less conventionally looking wind turbines have a number of advantages over the horizontal-axis machines. Their efficiency is unimpaired by the direction of the wind and, as the rotor rotates horizontally, there is less gravitational load and so fatigue is less and smaller, cheaper structures can be used.

There are three main designs of vertical-axis machine, with many variants: the Savonius Rotor, the Darrieus Rotor and the Variable Geometry Rotor.

The *Savonius Rotor*. A common version has two blades which, when looked at from above, together form an S shape. It is cheap to construct but relatively inefficient and so its uses are largely confined to water pumping and DIY applications.
The *Darrieus Rotor*. This has a catenary shape like the blades of an egg-whisk to minimise bending stresses and is being developed mainly in the U.S.A. and Canada. Both small versions (up to 40 kW) and large models (up to 0.5 MW) are being tested and in Canada a 4 MW machine, Eole-1, is on the drawing-board.
The *Variable Geometry Rotor*. Developed by Dr Peter Musgrove of Reading University, this has an H-shaped rotor, the upright sections of which are adjustable. At low wind speeds the sections are vertical but as the wind force increases they pivot into an outward-facing arrow shape to reduce the aerodynamic stresses.

Future Potential of Wind Power

Apart from the obvious environmental advantages of using wind power to generate electricity and to save fossil fuels, the main considerations in any wind power development scheme are economic.

Since wind turbines cannot provide a firm source of power but must either be used with wind available or incorporate storage or a back-up system, the value of wind energy must be estimated in terms of cost of fuel saved. The Cost of Energy Delivered (COE) may be calculated by multiplying the capital cost of a turbine by the annual charge rate and dividing by the annual energy output. This cost must then be compared with the cost of fuel saved. In 1981 Musgrove compared the cost of wind generated electricity (using the U.S. MOD-2 as a model) with the coal-generated electricity that supplied 76 per cent of the CEGB's load.[11] He concluded that land-based large wind turbines could deliver electricity at the cost of 1.66 pence/kWh where the average wind speed is 5.4 m/s, compared with an average fuel cost of electricity from coal-fired stations of 2.30 pence/kWh (allowing 3 per cent per year increase in coal costs from 1980 until the end of the century). Thus there was already a strong economic case for land-based wind energy systems. The economics for grid systems improve further in areas where the average wind speed is more than 6 m/s and there will be further advantages with the cost reductions expected in the next generation of large wind machines. System integration studies show that utility grid systems can absorb up to 20 per cent of their needs from intermittent wind sources.

Offshore Wind Power Systems (OWPS)

As has been mentioned, these were first proposed in the U.S.A. in 1973. Current aerogenerator technology together with Dr Musgrove's work on vertical-axis systems have led to the possibility of clustering hundreds of offshore windmills in suitable U.K. locations (such as the Wash) as previously suggested by Heronemus. Seventeen British offshore sites with a low-tide water depth of no more than 20 m have been considered. Preliminary calculations show that there are enough shallow areas off the U.K. coast to site windmills capable of an output exceeding 20 per cent of the installed capacity of the CEGB.[12] Wind speeds are higher than on the nearby land and there are fewer environmental objections. Since the U.K. has experience of both high-speed winds and offshore construction, technology in this form of wind energy is a possibility for the future. Although capital costs could be high and transmission losses would occur before the electricity could be delivered to the mainland, Musgrove has pointed out that escalating conventional fuel costs in the 1980s will quickly close the price gap between OWPS and other forms of electricity generation. In Sweden, also, variations on the Darrieus-type vertical-axis machines have been developed for use in offshore electricity generation. Notable is a 180-m diameter, 20 MW machine known as the Poseidon L-180, designed by Dr Olle Ljungstrom of the Swedish Aeronautical Research Institute.

Advantages and Disadvantages of Wind Power

Wind energy is a more concentrated form of renewable energy than direct solar radiation. Wind-powered water pumps operating on over a 3 m/s wind can cost as little as one-seventh the amount of a solar-powered pump with the same output.[13] Any increase in wind speed would improve the economics further.

Wind energy can also be used in many applications directly at the point at which it is required. It is a decentralised power source which makes it exceptionally attractive for remote and inaccessible consumers not linked to a supply grid, such as many islands or rural villages in developing countries. It eliminates the need for

an expensive centralised plant and transmission lines. It can also supply a centralised electricity-supply source either from small private generators or especially
when concentrated in wind farms and offshore wind power systems. It is ideal for
use in many hybrid renewable-energy systems as this saves on storage costs. Above
all, of course, it is a fuel saver – essential as fossil fuels become depleted.

The main disadvantages of wind power are that it is an intermittent supply
source and windmills are extremely site-sensitive. There is still insufficient knowledge of wind energy availability at a local level and much more detailed local data
are required to facilitate a more widespread use. In the case of small stand-alone
units, the problem of fluctuating supply and the need for storage or back-up
systems adds to the expense. Battery storage or conversion to hydrogen are possible
means of overcoming the problem but, in common with other forms of renewable
energy, storage is an expensive and limiting factor since the wind is not continuously
available as a high-energy source. This makes it unsuitable for many industrial processes. In direct link-up with the grid, however, this problem is irrelevant and the
energy cheap.

Often, it is difficult to find appropriate sites. The wind cannot be used in towns,
built-up areas or where other obstacles such as woods, impede the air flow. There
are many areas with little wind, little space or an unsuitable environment where
there is simply no possibility of installing a windmill. Also technical expertise is
needed to build the more sophisticated types of windmill, because the siting, installation and connection to the load are as important as the correct rating. This factor
and the initial cost makes them unsuitable for many developing countries. Other
objections are that wind generators can be regarded as visually unattractive and
may create a certain amount of noise, although this is rarely serious.

Despite the known limitations of wind power, it is clear from estimates of the
potential availability of wind in many countries that it can make a significant contribution to energy supply. In countries such as the U.S.A. this resource is likely to
be making an impact by the late-1980s or early-1990s. For example, the Lockheed
Company stated that by 1995 wind could be saving the U.S.A. 2200 million barrels
of oil a year – equivalent to total U.S. oil imports for 1975. By then 18 per cent of
U.S. electricity could be wind generated from large aerogenerators attached to the
national grid. Improvements in energy-storage technology will also greatly enhance
the future prospects of small-scale free-standing wind power units as another energy
option. In the U.K., the cautiously modest funding policy of the government makes
rapid development somewhat unlikely, although Dr Musgrove has estimated that
with government funding of £5 million per year until the late-1980s, wind power
could supply 20 per cent of the U.K.'s electricity needs by the turn of the century.
This is equivalent to the estimated contribution of nuclear power without the
attendant hazards. For many of the developing countries, small-scale wind power is
likely to become an economic resource and feasibility studies completed by the end
of the 1970s were distinctly optimistic. With the considerable number of major
systems in the 1–5 MW range under construction or being tested by 1980, it appears
that in many different countries of the world the windmill is once again taking its
place as an economic and environmentally benign source of power.

References

A. GEOTHERMAL

1. Ellis, A. J., Geothermal Energy Utilization and the Environment, *Mazingira*, Vol. 5, No. 1, 1981
2. Nunz, G. J., *The Hot Dry Rock Geothermal Energy Development Program: an Overview*, Affirmative Action/Equal Opportunity Employer, LASL 79-79, November 1979
3. *R.E. News*, Department of Energy, London, May 1980
4. Moss, J., 'Geothermal Go-ahead Expected in Grimsby', *Electrical Review*, Vol. 210, No. 18, 7 May 1982
5. *Department of Energy Press Notice No. 75*, Department of Energy, London, 28 June 1983
6. Edmunds, S. W., 'Thermal Uses of Geothermal Energy', *International Journal of Ambient Energy*, 1,1, 1980, pp. 63-7

B. WATER POWER

7. McGuigan, D., *Water Power, Natural Energy and Living*, NEC, London, 1978
8. Merrill, R. and Gage, T. (eds), *Energy Primer – Solar, Water, Wind and Biofuels,* Dell, New York, 1978
9. Simon, R., 'Small Hydro Makes a Big Splash', *Soft Energy Notes*, Sept./Oct. 1982, pp. 108-9

C. WIND POWER

10. Lipman, Professor N. H., 'Reading University, Blowing Hot and Cold on British Windmills', *The Guardian*, August 1980
11. Musgrove, Dr P. J., 'Wind Energy', *Journal of the Royal Society of Arts*, No. 5301, Vol. CXXIX, August 1981
12. For a discussion of the features to be taken into account in assessing OSWP *see* Moore, D. J., *Offshore Wind Data*, Central Electricity Research Laboratories, Leatherhead
13. Earthscan, *New and Renewable Energies 2*, IIED, London, 1981, p. 34.10

Further Reading

Butler, E. W. and Pick, J. B., *Geothermal Energy Development*, Plenum, New York, 1979
Noyes, R. (ed.), *Offshore and Underground Power Plants*, Energy Technology Review No. 19, Noyes Data Corporation, Park Ridge, New Jersey, 1977
Wahl, E. F., *Geothermal Energy Utilization*, Wiley, London, 1977
Edwards, L. M., Chilingar, G. V., Rieke, H. H. III and Ferth, W. H., *Handbook of Geothermal Energy*, Gulf Publishing, Houston, Texas, 1982
Christopher, M. and Armstead, H., *Geothermal Energy*, Spon, London, 1983
Grant, M., Donaldson, I. G. and Bixley, P. F., *Geothermal Reservoir Engineering*, Academic Press, London, 1982
McGuigan, D., *Small Scale Water Power*, Prism Press, Dorchester, Dorset, 1979

Simeons, C., *Hydro Power*, Pergamon, Oxford, 1980

Watt, S. B., *A Manual on the Hydraulic Ram for Pumping Water*, Intermediate Technology Publications, London, 1978

Keoppl, G. W., *Putman's Power from the Wind*, 2nd edn, Van Nostrand Reinhold, London, 1982

Kovarick, Pipher and Hurst, *Wind Energy*, Prism Press, Dorchester, Dorset

Workshop on Alternative Energy Strategies, *Energy: Global Prospects 1985–2000*, McGraw-Hill, New York, 1977

Flood, M., *Solar Prospects: The Potential for Renewable Energy*, Friends of the Earth/Wildwood House, London, 1983

Dawson, J. K., 'The Renewable Sources of Energy in the United Kingdom', *Proceedings of the Institution of Civil Engineers*, Part I, Vol. 74, February 1983

Golding, E. W., *The Generation of Electricity by Wind Power*, Spon, London, 1955

Lipman, N. H., Musgrove, P. J. and Pontin, G. W. (eds), *Wind Energy for the Eighties*, Peter Peregrinus, London, 1982

Selzer, H., 'Assessment of the Technical/Economic Prospects for Wind Energy in the European Countries', in Palz, W. and Schnell, W. (eds), *Wind Energy R & D in the European Community*, Series G, Vol. 1, Reidel, Dordrecht, 1983

10

Conservation

'In the short term, energy conservation must be the cornerstone of our policy; the potential for saving is immense . . . it is now certain that if we do not change our ways while there is still time our society will risk dislocation and eventual collapse'
Roy Jenkins as
President of the E.E.C.

Changing Traditions

Conservation was practised throughout history until the second half of the twentieth century. Most civilisations reclaimed, recycled and re-used materials. Old ships' timbers were preserved to build houses and barns; rags were used in the paper industry; organic wastes were spread on the land as fertiliser. As industrialisation progressed with cheap and abundant energy and materials, conservationist habits were gradually abandoned in favour of manufacture using virgin raw materials. These were often imported relatively cheaply and avoided the more expensive labour costs incurred by reclaiming and sorting used materials before recycling. Increasingly too, as both social and industrial structures developed towards specialisation, they became more energy-intensive and the overall use of energy became more profligate and more prone to promoting wasteful practices. Also, after the 1950s accelerating changes in new product design and concepts such as built-in obsolescence to guarantee continuous markets for replacements were introduced to stimulate demand and increase production. The increasing throughput of materials, in turn, led to higher overall energy consumption, the creation of additional waste and further pressure on reserves. Before the 1973 oil crisis some people were becoming alarmed by these runaway trends in energy consumption and waste. Some of the 'prophets of doom' as they were then labelled, came from within the oil companies themselves but, on the whole, up to the 1970s such critics were associated with the antipollution or anti-industry movements.

Gradually, with an increasing public awareness of pollution and its environmental hazards, and recognition of the fragility of long-term supplies of some fuels and raw materials, conservation was recognised as an alternative to wasteful consumption and production. There were frustrations that the public could do nothing to prevent an oil embargo, the effects of a miners' strike or controversial decisions about the adoption of a nuclear-power programme. Thus a conservation movement grew up as a 'grass-roots' or 'alternative' movement, originating on the West Coast of America with the 'flower people' and rapidly spreading to other industrialised nations, including West Germany and the U.K. Conservation was no longer concerned only with wild-life and an unpolluted environment but with questions of scale, energy and wasted resources. The grass-roots conservation groups gradually gained momentum and had to be recognised, if only as 'fringe' movements. Their literature and demonstrations proliferated, but since their concern for conservation was usually expressed in concepts such as 'self-sufficiency', a return to the land or decentralised communities, their message had interest rather than impact in a highly centralised and interdependent society.

Then in 1972, before the Yom Kippur War and Middle East oil crisis, came a very controversial semi-official report based on studies by the Massachusetts Institute of Technology. This was the Club of Rome's *Limits to Growth*.[1] It struck at the very root of the industrialised countries' unspoken assumptions about the value of indiscriminate industrial 'progress'. It forecast that, unless major changes in policy were forthcoming, at the most 100 years separated mankind from catastrophe. Initially the report was ill-received in established circles. The *Economist* magazine dubbed it as 'the highwater mark of old-fashioned nonsense' and it was widely criticised by economists, industrialists and technical specialists. Yet the controversy itself made many people uneasy and provided fuel for the conservation movement. Gradually, it was appreciated that economic growth might not, after all, be infinite. It might be limited not only by bad management or poor productivity but because of resource scarcity, not least of energy. Conservation was slowly becoming respectable since it was seen to be necessary. Indeed the U.K. government's Green Paper *War on Waste*, published in September 1974, stated that 'the government believes that there should be a new national effort to conserve and reclaim scarce resources − a war on waste involving all sections of the community. We all instinctively feel that there is something wrong in a society which wastes and discards resources on the scale we do today.' In the same year the government introduced a Control of Pollution Act, pollution being closely linked with the wasteful squandering of energy and materials.

Since that time official forecasts and events themselves have made it plain that pressures on resources will increase worldwide; fuel and raw materials will become more expensive; and certain products, such as oil, will run out in the foreseeable future. The *Global 2000 Report to the President of the United States* (1980), commissioned by President Carter, further depicted the horrifying economic and environmental implications of the world continuing to consume energy at present rates and by current methods.[2] An excellent analysis of these issues is to be found in *Environmental Impacts of Production and Use of Energy*, by E. E. El-Hinnawi[3] sponsored by the United Nations Environmental Programme.

In nearly every industrial country conservation has been seen as a necessary and inexpensive way of buying time to develop and instal alternative energy forms and of extending the lives of material and energy reserves. It even came to be regarded for the first time as a major potential energy source in its own right. In developing and Third World countries also, conservation was rapidly becoming a necessity, not only because of the scarcity of fuel-wood and the higher cost of imported energy but in order to avoid some of the environmental problems experienced by the energy-squandering 'developed' nations.

The question of how much energy is required and how it should be conserved to meet essential needs depends both on the end-use of the energy consumed, and on how much is wasted *en route*. An overall view is essential, as all conservation policies involve choices from the national level through to the individual.

Despite the numerous unpromising growth and energy-depletion statistics published during the 1973-83 period, most governments and industries took surprisingly long lead-times to introduce some comparatively simple energy-saving measures. In the U.K. this was partly due to the availability of North Sea oil and gas and indigenous coal deposits which made the problem, though acute, appear less imminent. Also with centralised power-supply industries in active competition with one another, the emphasis was for many years on consumption rather than conservation. Nor were there equivalent agencies or institutions to provide or market energy conservation.

Energy-conservation Measures in the U.K.

Since the 1973 oil crisis, successive U.K. Governments implemented a number of *ad hoc* measures to encourage conservation. In 1974 Eric Varley, then Secretary of State for Energy, announced a 12 point package to assist conservation in buildings. These initial steps were augmented in 1977 when his successor, Tony Benn, presented a 4-year £350 million programme for conservation in public sector buildings, housing insulation and energy management and training.

In 1978 the government formally stated its response to the changed energy situation in a Green Paper entitled *Energy Policy – a Consultative Document.*[4] This based its approach to conservation on the assumption that 'improving the efficiency with which energy is used brings benefit both to the individual and to the nation.' It went on to spell out the main areas of government energy-conservation policy as follows:

(i) Energy prices need to reflect at least the cost of supply. . .
(ii) Consumers of energy need to be in a position to take decisions in the light of adequate information about energy costs and about the ways in which energy can be more efficiently used . . . The government regards its role as ensuring that the available information is comprehensive, if necessary by filling gaps itself. In appropriate cases it may be necessary to ensure by legislation the provision of comparative information . . . In addition the government has a particular responsibility for securing recognition of the national importance of energy conservation.
(iii) Public authorities are themselves responsible for about 6 per cent of energy use

and the government has a particular responsibility for seeing that potential reductions in this consumption are realised.

(iv) Public sector housing, which accounts for about 9 per cent of total energy use, is another area where no substantial progress can be expected without major public expenditure.

 (v) The government is identifying the areas where research and development could lead to a significant improvement in energy use. Government financed research will supplement the substantial amount of research relevant to energy conservation which is already taking place in the energy industries, in manufacturing industry and elsewhere.

(vi) In certain cases mandatory measures to promote energy conservation are appropriate. . .'

The Green Paper continued by saying that these policies needed to be reinforced by the adoption of a mixture of three courses of action designed to maximise conservation by raising energy prices to the consumer: by taxation or other means; by reinforcing or extending mandatory measures; and by encouraging energy-saving investment through grants and tax allowances. Unfortunately the U.K. government did not implement the necessary measures with the same speed and vigour as other European countries that lacked the same advantageous fuel reserves. However, it has been accepted that measures available to government to stimulate conservation include energy pricing, information and motivation, setting a good example in the public sector, specialised advice and training, standards, research, development and demonstration, mandatory measures and fiscal incentives.[5]

U.K. Government Policy since 1979 – The Fifth Fuel

Following the general election in May 1979 there was a definite philosophical shift in energy policy and many of the previous government's modest conservation measures were scrapped in favour of pricing policies. True to Conservative principles the new government centred on an increased reliance on voluntary effort, as opposed to mandatory controls or grants, and argued that the key to conservation lay in increasing prices, especially for industrial users. In the words of the Secretary of State for Energy 'we would have to be very selective (in providing grants) and need to be persuaded about any particular scheme. The greater the incentive for profitable firms to reduce energy, use it more efficiently and introduce energy-saving equipment in their premises, the harder it is to make the case that money should be diverted from other vital incentive schemes for conservation in the domestic sectors.'[6] The price of gas was accordingly artificially raised in January 1980 and again in late-1983 – to discourage people from switching from more expensive fuels to this increasingly scarce resource. The price of electricity and coal continued to be increased as did the excise duty on petrol. Price controls were also abolished on paraffin. Symbolically perhaps the Department of Energy halved its staff on energy conservation. These cuts were denounced in an editorial in *The Observer* newspaper as 'the latest step in the demolition of a coherent energy-saving policy' (January 1980). Yet pricing alone was unlikely to produce rapid results. It provided an incentive but not the means to undertake the necessary investment required for more efficient energy use. It provoked cut-backs and lowering of standards rather than positive new conservation measures.

In the U.S.A. it was estimated that between 1973 and 1978 efficiency improvements had added 2.5 times as much 'energy service' to the economy as combined increases in nuclear power, imported oil and coal extraction.[7] In the U.K. final energy consumption between 1979 and 1982 fell by 12 per cent whereas GDP in the same period fell by only 1.5 per cent. David Howell, when Secretary of State for Energy had openly acknowledged conservation as a 'Fifth Fuel' capable of contributing 25 per cent or even more to the U.K.'s energy needs — coal, oil, gas and nuclear power being the other four. At the same time the projected energy requirements for the next 20 years were revised downwards.[8] Yet despite the fact that estimates of the potential of energy conservation have been growing from year to year, U.K. Government expenditure by 1984 was pitifully small by comparison with most other industrial countries. Although an Energy Efficiency Office had been established and a more vigorous conservation campaign promised 'to make Britain the most energy efficient country in the world,' further cuts in spending on renewables were envisaged (according to Sir Martin Ryle, Astronomer Royal, the U.K. spends less on R & D for renewable-energy sources than does Nicaragua!) The total U.K. expenditure on conservation, as allocated in various government departments, declined from £163.6 million in 1980/81 to £149 million in 1981/82. Of the 1981/82 figure some £56 million was 'not specifically allocated to energy' but was 'available for use within the overall block of prescribed expenditure.' Such provisions, most of which fall outside direct Department of Energy funding, make the real allocation of funds for U.K. conservation extremely difficult to assess.

An ENDS report (January 1983) aptly commented: 'these policies have flown in the face of independent assessments which have chipped away at two of the principal articles of faith in the Government's energy policy — its dogged adherence to economic pricing as a tool for ensuring rational energy usage (a phrase common in government circles from the early 1980s) and its support for major investments in energy supply such as the proposed nuclear reactor at Sizewell.' In 1981 the Select Committee on Energy had criticised the Department of Energy for 'lacking any clear idea of whether investing £1300 million in a single nuclear plant is as cost-effective as spending a similar sum to promote energy conservation,' and by 1982 had itself concluded that many conservation measures were in fact more cost-effective than most energy-supply investment. Yet despite this and other reports, such as the Armitage–Norton report on conservation in industry and the Earth Resources Research (ERR) scenario *Energy Efficient Futures* (January 1983), 'economic' pricing has remained the basis of government energy policy, including conservation. As Peter Walker, Secretary of State for Energy, himself pointed out in 1984 — the U.K. spends £100 million a day on energy, of which £20 million is wasted.

ENERGY: CRISIS OR OPPORTUNITY

U.K. Conservation Initiatives by Sector

With the growing necessity of energy conservation, initiatives have steadily spread. In the U.K. in 1982 the largest energy-consuming sector was Domestic and Public Administration Buildings (34 per cent), followed by Industry (32 per cent) and Transport (25 per cent). Energy-conservation initiatives have tended to mirror these consumption patterns. Table 10.1 suggests the estimated potential percentage reduction in energy demand per unit of activity in the various sectors.

Table 10.1   *Energy savings estimates, U.K. Percentage reduction in energy demand per unit of activity*

| | Shell–Western Europe | | |
	Technical potential over 1973	1973 to 2000 projection	Leach – U.K. 1976 to 2010
Industry	20–35	15–30	22–35
Domestic	40–60	20–40	37–42
Non-domestic buildings	40–50	15–35	30–60
Transport	20–35	10–25	15–40

Constructed from: Shell Briefing Service, *Improved Energy Efficiency*, 1979; G. Leach *et al.*, *A Low Energy Strategy for the United Kingdom*, IIED Report, Science Reviews, IIED, London, 1979.

(a) Domestic and Public Buildings

These play a vital role in any conservation policy since building stock is more permanent than industrial machinery or transport which have to be replaced every 10-20 years, and since buildings consume such a high proportion of total energy supplies. Figure 10.1 illustrates the pattern of consumption in domestic buildings. A domestic dwelling with poor insulation can lose up to 85 per cent of its heat through lofts, windows, doors, walls and floor. Moreover in Western Europe every extra °C of indoor temperature accounts for a further 5-10 per cent of space-heat consumption. Modern architecture with glass-sided office blocks and factories which cause enormous heat losses require more heating in winter and air conditioning in summer. These hardly conserve our fuel resources. An outstanding example built in the post-1973 era is the Pompidou Centre in Paris. This mecca of modern French culture, constructed mainly of glass and steel with most of its pipework on the outside of the building, must surely win an accolade for energy-squandering architecture. Apathy towards suitably insulating buildings is partly explained by the fact that one cannot *see* heat loss. The technique of thermomorphic surveys may help to demonstrate the extent of energy loss in this sector.

Many studies point to the considerable energy savings to be made by better designed houses and increased insulation standards. The substantial energy savings thus achieved in Sweden, Canada and the U.S.A. serve as illustrations. In the U.K. in

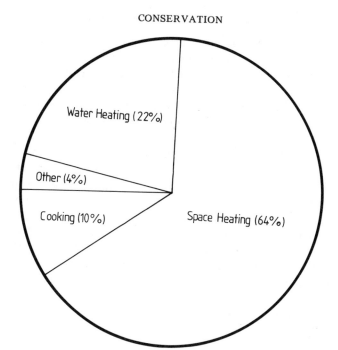

Figure 10.1 *Patterns of energy consumption in domestic buildings*

1979, ETSU estimated that the technical potential for conservation in buildings was 45 per cent. The Building Research Establishment also suggested that in specially designed low-energy houses up to 70 per cent of the current average domestic energy consumption could be saved. For non-domestic buildings the British Gas Corporation estimated a 20 per cent potential cost saving. Even more significant savings were postulated in 1979 by the American Society of Heating, Refrigerating and Air Conditioning Engineers (ASHRAE) as a result of intensive research,[9] and by Gerald Leach in the IIED Report *A Low Energy Strategy for the United Kingdom*.[10] Leach contended that well-designed buildings should be able to reduce space-heating loads and appliances by half, while maintaining current heating levels. Such estimates were initially treated with caution, yet in the same year a specially designed experimental house built by Philips at Aachen in West Germany actually achieved a 90 per cent reduction in the simulated average energy consumption for a family of four. This demonstrated the art of the possible, given high building and insulation standards and the incorporation of heat pumps, solar panels, and similar energy-saving devices. The Aachen house set a lead for future low-energy building design in northern Europe as did the Bronwell low energy home, and other U.K. experiments in Milton Keynes together with many experiments in the U.S.A., such as Brownell's low energy house in New York State. Those countries represented at the Tokyo Summit meeting on energy in May 1979 were pledged to a 5 per cent annual reduction in oil consumption. The U.K. government assessed that this could be achieved mainly by reducing energy consumption in buildings. Measures to improve insulation in houses had already been reinforced in 1975 by a Home Insulation Act, which doubled the previous insulation requirements for new dwellings, providing for a minimum loft insulation of 3 inches (or 75 mm). In a further Home Insulation Act (1978) very modest financial incentives were given to private householders to instal basic insulation where none already existed. In 1981 an Energy Conservation Act empowered the government to set mandatory efficiency standards for space-heating, water heating and gas appliances, and improvements were made to the Building Regulations to provide for higher conservation standards. In addition, a programme of insulation for local-authority buildings was set up together with a longer-term project for introducing new insulation standards and heating controls in the National Health Service, educational establishments and other public buildings under the government's Property Services Agency. In fact by 1981 the Department of Environment's Property Services Agency had already achieved a 40 per cent reduction in the energy used in its civil estate compared with 1972/73.

Officially sponsored research into building conservation was also carried out within the Department of the Environment's Research, Development and Demonstration Programme at the Building Research Establishment and within the Housing Development Directorate. Priority was given to: improved methods of ventilation control; study of the efficiency of heating controls; field trials of methods of insulating solid walls; assessment of problems associated with insulation; the study of heat-recovery technologies and heat pumps in buildings; new district-heating installations; and heat metering and charging. Surveys of existing building stock and estimates of the potential of passive solar heating in new buildings were undertaken to establish a further range of possibilities for energy saving. On the other hand, mandatory measures to improve conservation remained extremely limited. The government was unwilling to introduce further legislation to improve insulation standards (as had been done in other countries), or to set more stringent limits for domestic and commercial space-heating. Even by 1984 the minimum standards set by the Home Insulation Acts were well below the requirements necessary to achieve effective energy conservation in the majority of buildings and there were still in the U.K. over 6 million homes including many council houses without any loft or hot-water insulation. At local government level, while some authorities had put conservation measures into practice, saving up to 20 per cent of their fuel bills, others did not even know how much they were spending on fuel, despite the great opportunities for conservation in a wide number of areas from council housing to the replacement of less efficient street lighting. In the private building sector also, the voluntary nature of government policy towards energy conservation had not produced the necessary results. One reason lay in the slow rate of replacement of buildings and the long construction times. Another was because of the very large numbers of private householders to be persuaded and the high cost of building and heavy mortgage rates, especially in the economic recession; building societies did not offer the necessary incentives for such improvements.

There were also difficulties in persuading construction companies to insulate new buildings to efficient standards if this meant increasing capital costs, since market conditions dictated cheaper and not necessarily energy-efficient housing. Yet in order to maximise conservation, the running costs as well as the capital outlay have to be taken into consideration at the design stage. Another problem was that the 75,000 or so registered town planners, architects, engineers and quantity surveyors were often ill-equipped to advise on the relatively novel energy-saving aspects of building design.

In June 1982, the U.K. House of Commons Select Committee on Energy itself was stimulated to write a report on *Energy Conservation in Buildings*.[11] The Committee concluded that savings in the order of 30 per cent of present delivered energy consumption in both domestic and non-domestic buildings could be achieved using existing technology. It blamed failure to invest in building energy-conservation schemes on ignorance, other investment priorities, unsuitable accounting practices, uncertainty about future fuel prices, the existing leasing structure for buildings, the lack of co-ordination in the conservation industry and 'the fragmentation of responsibility among government bodies and above all, the lack of political will at the heart of Government which smothers the efforts of the Department of Energy's Conservation Division' (para. 64c). The Committee recommended 'a more active government policy to overcome the obstacles to cost-effective investment in energy conservation including increased expenditure and government grants, a wide-ranging programme of house insulation, the review of standards, codes of practice, regulations and legislation affecting conservation in buildings, and the setting up of a Conservation Committee or Agency. To increase the energy contribution in this essential area such recommendations should be implemented as quickly as possible, thus extending the life of fossil fuels until viable replacements have been developed. However, since in 1983 the U.K. was a net exporter of energy, with production about 20 per cent above consumption, it was difficult to inculcate a sustained sense of urgency. Nevertheless in 1984 *The Tenth Report of the Royal Commission on Environmental Pollution* strongly criticised government's lack of attention to energy conservation, pointing out that conservation was 'prevention rather than cure' and the cheapest, safest solution to the acid rain/nuclear waste dilemma.

(b) Manufacturing Industry

There have been increasing incentives since 1973 directed towards encouraging energy conservation in industry, including retrofitting where possible. A limited two-year government scheme (1978–80) made £25 million available to industry for replacing or improving inefficient boiler plant, insulating premises and improving combined heat and power (CHP) systems. Cash grants were given to firms willing to demonstrate energy-conservation projects and an initial subsidy was provided towards the cost of employing an accredited fuel consultant to survey energy-conservation possibilities. A 100 per cent tax allowance for the first year was granted for expenditure on the insulation of industrial buildings. £18 million was also made available over the four-year period 1979–83 for research and development into energy conservation in industry. In addition there were attempts to stimulate interest in conservation by the establishment of regional energy-conservation officers.

Their function included endeavours to persuade industries to improve energy conservation by voluntary means including appointing their own energy managers. By November 1981 there were over 5000 energy managers divided into 70 regional groups. They organised courses and seminars to exchange information on this relatively new topic of concern and also carried out energy audits with the help of ETSU and the research associations of the various industries concerned.

While government conservation policy took some considerable time to get under way in the U.K. most industries were, of necessity, increasing their own energy efficiency. Between 1973 and 1983, total energy consumption in industry fell by over 31 per cent, while industrial production fell by only 9 per cent. Energy-conservation measures together with the 1973–5 and post-1979 recessions were mainly responsible. The potential for conservation in industry is extremely high, as may be seen by the JEL low-energy factory at Stockport which was opened by Prince Charles in December 1983, and consumes 40 per cent less energy than a similar conventionally built factory.

(c) The Fuel-supply Industries

From the late-1970s most of the fuel-supply organisations, including the Gas Council, the Electricity Council, Mobil, Esso and Shell, offered free energy-saving advice in an attempt to stimulate public awareness. (Over 50,000 people visited the Gas Council's first energy-advice centre in Birmingham in its first year 1979/80.) Esso made available information on the techniques of improving boiler efficiencies used in its own plant-maintenance programmes. Shell's briefing reports such as *Energy Efficiency* and *Improved Energy Efficiency* repeatedly make the point that conservation and more efficient energy use is a better response to the oil crisis than switching to other forms of fuel.

In 1982 14 per cent of U.K. delivered energy was supplied by the *Electricity Industry*. About 75 per cent of all primary fuel was used by the fuel industries themselves during the conversion process to secondary fuels. Thus about 4 units of primary fuel were needed to produce 1 unit of electricity since energy is consumed during the mining of coal, the refining of oil, the production and transport of gas and the generation and transport of electricity from whatever energy source.

Paradoxically, while the U.K. consumer was being encouraged to switch to electricity even for low-grade heat (hardly a wise conservation move in terms of overall energy efficiency), the electricity industry itself was taking a number of steps to improve its own conversion rates and minimise waste. Power stations were gradually modernised and average efficiencies rose from 27.9 per cent in 1967 to 34.1 per cent in 1981/82, with a thermal efficiency of 37.87 per cent obtained by the most efficient station. However, with dependence on fewer but larger power stations, a greater proportion of energy is used in transmission and a greater capacity must be provided to guard against possible breakdown. This over-capacity itself was 35 per cent in 1980, the increased margin often cancelling out the energy savings gained through increased efficiency.

As well as implementing conservation measures through better insulation and increased thermal efficiencies, one of the more publicised early contributions of the *gas industry* to energy saving was the Gas Energy Management Awards Scheme instituted in 1975.

The *oil industry* took major steps to reduce its own energy consumption and to curtail waste. To give an example, in 1973 it was estimated that 'flaring' burned up some 4.5 billion ft³ of natural gas each year. Shell and Esso were consequently ordered by the U.K. government to make substantial reductions in the amount of gas being flared. According to the oil companies, this reduction in flaring meant a substantial reduction in the oil output. Esso estimated that this might be as great as 10 per cent in the Brent field alone.[12] Nevertheless great progress was made to reduce such forms of waste, as has already been mentioned in chapter 3. To give another example, BP devoted considerable resources to reducing its own energy costs (which in the mid-1970s accounted for about 30 per cent of operating costs). The company launched an extensive energy-conservation investment programme and by 1977 had already reached its energy-saving targets for 1980. By that year savings amounted to 17.5 per cent over 1973 rates, involving a saving of £178 million, and the company determined to reduce energy consumption by a further 10 per cent over the period 1980–85. Shell adopted similar policies.

The *National Coal Board* (NCB) saved 10 per cent of its energy bill in the two years 1976/77 mainly by reducing its own coal consumption. The NCB, conscious of its role as the main energy producer, set up an energy-conserving group at each colliery to investigate new conservation techniques.

(d) The Petrochemicals Industry

Like the oil refineries and power stations, the petrochemicals industry uses oil both as a raw material and as an energy source. By 1980 crude oil accounted for some 90 per cent of raw material consumed. In the event of an oil shortage, this industry will obviously decline unless alternative fuels with similar properties, such as oil from coal, can be found. The need to conserve energy both through increased efficiencies and reduced dependence on scarce energy forms is therefore of paramount importance during the transition period. The potential for conservation was amply illustrated by a Department of Industry report on the Soap and Detergents Industry (1979) which estimated possible energy savings of 22,000 tonnes of fuel oil per year in this sector alone through better insulation of process plant, heat recovery, lagging, improved boiler efficiency and better space-heating control. Experiments in the plastics industry similarly aimed at decreasing energy inputs and also at producing plastics from alternative energy sources such as biomass.

(e) Transport

About 45 per cent of oil consumption is used in the transportation sectors of the world outside the Communist areas. About three-quarters of this is used in road transport. Within the U.K. road transport consumes 20 per cent of the country's total fuel, air transport 3 per cent and railways and water transport 1 per cent each. Private cars take up around 50 per cent of transport fuel; other road transport and commercial vehicles 25 per cent. Transport-usage patterns and motor vehicle and engine design therefore have an important part to play in any energy policy. Up until 1973 most engine designs had taken very little account of fuel economy with the number of miles per hour a more important performance factor than litres/gallons per mile. By 1983 fuel consumption had become a crucial design factor.

The IIED report of 1979[10] quoting the U.S. Environmental Protection Agency gave an 'energy flow map' for a typical car at that time. This showed that 33 per cent of the fuel entering the car was diffused through the exhaust, 29 per cent in cylinder cooling, 6 per cent in air pumping, 8 per cent in engine friction, 6 per cent in air drag and 5.5 per cent at the wheels. Only just over 12 per cent of the fuel performed useful work at the wheels. Leach estimated that by improving stream-lining, reducing speed limits, switching to steel-cased radial-ply tyres, ensuring proper tyre inflation, reducing vehicle weight, making better use of low-friction lubricants, improving tuning of carburettors and ignition systems, using thermo-statically controlled fans and continuously variable transmission systems and, above all, improving engine performance and smooth driving, there could be a 46–57 per cent fuel saving.

Up to 1984 no government measures were taken in the U.K. to compel manu-facturers or motorists to save on fuel consumption although mpg performance had to be displayed on all new cars and there were road speed restrictions to encourage the more economic use of fuel. The government has also set a voluntary 10 per cent target in engine-efficiency improvement by 1985. Owing to initiatives taken by car manufacturers spurred on by the ever-increasing cost of petroleum to the customer, it seems probable that these targets will be achieved or even exceeded ahead of schedule. Most European car manufacturers have committed themselves to a 25 per cent increase in the energy efficiency of their products during the 1980s. In the U.S.A., with its larger cars and far higher *per capita* fuel consumption, legislation was introduced to coerce manufacturers to increase their mpg ratios annually.

Another approach to conservation in the transport sector would be to encourage rail passenger and freight travel instead of road transport since on average the latter uses some two-and-a-half times as much primary energy per passenger mile as the railways. However, this again would require government intervention in the form of financial incentives.

A renewed use of canals for freight transport would be an even greater fuel economy, the savings per tonne/mile in this case being in the order of 20:1. Yet in 1973 Britain transported only 1 per cent of her freight by canal compared with 51 per cent in France, 33 per cent in West Germany and 66 per cent in the Netherlands. It is true that Britain's canals are narrow compared with those on the Continent and would need developing, but the British Waterways Board has a relatively low financial allocation for the maintenance and development of canals compared with other EEC countries, most of which are rapidly expanding this transport network. Of all forms of conventional freight transport, canals are the least energy-consuming.[13]

(f) Agriculture and the Food-processing Industries

Western agriculture is highly capitally intensive but the direct use of fuel accounts for only about 5 per cent of farmers' costs. Modern mechanised agricultural production is, however, dependent on large energy inputs for machinery, irrigation, fertiliser, food processing and transport. It has been estimated by Chapman that in the U.S.A. 20 per cent of primary energy is consumed in the 'food chain', and in the U.K. about 17.5 per cent.[14] These percentages are expected to rise and be reflected worldwide as other countries adopt modern Western agricultural techniques without calculating how this extra energy will be provided in the long term. A crucial concept is again the 'energy ratio', which is the ratio between the amount of energy gained from the food eaten and the amount of human and mechanical energy consumed during its production. In primitive farming the ratio is about 15–20:1. In mechanised industrial farming, on the other hand, the ratio is barely 2:1 and for certain products it is even negative. There are indications that some modern forms of food production and transport to the consumer require as much as 10–30 times the total energy per unit output as primitive agriculture. Most of this extra energy is in the form of oil and its by-products. Intensive farming methods, including some of the 'biotechnologies' such as hydroponics and advanced fermentation technology, are likely to increase the energy intensity of farming still further.

With the prospect of oil shortages and world population increases, it is obviously highly desirable to reduce the dependence of agriculture on energy, both by improving fuel efficiencies at each step in the 'food chain' and by reverting to less energy-intensive methods of agriculture. On a world scale, only the latter course appears feasible. As early as 1976 it was suggested that if it were attempted to feed mankind using the agricultural technologies of the U.S.A., all known oil reserves would be consumed by agricultural requirements alone within 30 years.[15] Clearly alternative ways of fertilising crops will have to be found, and food will have to be processed and, if necessary, packaged and transported more efficiently. As far as the fuel efficiencies of the food-processing industries are concerned, it has been estimated that 35 per cent savings could be achieved by the year 2010.[10]

Energy Conservation through Reclamation, Re-use and Recycling

So far we have examined two areas of fuel conservation, namely government measures and initiatives taken by industry or suggested by other commercial organisations to improve energy efficiency at the point of use. These would aim also to eliminate the mismatches of supplying high-grade energy such as electricity to meet low-grade requirements. A third approach is to secure energy conservation through re-use and recycling of used materials. Although normally known as materials conservation, reclamation, recycling and re-use can contribute considerably to overall energy saving by short-cutting some energy-consuming steps in the manufacturing process, lengthening durability and product life so reducing consumption, and in some cases recycling or returning the spent product as fuel.

Reclamation and recycling also incur less pollution than production processes using virgin materials. The two main disadvantages of recycling are the time, effort and expense involved in waste recovery and sorting, and the fact that in some cases such as the recycling of paper and glass a deterioration in quality (as opposed to using raw materials) cannot be avoided.

In the U.K. in 1978/79 over 380 million tonnes of waste were jettisoned annually. As figure 10.2 shows, agriculture accounted for 200 million tonnes; mining and quarrying produced 110 million tonnes of waste; building and power stations 15 million tonnes; and other industry 27 million tonnes. Domestic and commercial wastes, together with household waste brought to dump sites (known as 'collected waste'), amounted to 18.3 million tonnes a year. This included 3.6 million tonnes of paper and board and around 1 million tonnes each of metal and glass. Some of these materials are in the form of discarded packaging, which accounts for about 28 per cent by weight[16] of domestic dustbin contents (11.25 million tonnes). The weight of waste for disposal has stayed surprisingly constant since 1930, although it has become bulkier. Around 85 per cent is still disposed to landfill, although sites conveniently near to conurbations are becoming difficult to find. Disposal costs in 1983 were as high as £20 per tonne in London. Moreover, Warmth and Energy from Rubbish (WARMER), launched in the U.K. in 1984, estimated that these figures would rise and that the U.K. would be dumping over 30 million tonnes that year with a calorific value equivalent to 12 million tonnes of coal.

The advantages of refuse reclamation and recycling are therefore considerable on environmental, resource and energy-conservation grounds. Yet by 1980 only about 2 per cent of domestic waste was, in fact, being reclaimed in the U.K. In addition there is the energy and material saving already obtained through cars and other steel products recovered from scrap. Steel produced from scrap saves about 30 per cent of the energy of that produced from iron ore. Similarly, 1 tonne of copper produced in the U.S.A. from ore mined from an open pit requires 25,000 kWh of energy, but only 2500 kWh if obtained from municipal or other scrap.[17]

Recycling is obviously easiest within the manufacturing industries themselves as this eliminates the need for collection and sorting. In the U.K. by 1980 some 66 per cent of the lead, 50 per cent of crude steel and 40 per cent of copper used in manufacture were being recycled, together with substantial amounts of paper and board. Prospects for recycling are best for ferrous metals, paper, glass, aluminium and biological refuse. With the latter, Britain's performance by 1980 was worse (*per capita*) than any other EEC country, although it was better in respect to paper and board. Recycling technology, though developing in its applications, is not new and could well be extended, even adapting proven methods to conserve energy and scarce or imported raw materials. For example, the City of Rome uses 800,000 tonnes of mixed refuse annually, separating it at one of four main plants and producing out of it large quantities of paper pulp, animal fodder, ferrous bales and compost. The Milan trolley-bus network has run on electricity generated by the combustion of urban refuse for many years. Many countries provide financial incentives to individuals taking re-usable waste to a collection point. In Phoenix, Arizona, for example, 'Golden Goats' were installed in shopping precincts and park-

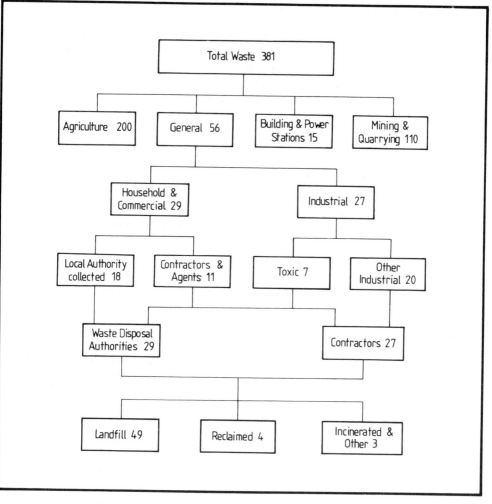

Figure 10.2   *Solid waste in the U.K. (in million tonnes) 1978/79*

ing areas in 1978. These machines sorted, weighed and separated aluminium cans at a return to the public of 13 cents per lb. An even more unconventional if less universally applicable method of saving energy in refuse collection was suggested by the President of the Asian Recycling Association in 1980. He claimed that a few tonnes of *Lumbricus Rubellus* (the red wriggler worm) could eat up to 150 tonnes of wet domestic rubbish per day and convert it into organic fertiliser and still retain enough energy to reproduce five times annually for up to six years! This light-hearted example merely illustrates the diversity of recycling methods at every level of agricultural, urban and industrial development if harnessed to appropriate uses by man's ingenuity or necessity.

Aluminium and Tinplate Industries[18]

Energy savings can be particularly significant when tinplate and aluminium are re-cycled. In the U.K. some 9000 million food and drink cans are used each year, consuming around 700,000 tonnes of tinplate and 40,000 tons of aluminium. Aluminium requires enormous energy when produced from bauxite (91,000 kWh per tonne) but only 3000 kWh when made from scrap — a saving of 96 per cent. In the case of tinplate some 50 per cent of energy used is consumed in iron-making which becomes unnecessary when tinplate is manufactured from scrap. Consequently increasing efforts are being made to increase recycling in this area. By 1984/85 can recycling is expected to have expanded from its present 1700 million cans to 2600 million. Although the savings in energy terms would be significant, even allowing for collection, sorting and processing, the widespread adoption of such schemes depends very greatly on co-operation from local councils and public participation promoted by publicity campaigns or other incentives. A line drawing of the main components and flow lines of a typical metal-recovery plant is shown in figure 10.3. Contrary to arguments by some manufacturers, retailers and local authorities in the U.K., a survey of 1350 people by the Manchester Business School in 1980 concluded that consumers preferred to participate in waste-recovery and recycling schemes, even where these involved more work.

The Glass Industry

Energy savings in the recycling of glass and glass containers have been the subject of several studies in the U.S.A. and the U.K. In the U.K. the Glass Manufacturers' Federation, representing the views of many in the industry, point out that recycling glass saves energy in two ways. The first way is in the acquisition, preparation and transport of raw materials used to make glass, for example, in the manufacture of soda ash. While 4258 MJ per tonne is needed for virgin raw materials, only 941 MJ is needed when cullet (broken scrap glass) is used, a saving of 78 per cent. Second, there is a 2 per cent energy saving in the smelting for every 10 per cent of cullet used. Also chemical losses during smelting are less using cullet so that 1 tonne of cullet can replace 1.2 tonnes of raw materials. Glass recycling also saves in energy in that fewer of the 7000 million bottles and jars produced annually in the U.K. (weighing 2 million tonnes and accounting for some 9 per cent of all domestic waste) have to be collected and disposed of by the local authorities. Because of these advantages, increasing efforts have been made to recover glass in the U.K. through 'Bottle Bank' schemes operated by the glass manufacturers in conjunction with local authorities. By 1983, local authorities had introduced more than 1650 'Bottle Banks', recovering approximately 122,000 tonnes of glass, and major recycling plants had been built in Knottingley (Yorkshire), Alloa (Scotland) and Harlow (Essex).

Another energy-conservation possibility involving the glass and other industries is the use of refillable returnable containers, as exemplified by the U.K. milk-delivery service. The resource costs for returnable systems are generally lower than for non-returnable containers with an energy saving of some 20 per cent.[19] Again there is the added advantage of fewer refuse-disposal requirements for local authori-

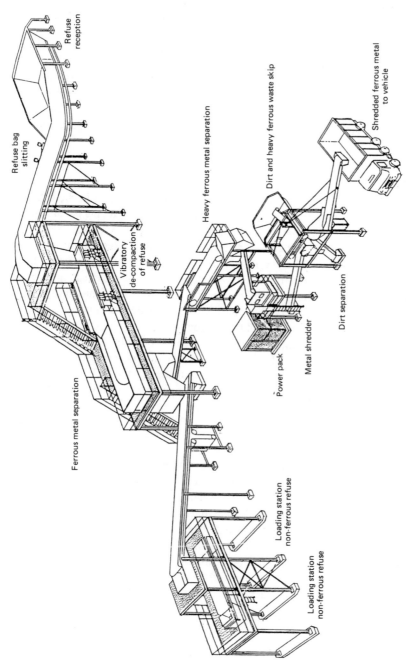

Refuse reception

Refuse bag slitting

Heavy ferrous metal separation

Dirt and heavy ferrous waste skip

Shredded ferrous metal to vehicle

Vibratory de-compaction of refuse

Ferrous metal separation

Dirt separation

Power pack

Metal shredder

Loading station non-ferrous refuse

Loading station non-ferrous refuse

Figure 10.3 *Diagram of a typical metal-recovery plant* (source: Metal Recovery Ltd)

ties. Estimates differ as to whether a return to such a system would create more jobs in the collection, sorting, cleaning and redistribution process (argued for example in a Friends of the Earth study in 1980) or would in fact increase unemployment owing to a reduction in the demand for disposable containers. Undoubtedly the system works in those countries where it has already been introduced, such as in West Germany and France. It is also possible that those sections of the industry that consume energy and raw materials on a large scale for ultimate waste disposal may find that they can use their expertise and equipment to produce items of longer-term value to society. In the long term, as the supply of both energy and raw materials becomes critical, recycling is likely to become increasingly economic and 'waste' will be regarded as a misplaced resource.

The Paper Industries

Paper recycling also brings significant energy and materials savings. In the U.S.A. studies have shown that 50 per cent of the energy used in producing paper can be conserved by using recycled pulp. In the U.K. in 1982 waste paper was used as a raw material for about 58 per cent of all paper and board production out of a possible recoverable total of some 70 per cent (the remainder being consumed, degraded or contaminated). Alternatively waste paper can be recycled and processed, sometimes with plastics and other wastes, for use as fuel. Studies have shown that paper waste has a calorific value of 10.3 MJ/kg, a little over a third of that of industrial coal. A tonne of paper yields 0.23 tonnes of oil equivalent when burned. On the other hand when manufactured from virgin wood pulp it consumes 1.6 tonnes of oil equivalent. Because of the importance of conserving forests for essential purposes such as food and fuel production (see chapters 7 and 11) paper conservation and recycling should perhaps be regarded as one of the highest energy-conservation priorities.

Lubricating Oil

This oil may be easily and cheaply recycled to save energy. In the U.K. government's Green Paper *War on Waste* (1974) it was stated that over 30 per cent of total consumption could be reclaimed, refined or burned as fuel. Yet by 1983 measures to this end had been taken by only a few individuals and organisations, such as British Rail. In the U.S.A., in several states such as Minnesota, law requires all retailers of motor oil either to provide a collection tank for the deposit and collection of used oil or to indicate the nearest oil-collection site. In West Germany a government-subsidised oil-recycling programme recovers over 70 per cent of used motor oil. By 1984 there were no similar schemes in the U.K.

Other Industries

There is considerable potential for energy saving through recycling in most manufacturing processes. Fibres may be reclaimed from used textiles. Rubber, as used in retread tyres and road surfacing, can be converted to a light fuel oil by destructive distillation. One by-product of this process is a residue with the same properties as coke which may be used as a pulverised solid fuel, especially in industrial burners. The EEC Commission estimated that in 1980 around 140,000 tonnes of used tyres were recycled annually, saving some 100,000 tonnes of crude oil imports. Recycled non-ferrous metals such as lead, zinc and copper also offer large energy savings. In 1981 60 per cent of lead, 30 per cent of zinc and 18 per cent of copper produced in the U.K. was from recycled scrap.

A recent ETSU report (R13, Vol. 11) states:

'Most forms of collected refuse are cost effective now against national investment criteria; . . . the use of collected refuse is now at the deployment stage and the main problems are to demonstrate this cost effectiveness and to establish commercially reliable routes from supplier to user. The resource has the potential to make an important contribution to U.K. energy supplies'.

Fuel from Waste

While recycling and re-use of materials indirectly saves energy, it is possible to convert urban refuse directly into fuel or fuel additives and also to use waste heat from industrial or power-generation processes. Before 1973 such notions attracted little attention. There are by now many different methods of recovering energy from urban waste and industrial processes, some of which will now be described.

Landfill gas recovery. Organic municipal solid waste (MSW) if dumped as landfill and left untreated, converts eventually to gas. As the waste decomposes anaerobically the landfill becomes a natural gas field, producing considerable quantities of a methane–carbon dioxide mixture. One tonne of MSW can yield 200 m³ of landfill gas, comprising perhaps 50 per cent methane. Exploitation of MSW was initiated in California in 1974 and, although recovery is economic only from dumps of 1 million tonnes or more, there were 15 such schemes under way in the U.S.A. by 1982. In New York, for example, the Brooklyn Union Gas Company distributes enough gas derived from a garbage dump on Staten Island to heat 15,000 homes.
Incineration with heat recovery as steam. Incineration is a popular and well-established method of treating urban wastes as it sterilises the residue, and reduces their weight by about 60 per cent and the volume by 90 per cent. However, incinerating plants have high capital and running costs and produce large quantities of gas through combustion. By the addition of a boiler to the plant the waste heat can be recovered to raise steam for a variety of purposes. In the U.K. the National Coal Board set up a district-heating scheme in Nottingham using waste heat from Nottingham's incinerator to supply the base-load, supplemented by a back-up coal-fired plant for peak demand periods. Other examples of the 1970s include France, where a similar district-heating plant exists in Paris, and Hong Kong, where waste heat from incinerators is used to desalinate sea water. Many new projects of this type are currently under development worldwide with over 1000 incinerators in operation.
Incineration with heat recovery for electricity generation. This method uses the steam to generate electricity. The cost of the additional generating plant, however, is high in comparison with an incinerator without heat recovery. For example, in the GLC's 1300 tonne/day incinerator with combined generating plant at Edmonton, the high capital costs have not been recouped. However, because of the expense of using valuable agricultural land for waste disposal, a similar plant in Amsterdam was apparently operating viably by 1979.

Pyrolysis. In this process refuse is heated in closed containers in the absence of air; it was developed in the U.S.A. A combination of combustible gases, a substance resembling crude oil, water and a solid residue or 'char' is produced. Pyrolysis has the advantage of producing some fuels that can be burned in slightly adapted conventional appliances or stored. By changing the operation slightly the relative amounts of solid, gas and liquid may be controlled. 'Gas' pyrolysis maximises the production of gas; 'oil' pyrolysis, the production of oil; and so on. Since there is no air in the sealed furnace, metal residues other than aluminium can be recovered from the char; this would be impossible in an incinerator because the metals would be oxidised. Char itself may be used as a fuel in the same way as very low-grade coal.

Dry-fuel supplement or waste-derived fuel (WDF/RDF). In this method, untreated solid wastes undergo magnetic separation before being shredded or pulverised and then burned as supplementary fuel in a gas or coal-fired boiler, sometimes in the form of pellets. WDF is similar to peat in density, volume and heat release, with a calorific value of 10–20 MJ/kg. Simplicity is one advantage since boilers require minimum modification. Imperial Metal Industries Ltd provide over 30 per cent of the power for their plant at Witton, Birmingham, by using WDF. The firm collects up to 60,000 tonnes of untreated refuse annually from the West Midlands Metropolitan Council. Similar schemes are operated by the Blue Circle Group in agreement with the Wiltshire County Council, and by Newcastle-upon-Tyne and Doncaster Councils. The Tyne and Wear County Council's Byker Reclamation Plant in Newcastle-upon-Tyne was Europe's first municipal WDF plant aimed at processing Newcastle's domestic and commercial refuse. The plant initially reduced waste-disposal costs from £10 to £7 per tonne per year and in addition supplied district-heating to local homes, schools and community centres. Currently it produces 6–7 tonnes/h of fuel pellets with a weekly output of 150–250 tonnes/week. Other schemes followed.

Fluid-bed refuse incineration. This is a low-cost incinerator process first developed in the U.S.A. The refuse is shredded and fed on to a bed of sand. Air under pressure is passed under it, throwing the rubbish and sand into the air. The result is a fluid-like mixture of refuse and sand in constant motion. At a critical temperature the refuse ignites, undergoing a high degree of combustion as the oxygen easily permeates the material. Advantages are that the fluidised-bed boiler has excellent heat-transfer properties, is cheap to construct and, as burning is complete, emitted waste gases are minimal. A plant in Menlo Park, California, uses such waste-derived heat to produce 1,000 kW from 100 tonnes of waste. In 1973, a study was undertaken for the State of Connecticut, U.S.A., to compare the energy-conversion efficiencies of nine energy-recovery techniques, tabulating the energy content of the original refuse and the energy output and conversion efficiency of each. The most efficient method in the analysis was the dry-fuel supplement with ferrous metal reclamation, which had a conversion efficiency of 95 per cent. If the reclamation of materials were improved by combining glass and aluminium recovery with this method, the energy-conversion efficiency would be reduced to 65–70 per cent. Even this is a higher efficiency than that achieved by pyrolysis or incineration. Obviously methods must always be adapted to maximise overall conservation under local conditions.

Combined Heat and Power (CHP)

A much neglected form of energy conservation with enormous potential, which is currently attracting growing interest, is the use of heat rejected as waste from power-generating schemes or industrial processes. In a typical power station, for example, only 30–35 per cent of the primary energy supplied is converted to electricity, the remainder being dispersed in the conversion process. Of this nearly three-quarters is currently emitted as waste heat which is dispersed in cooling towers. The process makes heavy demands on local water supplies which are required to draw off the waste heat and are then discharged into lakes, rivers and sea. This waste heat can nevertheless be utilised in many ways such as for pre-heating water supplies for industrial processes; for district-heating (DH) schemes in which neighbouring buildings are provided with central heating and hot water; or for such purposes as agriculture, the heating of greenhouses or fish farming. A power station that uses its waste heat constructively provides combined heat and power (CHP). In energy-conservation terms it has been calculated that there is a saving of 50 per cent when heat and electricity are provided from a CHP plant compared with the same amount of heat and electricity supplied by separate boiler and power-station facilities.[20] On the other hand, because new district-heating schemes using CHP supplied by a generating station or by some other centrally located boiler require expensive additional pipeline systems for heat distribution, they are most economical where the networks are large, where there is a high density of buildings close to the plant and connected to the system, and when these buildings are specially designed to be an integral part of that system. In the U.K., CHP-linked district-heating schemes are economically competitive with oil-fired central heating schemes involving more than 100 houses and have a very high efficiency on runs up to 10 km.

One problem is that to maximise the overall energy efficiency of CHP the generating stations should both be located near to major residential areas and be operated so that waste heat is continuously available. Unfortunately, in the U.K. at present, the large power stations that provide a continuous base-load are often remote from population centres, while the nearer, smaller stations are frequently required only for peak supply periods. Many of these smaller power stations in urban areas are reaching the end of their working lives and are currently being decommissioned. If, however, diesel engines and gas turbines are employed in smaller-scale combined heat and electricity generation schemes, they can still form part of a 'total energy conservation' concept.

Although the energy potential of CHP schemes has been recognised for some time, interest in the U.K. has been slow to gather momentum. It was only in 1977 that a 22 member committee was eventually set up under the chairmanship of Sir Walter Marshall to review the possibilities of CHP systems. Its conclusions were that such systems were not economic at current prices but that they could become a more attractive proposition by the year 2000. Prompted by the committee's findings the Department of Energy in its 1979 energy projections forecast that CHP schemes could save the U.K. about 2.5 mtce per year by 2000. Up to 20 per cent of space and water-heating requirements could be met by CHP if schemes were introduced systematically.

In April, 1980, the Secretary of State for Energy initiated a two-stage programme to prepare proposals for CHP/DH schemes in one or more lead cities in conjunction with interested local authorities and the electricity-supply industries. The Summary Report and Recommendations (published in July 1982)[21] summarised the finding of the stage I studies. It is supported by a series of supplementary reports describing in full the work carried out and the assumptions made. Following the presentation of an Interim Report by the lead consultants, W. S. Atkins and Partners, in December 1980, the following nine cities were designated for detailed feasibility study: Belfast, Edinburgh, Glasgow, Leicester, Liverpool, London, Manchester, Sheffield and Newcastle-upon-Tyne.

After consultation with local authorities, areas of each city were selected that were likely to be suitable for CHP/DH implementation. Surveys were conducted to assess potential heat loads, distribution-system practicability and cost, and the penetration of the potential market for heat under conditions of consumer freedom of choice. CHP/DH schemes to meet the assessed heat loads were defined for each city using established engineering technology. The schemes range in size from approximately 400 MW(h) to 1100 MW(h) of peak heat demand.

A phased implementation programme, taking into account future inner city improvement and development plans, was drawn up for each city. Each scheme consists of a series of district-heating areas which would be connected by an arterial mains system to a power station at a time calculated to maximise early revenue and delay major capital investment. District-heating development would commence using coal-fired heat-only boilers. The CHP station would come on stream in about the seventh year of a total development period of 21 years. The engineering proposals have been designed to enable district-heating areas to operate independently should this become necessary because of curtailed development.

Three fuel-price scenarios were adopted for evaluation purposes. The lowest price scenario, which is the most rigorous test of CHP/DH economic viability, was taken as the base case for evaluation. The profitability of each scheme was assessed by discounted cash flow. The price of heat was related to the lifetime cost of comparable alternative heating systems proposed by the electricity, gas and coal industries. All nine schemes could provide heat at 10 per cent below the cheapest alternative while showing rates of return about or above 5 per cent per year as required of new investment by nationalised industries.

Meanwhile, several local authorities had begun their own CHP projects on a limited scale and examples were already to be found in Basildon, Hereford, Peterborough and Nottingham. Other interesting pilot projects included the Drax nuclear power station which in 1979 began to supply waste heat for half an acre of glass houses for tomato growing and to heat lagoons for a nearby eel farm. In the same year, the British Steel Corporation sponsored a pilot project for the utilisation of waste heat from a central boiler in their Newport steelworks to supply a district-heating scheme. Nevertheless by 1983 the U.K. government was not supporting CHP/DH initiatives at a rate consistent with the demands of an energetic conservation policy. Among the various reasons given were the recession, the fact that the U.K. is substantially well-endowed with fossil fuels, and the typically British argument that CHP removes consumer choice. The dominant position of the powerful monopolistic energy-supply industries whose interests favour the expansion of supply as opposed to the curtailment of demand has also hindered CHP expansion.

One of the most ambitious CHP schemes planned in the U.K. by 1982 was in the London Borough of Southwark. Costing an estimated £602 million this 1000 MW proposal envisages purpose-built generating stations to supply low-grade heat by direct connection to the whole of the borough. Over the 26-year construction period the system could save 5.5 mtce and thereafter it would save $\frac{1}{3}$ mtce per year, saving an estimated £232 million and £71 million, respectively. This is equivalent to a 48 per cent reduction in the annual fuel bills of residents during the construction period and 80 per cent thereafter.[22] These are not the only benefits to local authorities and residents. CHP/DH investment stimulates direct employment generation, improves the quality of much housing stock, is beneficial to those without central heating and, in cases such as Southwark and Newcastle,[23] emphasises the local authorities' commitment to develop the inner city.

During the early-1980s there emerged several new reports and recommendations on CHP/DH, including that of the House of Commons Select Committee on Energy,[24] which proposed, among other things, that government should assess and finance any new projects on the same terms as other public-supply industries and the British Gas Corporation should give assistance to heat and electricity production and offer gas at a competitive tariff. The Committee also thought that new statutory provisions were required to ensure that Electricity Boards provide heat or assist in the provision of heat.

The Energy Act 1983 placed a duty on Electricity Boards to adopt and support CHP and district-heating schemes and to purchase the electricity from private schemes on fair terms. The Secretary of State for Energy acts as an arbitrator in any dispute between private generators and Electricity Boards. Several of the former statutory obstacles to CHP have therefore been removed since the Conservative Government (1983) saw an expansion of private electricity generation (especially by industry) as part of its goal of making the electricity-supply industry face greater competition. The development of private CHP is one way this aim could be achieved.

Other countries, such as France, Denmark, Sweden, West Germany, the U.S.A. and the U.S.S.R., have been operating CHP schemes on a much larger scale for some years. In 1974 for example, the Bolinden chemical works in Sweden became the first manufacturing concern to deliver waste heat for district-heating. The surplus steam from the production of sulphuric acid went into an energy centre where it was converted into electricity and heat. The water leaving the centre had a temperature of 90–110°C and, by means of heat exchangers, was used to heat water for the district-heating scheme. Twenty thousand homes were being heated by this scheme in 1980 and it was hoped eventually to double the number. In all, by 1983 CHP schemes supplied over 25 per cent of the total energy used in Sweden for hot water and space-heating.

Early CHP schemes developed in other countries include the 'Agrotherm' scheme in Dusseldorf, West Germany, which exploits the waste heat of a large power station by circulating it in an underground network of pipes at about 20°C to stimulate agricultural growth in an area of about 30 hectares of land. An associated project is the intensive breeding of edible fish. In the U.S.A. also, district-heating and cooling systems, generally using steam rather than hot water, are being developed by the private electricity utilities. New York's Consolidated Edison system is the largest, supplying more than 10 billion kg of steam for space-heating, air conditioning and manufacturing in the area (1983). The steam is obtained from the back-pressure turbines of five electricity-generating stations together with other peak-demand boilers, and is conveyed by an underground distribution grid over 70 miles in length. Government support of CHP has perhaps been most marked in Denmark. An Act on Heat Supply came into force in September 1979, so providing the framework for national and local authorities to develop power-plant heat in all built-up areas. Together with natural gas and renewable-energy sources, CHP and DH schemes form the main planks of Denmark's post-petroleum energy strategy.

The Conservation Option

As we have seen, improved technical efficiency in all fields of production together with the strenuous implementation of the policies and practices described can contribute very considerable energy savings. Professor Sir Martin Ryle (former Astronomer Royal), in his paper *The Energy Problem*[25] outlined a number of areas where the combined possible energy savings would be equivalent to three-quarters of the present energy supplied by oil and gas. As far as nuclear power was concerned, he pointed out that 'since the capital cost of a nuclear power station if spent instead on the saving of energy now wasted, would give some three times more energy than the station would produce in its lifetime, programmes for insulating buildings, improving the efficiency of industrial and other machinery, use of waste heat, etc., should have clear priority.' He also stressed the arguments formerly used by the 'grassroots' conservationists: 'The techniques involved are nearly all well-established and labour-intensive, and would provide employment over a wide range from the unskilled, to the electrical, machine-tool and other industries greatly in need of work, to sophisticated control engineering; they could be installed on a time scale shorter than any nuclear programme and could start saving energy *now*.'

Energy conservation, if viewed in this positive light, is not only one of the most necessary but also one of the least expensive energy options. It is also indigenous, clean and a supply source in its own right, while reducing energy inputs. Although new conservation techniques are constantly being developed, there are many that are proven and do not require large funding or government research grants. Possibly this fact makes them unglamorous and less attractive than some of the newer technologies. Yet any initial capital outlay is usually recovered in a far shorter period than investment in new energy developments. Many building-insulation measures, for example, have a pay-back period of under 5 years. It is for this reason that all proposed energy investment should be considered on an opportunity cost basis and compared in each case with the alternative possibility of spending the same amount of money on energy saving.

To summarise, energy conservation may be achieved in three main ways. The first is simply by reducing unnecessary demand at source and by a greater consciousness of the way in which energy is used. This reduction may be implemented by certain changes in lifestyle or habits at work or home. Devices may be switched off when not needed; 'normal' temperature levels in homes, public buildings and factories may have to be lowered and the unnecessary use of fuel for transport or certain manufacturing processes curtailed.

The second area of conservation is an extension of the first. It includes the introduction of techniques to increase the efficiency with which energy is used in each of its applications, and increased energy-recovery methods. These, in turn, reduce demand at source still further. Most of the energy savings currently achieved through conservation fall in this category.

The third way is concerned with recovering and re-using or recycling used manufactured products and waste. The total amount of energy used in the raw material–transport–production–marketing–consumption–waste disposal cycle has frequently been found to be less using recycled products than by using virgin raw materials. In some cases waste materials may be used directly as a fuel source.

Restraint in energy use, improved efficiency and the extension of recovery techniques are not in themselves long-term solutions to the energy problem. All these measures can do is to buy time by slowing down fuel-depletion rates, by relieving environmental problems and pollution, and by providing the necessary lead times for the research and development of other energy forms and technologies for using the remaining fossil fuels to better effect. Since the world population is expected to reach 6.3–6.5 billion by the year 2000 and the developing countries need to increase their share of energy resources, conservation can only ever be a partial short-term to medium-term solution.

Some Longer-term Considerations

There is another, more radical way of looking at energy conservation which goes far beyond merely reducing consumption. This fourth way, perhaps an extension of the third, entails using the appropriate energy source in the first place, examining the necessity for certain products and abandoning such concepts as built-in obsolescence, wasteful design and throw-away or disposable commodities. It also involves specifically matching the energy form and application to its end-use to avoid the wastage incurred by 'mismatches'. This last stage in materials and energy conservation is a difficult step to achieve in a society committed to growth. So far it has found little support, despite the major energy savings that can be obtained. The need for a correct end-use for each fuel, for example, can be well illustrated by the case for low-grade domestic heating where 100 units of primary fuel will produce about 25 units of heat in an electrically heated home, but about 70 units of heat in a house centrally heated using oil. For heating purposes therefore, oil has almost three times the primary-fuel efficiency of electricity. Electricity is inefficient but oil is scarce. One must evaluate alternatives. Since one-third of all fuel in the U.K. is used for low-grade space-heating, a conservation policy embracing the notion of right end-use for each fuel clearly has important merits. This is where coal or renewable sources might make a significant contribution.

Perhaps even more fundamental and dramatic energy economies could be achieved by examining what people really need to sustain a viable existence and by eliminating useless and unnecessary procedures. For example, what goods and people *need* to be transported? Energy costs of commercial freight transport account for some 30 per cent of total costs. Yet according to United Nations statistics in 1978, 60 per cent of total world trade in manufactured goods was traded between industrialised countries themselves, often in similar commodities. 40 per cent of world oil consumption was used by the transport industries, sometimes merely engaged in reciprocal 'dumping' of each country's goods! Any really effective conservation policy in this sector must therefore include a revaluation and restructuring of fundamental trade and usage patterns, since such unnecessary trafficking of goods hardly constitutes an improved standard of living or a concern for energy conservation. By 1984 there were no signs of such changes being effected except as indirect results of the world recession.

Finally, there are those who believe that a conservation policy should encourage people to make absolute cuts in energy consumption by changing their lifestyles. Two main arguments are put forward. The first is that levels of material consumption in the industrialised countries are already much too high, both in relation to the poverty of the remaining two-thirds of the world, and in terms of the relevance to man's deeper spiritual, physical and psychological health. Consumption should be reduced by cutting back, doing without and embracing a fall in living standards (which would nevertheless be insignificant compared with standards in many Third World countries).

The second argument is put forward by those who defend current living standards in the industrialised world but who believe that present patterns of consumption are inefficient and wasteful. They would, for example, advocate policies that re-integrate living and working accommodation, as is usually the case in less-developed countries, so as to eliminate the energy consumed in travel. They advocate more labour-intensive technology both to increase employment and to save energy used in highly automated machinery. These people point to the fact that many of the most highly energy-intensive processes and products have only been introduced in the 1960s and 1970s without much appreciable difference to the qualitative standard of life. They also question the necessity of manufacturing methods that rely on the entire replacement of an appliance or component because of a minor fault.

So far, conservation practice has been confined to certain limited attempts at pricing policies, techniques to improve energy efficiency and the introduction of certain energy and material-recovery processes. The conservation option, if it is to provide a really significant contribution, must be seen in a broader perspective. The curtailment of unnecessary and wasteful energy consumption in the form of fuels and goods involves a fundamental review of lifestyles. Governments, industries and individuals might ask what constitutes an 'improved standard of life'? In the developing countries it is often merely food, shelter, clean water, work and the bare essentials for survival. In our industrialised energy-consuming countries it is assumed to be more gadgets and more comfort for more people, which in turn leads to a higher consumption of energy and raw materials. Towards these ends there have been massive cuts in all the major public services. Yet, to many people, beyond the basic necessities of life the 'standard of living' has more intangible connotations such as a reliable health service, an unpolluted environment, a good educational system, law, order and security. These cannot be easily produced through increased energy consumption and sometimes the reverse is true. Many years ago, Ivan Illich pointed out that the energy used above a certain level tends to pollute and corrupt, and that that level had long ago been reached in England. Barbara Ward in her Foreword to Lovin's book *Soft Energy Paths*[26] also stressed the potentially disastrous consequences of 'cataclysmic wealth' on individuals, institutions or whole societies. Such a sudden wealth creates 'an irrational lack of care about usefulness or waste. The process develops habits which accustom them to operating on the basis of excess and wastefulness, and, although different episodes have different endings, one prospect sees the affected groups, long after the cloudburst of wealth has passed, trying every kind of expedient — borrowing, sponging, speculating — to try to ensure that the private habits or public institutions of excess and waste are maintained. The result is at best a measure of social disintegration, at worst, collapse.'

Amory Lovins in *Soft Energy Paths*[26] forcefully points out that cutting back on energy consumption does not necessarily mean a deterioration in the quality of life but rather changing some of our habits and eliminating waste, inefficiency and pointless planned obsolescence. Energy conservation, therefore, apart from its ecological soundness, is one sure way of keeping our energy options open for the future. While we have the choice to conserve we still have excess. In the past, conservation has been a neglected and wasted energy option. In calculating the costs of conservation we must always bear in mind that, in the long term, no energy is as expensive as wasted energy — except perhaps *no* energy.

References

1. Meadows, D. H., Meadows, D. L., Randers, J. and Behrens III, W. W., *The Limits to Growth*, Universe Books, New York, 1972
2. Barny, G. O., *The Global 2000 Report to the President of the United States*, Pergamon, Oxford, 1980
3. El-Hinnawi, E. E., *Environmental Impacts of Production and Use of Energy*, Tycooly International, Dublin, 1981
4. *Energy Policy — a Consultative Document*, Cmnd 7101, HMSO, London, February 1978, pp. 20, 21
5. *An Interdepartmental Report by Officials*, Energy Paper 33, HMSO, London, 1979
6. *Hansard*, 29 October 1979, Col. 960
7. Taylor, Dr V., *The Easy Path Energy Plan*, Union of Concerned Scientists, U.S.A., 1979
8. Mr David Howell, Secretary of State for Energy, in address to Parliamentary Liaison Group for Alternative Energy Strategies, 16 June 1980

9. ASHRAE Standard, *Energy Conservation: Existing Building – Commercial*, 100-31, August 1979, ASHRAE, New York
10. Leach, G. *et al.*, *A Low Energy Strategy for the United Kingdom*, IIED Report, Science Reviews, IIED, London, 1979
11. House of Commons Select Committee on Energy, *Energy Conservation in Buildings*, RC401-1, HMSO, London, 1982
12. Environmental Data Services (ENDS), *Report No. 38*, ENDS, London, November 1979, p. 20
13. Weighell, Sidney, former General Secretary N.U.R., letter to *The Guardian*, 27 December 1979
14. Chapman, P., *Fuels Paradise*, Penguin, London, 1975, p. 6
15. Schumacher, Dr E. F., *Toward 2024 AD*, UOP Inc., Feltham, Middlesex, 1975, p. 48
16. *Figuring Out Rubbish*, Industry Committee for Packaging and the Environment, London, April 1981
17. Thomas, C., *Material Gains*, Earth Resources Research, London, 1979
18. For the sections on Aluminium and Tinplate, Glass and Power, frequent reference has been made to: *Recycling*, Industry Committee for Packaging and the Environment, London, April 1981, and supporting papers
19. Waste Management Advisory Council, *Study of Returnable and Non-Returnable Containers*, HMSO, London, 1981
20. Orchard, W. R. H., *Combined Heat, Energy and Power, 'The Southwark Scene'*, paper presented to Polytechnic of the South Bank, Energy Costs and Conservation Course, December 1981
21. W. S. Atkins and Partners, *CHP/DH Feasibility Programme: Stage 1, Summary Report and Recommendations for the Department of Energy*, Epsom, July 1982
22. *Combined Heat and Power in the London Borough of Southwark*, London Borough of Southwark Housing Department, 1981
23. Marsh, J., paper presented to South Bank Polytechnic, London, December 1981
24. House of Commons Select Committee on Energy, Combined Heat and Power, *HC 314-1* and *HC 314-2*, HMSO, London, 1983
25. Ryle, Professor Sir Martin, 'The Energy Problem', *Resurgence Magazine*, May/June 1980, p. 6
26. Lovins, A. B., *Soft Energy Paths*, Penguin, London, 1977

Further Reading

Gabor, D., Colombo, U., King, A. and Galli, R., *Beyond the Age of Waste*, 2nd edn, Club of Rome Report, Pergamon, Oxford, 1981
English, Bohm and Clinard, *Proceedings of the International Energy Symposium III*, Harper and Row, New York, 1982
Kent, D., *Dictionary of Applied Energy Conservation*, Kogan Page, London, 1982
McGuigan, D. and McGuigan, A., *Heat pumps. An efficient heating and cooling alternative*, Garden Way Publishing, Carlotte, Vermont, 1981
Orchard, W. H. R. and Sherratt, C. A. F. C., *Combined Heat and Power – Whole City Heating – Planning Tomorrow's Energy Economy*, George Godwin, London, 1980
Dryden, I. G. C. (ed.), *The Efficient Use of Energy*, 2nd edn, Butterworths, London, in collaboration with the Institute of Energy Action on behalf of the U.K. Department of Energy, 1982

Armor, M., *Heat Pumps and Houses*, Prism Press, Dorchester, Dorset, 1981

Payne, G. A., *The Energy Managers Handbook*, 2nd edn, Westbury House, Guildford, 1980

Department of Energy, Combined Heat and Power Group, *District Heating Combined with Electricity Generation in the U.K.*, Energy Paper 20, HMSO, London, 1977

Department of Energy, *Combined Heat and Power Group, Heat Loads in British Cities*, Energy Paper 34, HMSO, London, 1979; *Combined Heat and Electricity Power Generation in the U.K.*, Energy Paper 35, HMSO, London, 1979

Atkins, W. S., *Greater London Heat Density Survey*, Department of Energy, London, 1978

Pearce, D. *et al.*, *Decision Making for Energy Futures*, Social Science Research Council, Macmillan, London, 1979

Lucas, N. J. D. (ed.), *Local Energy Centres*, Applied Science Publishers, London, 1978

11

Energy Options for the Third World

'The world is a unity and we must begin to act,
as members of it who depend on each other'

Brandt Commission Report of 1980

By far the most critical energy problems of today are those that confront certain countries of the Third World, this being the generic term covering the world's poorest nations. It includes countries of Latin America and the more recently independent states of Africa and Asia. The other two 'Worlds' are the Western World (Western Europe, the U.S.A., Canada, Japan, Australia, New Zealand and South Africa), and the Communist World which includes those countries with centrally planned economies such as the Warsaw Pact nations and China.

In 1960, 24 countries in the Western World established the Organisation for Economic Co-operation and Development (OECD). The economic cooperation and development achieved by OECD benefited its members, but also served to widen the gap between rich and poor. By the end of the 1970s the world's 25 wealthiest nations, which included 16 OECD members, had *per capita* incomes of $7000 per head and over, as opposed to $300 and under in the poorest countries.

The wealthiest countries of the Western group and Communist Bloc are also referred to as 'industrialised' as opposed to the poorer 'developing' countries of the Third World; or 'North' as opposed to 'South', a concept stressed in the *Brandt Commission Report* of 1980.[1] The North, with just over 24 per cent of the world's population, commands 80 per cent of its total income and 95 per cent of its manufacturing capacity. The South consists very largely of Third World countries, many of them with large populations and few resources; or without the necessary knowledge and wealth to develop their potential material and energy resources. Table 11.1 groups each country of the world into one of five bands according to its *per capita* GNP and demonstrates that 124 countries, or 70 per cent of those listed, have *per*

268

Table 11.1 *The wealth of nations*

Income group	Number of countries	Population mid-1978 (millions)	GNP 1978 (U.S. $ billions)	Average GNP per capita 1978 (U.S. $)
Less than $300	36	2008	415	210
$300–699	38	493	232	470
$700–2999	50	571	841	1470
$3000–6999	29	536	2222	4150
$7000 and over	25	552	5083	9220

Source: World Bank.

capita incomes of under $3000 per year and may be classified as Third World or developing countries.[2] These contain over 3 billion people, out of the 1980 world population of 4.3 billion. Nearly two-thirds of them come into the category of 'rural poor', with some 1400 million (33 per cent of the world's population) in southern Asia, 320 million (7 per cent) in Africa and 110 million (3 per cent) in South America. These *least developed countries* (LDCs) also contain the great majority of the world's 800 million 'absolute poor'. They vary widely in their energy economies, and it is obvious that the energy problems of the different countries require different solutions.

Fuel Resources and Consumption within Third World Countries

Resources

There is considerable unevenness in the distribution of conventional fuel resources (oil, gas, heavy oil and shale, coal and hydropower) between the Third World countries although UN estimates in 1979 indicated that possibly only about one-third of the oil in these countries had been identified and there is still equal uncertainty about natural gas reserves.

Many of the countries do not possess adequate reserves of conventional fuels to meet subsistence let alone development needs, although in some cases exploration to date has not been extensive owing to lack of funds. A few Third World countries have been able to develop *indigenous energy resources*, a need which has become increasingly urgent since 1973. In addition to OPEC members, the most well-endowed countries include Botswana, Swaziland and Zimbabwe in Africa; India, China and Indonesia in Asia; Brazil, Mexico and Venezuela in Latin America; and also Turkey and Yugoslavia. Most others have to rely heavily on imported fuels to sustain their commercial and industrial requirements. However, the falling prices of their exports in relation to oil and their stagnant economies frequently deprive such countries of the foreign exchange necessary to pay for the energy required for their development.

Consumption

The magnitude of the energy problem in Third World countries that do not possess their own fuel resources is reflected in the extremely low consumption figures compared with the size of their populations and their development needs. While in 1979 the annual commercial energy consumption *per capita* in the U.S.A. was about 11 tonnes of coal equivalent and about 5 tonnes in Western Europe, it was only between 250 and 750 kg of coal equivalent for about half the population in Third World countries and less than 250 kg for the other half. According to Leach[3] the use of energy 'must increase considerably in the farms and villages of the developing world if rising populations are to be fed adequately, let alone better.'

In measuring energy consumption in developing countries, however, the distinction must be made between commercial energy and traditional (non-commercial) energy. Commercial energy is that which is sold in the course of commerce or provided by a public utility. It is broadly synonymous with the use of conventional energies such as oil, coal, gas, electricity from hydropower and nuclear energy. Traditional energy includes those fuels, such as wood, peat, dung and other organic residues, which are generally used in pre-industrial societies or in non-industrial rural areas. These sometimes supply over 90 per cent of rural needs.

(a) *Commercial fuels.* World commercial energy consumption in 1975–80 is shown in table 11.2.

This table demonstrates the enormous disparity in energy consumption between the industrialised countries and the developing countries. Nevertheless, it highlights the slowing down of consumption in nearly all countries as a consequence of the rising price of oil. In total, the developing countries (excluding China) consume only 12 per cent of the world's commercial energy, while growth rates have halved since 1975.

The use of commercial energy in the main economic sectors of developing

Table 11.2 *World commercial energy consumption 1975–80*
(million barrels a day of oil equivalent)

	1975	1980	Average annual growth (%)	
			1950–74	1975–80
World	122.1	137.8	5.0	2.5
Developing countries	13.9	16.7	6.9	3.7
Oil-importing, developing countries	10.4	12.4	6.9	3.6

Source: World Bank, *Energy in Developing Countries*, p. 2, August 1980.

countries is uneven. Typically less than 5 per cent is used in agriculture (mainly for fertilisers, irrigation and farm mechanisation). Industry consumes most energy, accounting for an overall average of 35 per cent of commercial energy consumption. The rate is higher in those countries with a large heavy industry base comprising energy-intensive processes such as steel, aluminium and cement production. Motorised transport accounts for 10–30 per cent of consumption, with higher figures in the middle-income countries, mainly attributable to increased road transport. In addition, most developing countries are embarking on electrification programmes using all conventional fuels including hydropower, and small amounts of nuclear power in some cases. In 1980, 25 per cent of all commercial energy used in developing countries was transformed into electricity.

Despite low overall energy consumption compared with industrialised countries there is a *predominance of oil* in Third World commercial-energy balances. In 1980, oil accounted for 55 per cent of total commercial-energy use in the developing countries (a higher proportion than in the industrialised countries), coal for 16 per cent and gas for 13 per cent. Nuclear energy accounted for only 0.6 per cent. Unfortunately, owing to lack of precise data, especially in rural areas, the use of traditional fuels has been vastly under-estimated. The abnormal dependence on oil as opposed to other commercial fuels of many developing countries is in part due to their economic development having taken off in the era of oil's relative cheapness and predominance as a fuel. This renders such countries even more vulnerable than the industrialised nations to any fluctuations in oil prices. Yet their oil consumption continues to increase while most industrialised countries are attempting to reduce their consumption.

Even before 1973 oil-importing Third World countries were incurring massive foreign-exchange bills for the fuel needed to sustain their economic growth. Since 1973 the gravity of their plight has intensified. By 1980, the cost of oil imports amounted to 49.3 per cent of total imports, having increased in volume terms over the previous decade by 9.6 per cent per year.

The sharply rising cost of oil imports has imposed severe strains on developing countries' economies and thus constrained their continued development. In 1980 oil-import expenditures accounted for a total of 3.3 per cent of their GNP and 26 per cent of their export earnings.[2] In individual cases this was as much as 60 per cent. As oil becomes an ever-scarcer commodity and the more wealthy industrial countries continue to use their superior purchasing power for their own preferential advantage, the developing countries will find it increasingly difficult and expensive to get the basic fuel supplies they need. Their situation is made worse because the export commodities on which they rely for foreign exchange to buy oil have been falling heavily in value since 1975. For example, in 1975 1 tonne of copper could purchase 115.4 barrels of oil. In June 1982 it bought only 38.1 barrels. A similar squeeze has hit other metals as well as food crops, as figure 11.1 shows.

Inevitably the very poor in rural areas are first to suffer when oil supplies begin to shrink and prices rise. Although their *per capita* energy consumption is very low, it is absolutely necessary for the bare essentials of life — for cooking, lighting, obtaining drinking water, shelter, essential transport and, once these have been

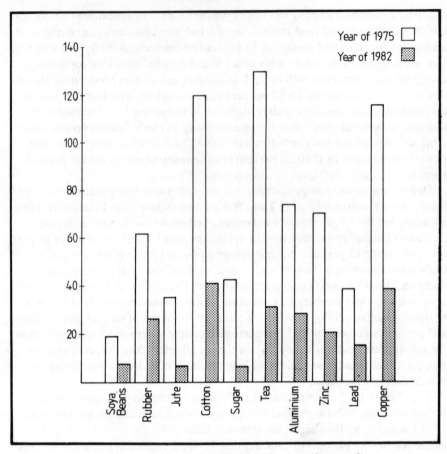

Figure 11.1 *Barrels of oil purchased by 1 tonne of commodity*
(data from *South Magazine*, August 1982)

achieved, for providing basic communal needs such as irrigation, fertilisers and fuel for village industry. As Leach commented 'rural energy deprivation of *all* kinds . . . is an insistent and potentially explosive fact cf the Third World.'[3]

The oil crisis has even hit the more prosperous, middle-income developing countries, where the rapid growth of cities, industries and transport based on oil-intensive technologies is already proceeding apace. These countries increased their dependence on oil by no less than 26.4 per cent per year between 1970 and 1980 and will be in a particularly vulnerable position as oil supplies begin to decline.

(b) *Non-commercial fuels.* While the urban and industrialised sectors of developing countries have been the main consumers of commercial fuels, the rural areas have always relied largely on the non-commercial energy sources. On a world scale total biomass sources probably account for some 15 per cent of annual fuel consumption or the estimated equivalent of 20 million barrels of oil per day. Since there is no means of quantifying this non-commercial energy, precise estimates are difficult to obtain and figures may vary. Biomass in different forms probably amounts to 15 GJ or the equivalent of 1 tonne of dry wood per person per year in developing areas of the Third World.[4] Usage naturally varies from country to country and within countries themselves according to climate, accessibility and location. Table 11.3 compares fuel-wood and charcoal consumption with the use of commercial fuels in different regions of the world. It will be appreciated, however, that while most LDCs are still in the 'wood age', many developed countries consume vast amounts of wood for non-essential purposes. A National Agriculture Lands Study in the U.S.A. in 1981 (quoted by Erik P. Eckholm in *Down to Earth — Environmental and Human Needs*) estimated that the average North American consumes about as much wood annually in the form of paper as does the average resident in many Third World countries in the form of cooking fuel.

It will be seen that in the least-developed countries of Africa and Asia, fuel-wood accounts for 86 and 64 per cent of total energy consumption, respectively. In some local regions up to 98 per cent of energy used is in the form of wood or charcoal while in the industrial economies fuel-wood generally accounts for a much lower

Table 11.3   *Fuel-wood and charcoal consumption 1978*

	Population (millions)	Fuel-wood and charcoal consumption per capita (m^3 of wood)	Energy equivalent of fuel-wood (millions GJ)	Fuel-wood as percentage of total energy use
North				
Market economies	775	0.07	508	0.3
Centrally planned economies	372	0.24	855	1.4
TOTAL	1147	0.13	1363	0.7
Africa				
Least developed countries	138	1.18	1532	85.7
TOTAL	415	0.85	3318	57.9
Asia				
Least developed countries	130	0.26	319	63.9
Centrally planned economies	1010	0.22	2068	7.9
TOTAL	2347	0.34	2387	16.6
Latin America				
TOTAL	349	0.78	2557	18.4
SOUTH TOTAL	3111	0.46	13353	20.6
WORLD TOTAL	4258	0.37	14720	5.4

Source: United Nations Conference on New and Renewable Sources of Energy, Nairobi, 1981.

percentage and can be as little as 0.3 per cent of the total. While wood and charcoal contribute 21 per cent of total energy consumption in developing countries (which includes oil-producing Third World countries) they contribute only 5 per cent of total world energy.

Fuel-wood is the preferred fuel of rural communities because of its local availability, suitability for decentralised energy systems and because until recently it has been cheap or even free. Unfortunately, owing to the rapid growth of Third World populations, to wasteful and inefficient practices and lack of replanting, the demand for fuel-wood in many LDCs is rapidly outstripping supply. Quite apart from the ecological, agricultural and climatic imbalances thus caused, threatening food production in certain areas on what the FAO terms 'an alarming scale', in some areas shortages of these essential fuels have become so serious as to be termed 'the Second Energy Crisis'. An FAO study for the UN 1981 Nairobi Conference on New and Renewable Sources of Energy revealed a situation of 'frightening proportions.' It was estimated that already more than 1000 million people were unable to obtain sufficient fuel-wood for even *minimum* requirements while another 1000 million were affected by lesser shortages. Unless rapid reforestation programmes take place and fuel-wood is used with greater efficiency, by the year 2000 over 2500 million people will need to replace wood by alternative cooking fuels.[5] This does not leave much time and will require massive investments of time and money to improve the supply, distribution and efficient use of fuel-wood.

The situation is in many ways more critical than shortages in oil supply: first because over half the world's population is directly affected; second because these people live at or near subsistence level with no margins for cut-backs in consumption and no means of switching directly to alternative fuels; and third because forests are themselves essential to the biosphere and to the preservation of ecological balance. As was shown in chapter 7, a lack of proper husbandry, forest conservation and reforestation programmes, together with overgrazing in certain areas, have already caused a steady erosion of woodlands and desert encroachment at a threatening pace in many parts of the world. Thus, unless rural communities can establish their own systems for planting new trees at rates sufficient to replace cutting (as is happening in some areas, such as in northern India), their fuel problems can only get worse. The inhabitants will have to journey increasingly further afield and spend more energy and time, or else more money, to procure their minimal daily needs. It is not unusual in some villages for women to spend the greater part of the day gathering fire-wood to cook the evening meal or for families to spend over 50 per cent of their earnings on purchasing fuel-wood where trees are scarce. Moreover, fire-wood competes with food production for land. Leach[3] quotes the Sahel in North Africa where fire-wood consumption is over 12 million tonnes per year (or 0.6 m^3 per person). It is estimated that 150,000–300,000 hectares/year must be planted in forests to meet energy needs by the year 2000, assuming 3–6 m^3 forest fuel per hectare on irrigated land. This is about 50 times greater than current forest planting. In Africa as a whole a fifteenfold increase is needed to meet these demands. On a world scale some 50 million hectares of fuel-wood plantation will be needed by the year 2000 to meet the *additional* domestic needs of developing countries for cooking and heating alone. Nor does this take into account the vast forest areas destroyed annually to satisfy the requirements of the industrial world for paper and other commodities. Meanwhile forests of the developing countries are being laid waste at a rate of about 1.3 per cent of the total or 10–15 million hectares per year. Unfortunately, the chances of even maintaining existing forests, let alone planting new ones, are problematic. Lack of skilled personnel, self-interested exploitation by wood contractors, clandestine cutting by poor villagers, damage caused by stray animals, neglected soil conditions and non-availability of common land in some areas are frequent obstacles to forest preservation or reforestation programmes. Where land is available it is often not properly irrigated or fenced. Most ominously the rate of deforestation is greatest in countries most heavily dependent on oil imports. Where no fire-wood exists families burn animal dung cakes and crop wastes rather than returning them to the land as fertiliser. This, in turn, also impoverishes the soil and leads to low agricultural yields and food crop failures.

Future Energy Options for the Third World

Two major facts dominate the energy scene for the Third World. First, it is clear that energy consumption must continue to grow if development is to proceed and if rural poverty and deprivation are to be alleviated. Second, it is equally evident that traditional and conventional approaches to energy supply have a limited future and need to be re-evaluated to ensure short-term survival and establish longer-term energy viability. Usually, rural development takes place with the inefficient use of traditional fuels such as wood, dung and charcoal gradually being replaced by the inefficient use of commercial fuels including kerosene, bottled gas and oil. If circumstances permit, these are ultimately replaced by inefficient rural electrification programmes which are frequently dependent on oil generators. Previous chapters have examined some of the alternatives. There is now a necessity, as never before, to maximise efficiencies at every stage of the energy transition and to find the most appropriate paths for Third World countries at each level of their development. Some possible conclusions, drawn from foregoing chapters, are given below.

(1) Fuel-wood Futures

Since in the present economic climate most impoverished rural areas of the Third World will not have the means of developing substitute fuel resources between now and the beginning of the next century, the most urgent priority is to improve the supply (including replanting), distribution and efficient combustion or distillation of fuel-wood. In 1981, the United Nations Conference on New and Renewable Sources of Energy in Nairobi (UNERG) identified six major needs, as follows:

(1) The improvement of present fuel-wood resources by better management of forest shrubland and by educating communities in elementary silviculture.
(2) The creation of new forest resources, large-scale plantations, community woodlands and individual plantings. (This will require considerable educational, financial and back-up support.)

(3) The organisation of distribution and better storage to ensure more equal access to fire-wood between urban and rural communities in the Third World. Towns and cities frequently denude the surrounding areas while remote regions may, in some cases, have surplus.
(4) The improvement of technology in the conversion of charcoal (which typically has only a 20 per cent efficiency) and the combustion of fire-wood (where efficiencies could be improved by up to 30 per cent). These could be achieved by improved design in charcoal burners and the introduction of wood-burning stoves, preferably manufactured locally, such as Lorena Stoves, to replace the traditional three-stone open fires.
(5) The reconstruction of land tenure with ownership and control of common land devolving on local families. This, in turn, would encourage replanting, better husbandry and also avoid unrestrained fire-wood gathering on common land.
(6) Finally, the gradual substitution of fossil fuels including coal, bottle or piped gas and kerosene in those urban areas still dependent on the biomass resources of the surrounding rural communities (presumably as an interim step before the transition to renewable-fuel economies).

However, the success of all such measures, as aptly summarised in a 1981 Earthscan report,[6] greatly depends on how they fit into existing social and economic patterns. Unless carefully introduced they could prove to be counter-productive. For example, an improvement in charcoal production techniques might have the effect of lowering the price of charcoal and hence of reducing the incentive for planting trees. In common with the enclosure movement in the U.K., the apportioning of common land might result in depriving the very poor of all access to fuel and, as now is the case in parts of India, they would have to pay for wood and dung which had hitherto been collected free. The introduction of expensive wood-burning stoves on a large scale could only replace traditional methods if safety or convenience factors were emphasised or if they were perceived to be superior.

Another suggestion that has been put forward as a means of conserving valuable energy in developing countries, particularly with respect to transport and agriculture (ploughing, threshing, grinding, pumping, etc.), is to improve and extend the use of draught animals in rural areas. In 1981 UNERG calculated that there were probably 280–300 million working draught animals (about 80 working animals for every one of the 2.7 million tractors) operating in the Third World. Compared with draught animals working in the U.S. or Europe up to the early part of the twentieth century, very little development had taken place in the breeding or efficient use of these animated energy converters. By simple design improvements in such things as harnesses or the wheel bearings of carts, both the life and output of draught animals can be increased.

(2) Fossil Fuels

It is by now evident that, in the long term, dependence on oil is increasingly prohibitive for developing countries with no indigenous oil reserves or significant export commodities. Even in 1980 current account deficits of non-oil-producing developing countries totalled $70 billion. Also, many developing countries have very limited amounts of other conventional energy forms; even where these exist, frequently they cannot be exploited quickly, since poor countries are unable to raise the immense capital sums necessary to build new coal mines, prospect for oil and gas, lay pipelines, extract oil from shale and sands, build refineries, electricity-generating stations and grids, and instal adequate hydroelectric schemes. Already the non-oil-producing developing countries have debts of over $300 billion, some 40 per cent of which is owed to private institutions. The World Bank has calculated that between 1960 and 1975 the developing countries had invested about $12 billion per year (1980 prices) in commercial energy production. This was about 5 per cent of their total investment and 1.3 per cent of GNP. They will have to invest at nearly double the previous rates to meet their energy commitments over the next few years. In addition to coal and oil which is currently being exploited where it exists and where funds and loans are available, substantial reserves of natural gas are known to exist in many Third World countries. This requires less capital investment and technical expertise for its development than do some other fossil fuels. However, natural gas, even when available, is limited by location. Unless it is used at the point of extraction, its distribution requires the expensive laying of pipe networks. In oil-producing Third World countries there is a move towards conserving and using gas previously wasted through flaring, and some other countries are attempting to locate reserves in hitherto unprospected regions. Nevertheless, the World Bank, which in 1980 supplied half of the public support for energy development, warned that the programmes necessary 'cannot be financed by the Bank.' It went on to state that 'it is by no means clear that the very large investments required for energy development in the oil-importing countries will be met.'[2]

(3) The Nuclear Option

Nuclear power is among the most centralised of all energy forms and the problems of introducing it on a large scale in the developing countries are particularly acute. One major problem was that by 1980 only 10 developing countries had electricity grid systems large enough to take a nuclear load. Another problem is that nuclear generation is particularly suitable for the *base-load* generation of electricity – the relatively small proportion of a country's needs that is fairly constant, perhaps in the order of 10 per cent of total requirements. Even if its output could be consistently used, the country would have to instal an equivalent capacity on stand-by in case of breakdown or for use during routine maintenance (which can account for up to one-third of running time), again a major diseconomy.[7]

Since few developing countries have indigenous nuclear expertise or uranium reserves, the development of a nuclear industry also increases local dependence on overseas industrialised interests as well as being a drain on foreign exchange reserves. Also, without the industrialised countries' long experience of working practices in the nuclear industry, safety standards are often less easy to maintain. For example, India, the country with the greatest nuclear expertise in the Third World, has frequently had to seek assistance from U.S. nuclear companies when problems have occurred. To illustrate, in 1979 when experts from the U.S. General Electric Corporation were called in to examine reactor defects, it was reported that 'one power station was so heavily contaminated . . . that it is impossible for maintenance jobs to be performed without the maintenance personnel exceeding the fortnightly radiation dose in a matter of minutes.'[8] Although it is of great financial interest to industrial nations such as the U.S.A., the U.K., France and West Germany to export their nuclear technologies to developing countries, the potential risk factors are correspondingly higher in countries without long-standing technological traditions.

(4) The Renewable-energy Option

Policies for the development of effective energy technologies for Third World countries must take into account the wider economic, social and cultural context in these countries. For example, large capital and energy-intensive projects tend to create very few new job opportunities and require many years to complete. It is in any case questionable whether the needs of the poor can best be met by high-grade energy and highly centralised systems — the industrialised countries' solution to energy demand in the twentieth century! The Third World's immediate need is most often for access to cheap, more autonomous sources of energy supply maintained by skills and materials locally available, and limited in dependence on costly imports and sophisticated technologies. Yet frequently these countries regard the small-scale 'soft energy' technologies as second-rate options already discarded by the industrialised nations.

The Third World's long-term energy requirements will ultimately have to be met by the development of local, indigenous resources. Although there must be immediate financial and technical help to overcome the short-term energy needs, it is likely that long-term and stable energy supplies will only be achieved with greater emphasis on self-sufficiency.

An alternative to fossil fuels and the nuclear option is therefore the renewable option which many people argue offers the best hope for social and economic development in the Third World. Apart from geothermal energy the renewable resources are solar-derived, including wind, wave, OTEC, hydroelectric and biomass, together with direct solar energy, with its mechanical, thermal and electrical applications. Any renewable-energy strategy for the developing countries would need to investigate the technical feasibility and costs of these energy forms, together with the end-use requirements of the particular village or urban setting, and the extent of their social and cultural acceptability. Many renewable-energy devices manufactured in industrialised countries for use in LDCs do not take such factors into account. Moreover they are frequently expensive and can be unreliable or difficult to adjust and maintain without trained technicians in the field. Each renewable-energy device or piece of equipment therefore should be precisely matched to the requirements or opportunities of a particular situation or community in order to make an effective contribution. Wherever possible it should be manufactured locally, not only for economic and political reasons, but also to increase the acceptability and understanding of such applications at a local level. Frequently the adoption of renewable-energy equipment in developing countries has been obstructed not only by expense, but because it has been entirely inappropriate in a given local context. In some ways, new and renewable energies have been 'technical solutions looking for appropriate problems.'[6]

Attempts have been made to assess the utility of both new and renewable energies in Third World (South) countries, comparing these to the industrial countries (North) with an indication of their cost-effectiveness. While some of the technologies are relatively new, others such as wind, water and hydroelectric power are already well-established and merely need extending. As has been shown in chapter 9, the potential for rural electrification through mini hydroelectric schemes is very great in many areas. For example, while Brazil in the 1970s planned to increase its generating capacity by 11 per cent per year until 2000, developing only sites larger than 10 MW, alternatively it has been suggested that the mini hydro-potential of the Amazon's many tributaries could completely and inexpensively replace all the oil-powered diesel generators that supply electricity to the river-bank villages.

Since *cooking* accounts for most of the fuel consumption in rural areas one possibility (mentioned in chapter 6) is to use solar energy for cooking. *Solar cooking*, however, presents problems. Although different designs have been tried under field conditions there are two major snags. The first is the initial cost to families who may be cooking with 'free' fuel-wood, at whatever cost to the environment and future energy supplies. The second is the social problem of persuading potential users to adopt different habits such as eating the main meal at mid-day and cooking in the open air to utilise the solar radiation. These factors may represent a major obstacle for communities that work in the fields during the day or in areas where hospitality to beggars and strangers is a compulsory religious requirement. Parts of India provide a good example despite the excellent incidence of solar radiation. Yet from a purely technical point of view, apart from initial costs which might even be met by government or foreign aid subsidies, there are no reasons why solar cookers should not gradually be substituted for fuel-wood burning in appropriate areas of the Third World.

Apart from the traditional fire-wood, crop residues and dung cakes, *biogas* offers the best current prospects for low-cost cooking. In addition it can provide heating and crop drying using renewable energy. As mentioned in chapter 7, *biogas* plants already operate in parts of the Third World, notably in India and parts of China where there are 8 million family-sized plants already installed (roughly 1 for every 120 people, taking China as a whole). Larger biogas units have been installed in India. There are, however, considerable problems connected with enlisting the co-operation of individual members of the community, particularly in societies where caste systems or social conventions prove to be obstacles. Again, India is an apt example. Even with small family biogas plants, there may be adverse side-effects such as accentuating the gulf between those who own animals to provide the wastes and the very poor. One consequence might be the monetising of dung which could no longer be collected and used by landless peasants. Also, although the biogas process is technically feasible, there have been many instances where families have had too few cattle to fuel the digesters, or where there have been failures because biogas digesters were introduced in areas where the ambient temperature was inadequate to maintain the level of gas production.

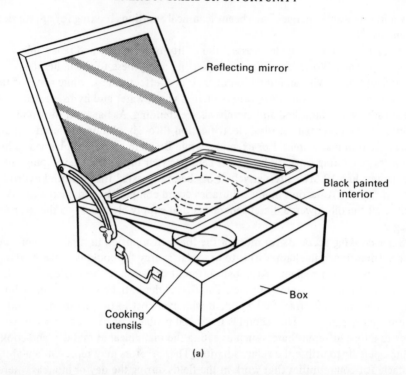

(a)

(b)

Figure 11.2 *Two simple ways of saving fuel-wood in cooking:*
(a) Solar Cooker, (b) Lorena Stove

An easier approach in many areas is to *improve the traditional cooking appliances* so that less energy is consumed per unit output. The traditional three-stone open fire has an extremely poor efficiency rate of some 5–9 per cent. This can easily be improved once the necessity is perceived. Because of the low efficiencies of open fires for cooking (and heating) the total amount of fuel used in rural areas of developing countries as well as the *per capita* cooking-fuel consumption rate vastly exceeds those of the industrialised countries. Estimates vary from country to country but cookery requirements are generally assumed to be 7–14 GJ/person/ year in LDCs compared with 1.5–2.0 GJ/person/year in industrialised countries.[3] The potential energy saving with improved cooking methods is thus considerable, provided that stoves are sufficiently inexpensive to be affordable by the poor, or that they can be hand-made using local materials.

Water-heating and pumping facilities can also be fuelled with decentralised low-cost renewable-energy sources. By the addition of a small boiler, biomass and solar thermal systems analogous to those used for domestic cooking can be used for the provision of hot water, although such systems still tend to be extremely expensive and obtainable only with government or foreign subsidies. Solar-powered water-pumping systems have nevertheless been installed in several developing regions, such as Kenya, Sudan, Niger, Upper Volta, Mali and Senegal.

Agriculture offers endless possibilities for appropriate decentralised renewable-energy applications. Solar thermal energy can be used for drying crops, for producing drinking water and to power desalination plants. To obtain drinking water in arid regions, various kinds of simple low-cost solar water stills can be made, as mentioned in chapter 6. Portable photovoltaic cells have also been used to provide power for crop spraying, thus saving human energy and fuels.

With *industrial and commercial uses* in mind, experiments in the use of solar power in various forms to replace other fuels in small industries have been undertaken in many countries, especially for low/medium-heat applications. In rare instances high-grade solar heat is used for local industry in Third World countries, but generally this is not economically feasible. Such experiments have demonstrated solar potential for individual projects in a local community.

Developments in solid-state low-power microwave technology have also opened the way for solar-powered telecommunications links, supplied by photovoltaic cells, especially in areas inaccessible to conventional energy forms. Examples exist in many developing countries such as Tanzania, Nigeria, Pakistan, Colombia, Saudi Arabia, and even within the Arctic Circle. These systems are reliable, durable, maintenance-free, transportable and non-polluting. Costs are relatively high but falling (*see* chapter 6), even though such devices are economically attractive only for remote or inaccessible sites where the cost of supplying fuel or maintaining batteries over a number of years is unacceptable.

Integrated village energy systems are the aim of attempts first made in the early-1970s, to develop integrated, hybrid, renewable-energy systems which could in combination supply the total energy needs of rural communities. In Israel, Iran, Mexico and India larger community energy systems able to supply up to 200 families are also being developed, based on the conversion of various renewable sources including biomass, wind, solar and water power, to mechanical, thermal or electrical supply. The Algerian government's solar village of Ain Hnache is an interesting prototype that aims to provide all the energy needs for 1500–2000 people. Energy systems are integrated on an appropriate scale with other activities such as food production. The interdisciplinary team of planners took into account every aspect of village life having first conducted a survey of the various requirements. All inhabitants receive training in how to use the 'new' technologies most effectively and are encouraged to develop renewable-energy techniques of their own. The Ain Hnache experiment, the first of many similar villages planned by the Algerian government (using revenues from oil and gas exports), represents an advance in living standards without large resource demands for poor rural areas. Although the advantages of such schemes are manifold, ranging from independence of centralised supply to improvements in local agriculture, the initial construction costs are high and may be out of the range of most developing countries.

Some *rural electrification* schemes based on solar cells are also being developed for certain remote tropical areas with a high incidence of sunshine but where no mains electricity exists or would be considered feasible. By 1984 these schemes could sometimes be competitive with the production of power by diesel engines in the range from 100 W to several kW. They are considerably cheaper than the prohibitive costs of linking up remote rural communities to a national grid (a fact that has led Egypt and India, for example, to adopt a policy of not connecting distant villages of less than 1000 people to the main transmission lines), but are still too expensive for widespread use in the poorer Third World countries.

Wind-power applications for the provision of village electricity supplies are another area of development in those locations with adequate wind regimes. In some cases it will be possible to integrate these with other forms of renewable-energy production such as solar or geothermal. However, apart from the initial expense of wind converters, insufficient precise information of local wind conditions and the additional cost of adequate storage facilities have so far curtailed the rapid expansion of wind-generated electricity for domestic supplies.

Energy Politics and Policies

By 1984 the governments of most Third World countries were sufficiently aware of the vital role of energy in their economies to have developed some form of domestic energy plan, even if this did not always constitute a policy. The emphasis, however, was usually confined to making conventional or traditional energy forms available for industrial development rather than introducing schemes and equipment to harness the renewable energies for rural areas. There are many reasons for this. First, it is easier to attract and allocate investments for conventional fuels since the energy contribution of some of the renewables is often uncertain. Second, the newer forms of equipment for harnessing renewable energy tend to be expensive and require trained technicians to adjust and maintain them. Such skills are not generally available locally, and thus if fossil fuels are not present the preference is to continue to use traditional fuels such as wood and dung by means of inefficient combustion. Also, energy consultants to developing countries are often drawn from the ranks of academics in industrial countries and frequently more accustomed to producing papers on power rather than power itself. Another reason is that overseas aid programmes have traditionally given priority to the more prestigious large-scale constructions. The organisations through which aid is normally distributed (the World Bank, UN agencies and so forth) tend themselves to be highly centralised institutions and as such unconsciously reflect their own centralised structures in the aid-administering bodies and schemes that they set up. Since the disparities between the North/South traditions and experiences are frequently as marked as their economies, the implementation of industrial energy-development concepts in rural areas has often proved highly unsuitable. Even the most efficient, effective and concerned institutions such as the World Bank also have obligations to investors whose wealth is frequently derived from the export of high technology and equipment to growing markets in the Third World.

At the same time there are signs that awareness of the advantages of developing traditional, alternative or renewable energies as a means of avoiding the 'oil trap' is growing even with respect to large-scale urban and industrial applications. Brazil is a case in point, pursuing policies that combine the development of fossil fuels and nuclear power with moves toward increased hydropower and liquid fuels from biomass. Yet such a radical recasting of energy supplies cannot take place overnight. There will have to be a controlled transition period during which simultaneous efforts are made: to diversify energy from oil, gas and coal; to develop alternative energies; to promote policies for maximum conservation; and to maintain and expand sufficient traditional energy supplies to prevent a severe short-term energy gap or economic recession.

To achieve the objectives outlined, substantially enhanced 'energy aid' will have to be made available to non-oil-producing Third World nations for some time to come. Even the short-term development of fossil fuels as well as the longer-term transition to renewable energy involves high initial capital costs. Following the 1973 oil crisis, the International Monetary Fund created new facilities to help countries with severe difficulties in their balance of payments caused by the higher oil prices. A common fund was established to stabilise the earnings of primary-commodity producers in these poorer countries. Together with agricultural-development funds and disaster-relief programmes came a new awareness of the vital part that energy plays in essential food production. By the end of the 1970s many international agencies were providing significant aid for the development of renewable-energy resources in Third World countries. There were various multilateral, bilateral and non-governmental agencies involved, the United Nations being the major multilateral organisation. Bilateral aid programmes to increase Third World energy supplies included those from industrialised countries such as the U.S.A., the U.K., France and West Germany, all of which had historical links with certain developing nations. The OPEC group also established the Arab Fund for Economic Development in Africa, giving particular support to Muslim countries. Many other energy aid programmes by non-governmental organisations were set up, but on a much smaller scale and usually oriented towards the human-resources aspects of development.

By far the largest source of support for energy development in Third World countries has been the World Bank. Together with the United Nations Development Programme a 60-country Energy Sector Assessment Programme was launched in 1980 and some 37 countries had been covered by 1983.[9] The objective was to assist countries in improving their energy situation within an integrated framework. Apart from its commitment to reforestation schemes, the policies being advocated by the World Bank in 1983 were broadly consistent with the need to augment conventional energies during a transitional period prior to the expansion of the 'renewables option'. Measures that were recommended included efforts to increase levels of oil exploration, the use of enhanced techniques to extract remaining indigenous oil, the building of more refineries in the developing countries, exploration for heavy oil and shale, and expanded investment in electricity generation. Similar policies were put forward in respect of natural gas and coal, including the production of synthetic fuel, coal gasification and liquefaction and the production of methanol from natural gas. Nuclear power was being supported in some countries for base-loads. All the fuels were to be used as efficiently as possible with substantial savings obtainable through conservation.[2]

Since 1973 several developing countries have begun to diversify their energy-supply sources and intensify exploration and research. Already by 1980 oil was being produced in Ghana, Guatemala, the Ivory Coast, and the Philippines — all of which had hitherto been thought not to have indigenous deposits. Oil or gas deposits had also been found in Chad, Pakistan, Thailand and Tanzania. Other countries, including India, Turkey, South Korea, and the Philippines, had taken steps to increase their coal and lignite production. Argentina, Brazil, Cameroon, the Ivory Coast, Uruguay and Sri Lanka were among those who had begun to increase their hydroelectric capacity. Over 30 developing countries were found to possess geo-thermal resources and some, notably Indonesia and the Philippines, are developing these, alongside other projects. The Philippines, for example, was developing wood-powered electricity generation and the use of gasifiers; and, as already mentioned, Brazil was taking the lead in alcohol production from biomass.

In all fuel-importing Third World countries, the overcoming of energy scarcity is a most urgent priority and the prerequisite of any form of sustainable development. The need is for the transition from a subsistence economy to a commercial economy based on indigenous fuels. Equipment must be appropriately matched to end-uses and be easy to maintain so that conservation methods will go hand in hand with increased supplies. Unfortunately, however, such ideal solutions are rare and conditions frequently necessitate difficult choices. The next 20 years are likely to see the rapid development of electricity in most Third World countries. Yet even extensive use of conservation technologies, together with the greater use of oil, gas, coal, peat and nuclear energy, will only buy time. Investments in capitally intensive, non-renewable fuels, while guaranteeing supplies in the short term, should be avoided if comparable investments in renewable supplies can satisfy long-term local demands. For example, even if reserves exist, natural gas may be an unattractive energy source for many poor Third World countries, given the high initial cost of laying down pipelines, if local biogas digesters can supply sufficient energy to meet demand.

Beyond these interim measures, there is a wide consensus that the Third World countries must look to local environmentally benign renewable resources to supply their future needs. The Brandt Commission's energy recommendations, for example, state that: 'An orderly transition is required from high dependence on increasingly scarce non-renewable energy sources. . . Prices which reflect long-term scarcities will play an important role in this transition. . . Special arrangements including financial assistance should be made to ensure supplies to the poorer developing countries. . . International and regional financial agencies must increase substantially their funding of exploration and development of energy sources including the development of renewable energy resources.'[1]

In pursuing policies such as these there are major practical, political, social and economic problems to be overcome before decentralised, locally based renewable-energy structures can be established in most countries. At a practical level there is the chronic lack of manpower and money to introduce even simple new technologies, especially in rural areas. Many of the renewable-energy technologies are still unproven in Third World environments. The infrastructure necessary to distribute, market or service new appliances is also lacking. There are enormous political obstacles, such as the reluctance of the governments of developing countries to settle for small-scale or simpler energy-harnessing equipment when it is produced by multinational companies based in developed countries that do not use the equipment themselves! There are social obstacles in persuading local people to accept and successfully use strange new devices and to abandon age-old traditions. Also the aid donors and major development agencies find it much more difficult to retain accountability when their funds and resources are deployed over wide sectors of the population in small amounts. Economically, too, the flow of aid to developing countries is declining in real terms owing to the world recession, widespread inflation and rising oil prices. As mentioned, the international purchasing power of Third World countries has been diminishing rapidly since 1973 while their own essential primary-export commodities fall in value relative to oil and other fuel import prices.

Consequently, any developments in renewable energies and aid sponsored by the industrial North to meet the energy needs of the South require extremely careful and sensitive planning. One way, which would necessitate funding and organisational skill, might be to form networks of local energy agencies with adequately trained personnel as a pre-condition of the wide dissemination of renewable-energy equipment for local energy systems. Such systems should, as far as possible, be based on a diversification of indigenous energy supplies so as not to be dependent on any single one. Education in energy use and conservation techniques could be promoted where there are local schools or radio stations. While the energy policies of industrialised countries may aim at 'growth' or improving the 'quality of life', the focus for Third World countries must be on satisfying basic needs to ensure survival. Equally, Third World countries should develop energy supplies with due regard to conservation in order to avoid some of the expensive mistakes made by the energy-squandering industrial countries during their evolution. Perhaps the only meagre advantage of their late development may be that they can draw on the experience and evolving energy consciousness of the energy-profligate nations. If the Third World countries were to consume non-renewable fuels at the same rate as the industrialised world, their energy requirements would greatly exceed the bounds of the possible. Therefore, since the energy use in most Third World countries is very small indeed compared with the U.S.A. or Western Europe, taking their energy needs into consideration 'would cost the U.S. and other developed countries relatively little but could yield significant benefits including making the world's use of energy more efficient and more secure for all nations, rich and poor.'[10]

Ultimately, the industrialised nations will need to reduce their consumption of fossil fuels while the developing countries need immediate access both to fossil fuels and to renewable-energy forms. If the industrialised countries' oil use, for example, were 15 per cent below present levels by the year 2000 and developing countries were allowed to increase their use by 5.5 per cent per year, world oil consumption would increase by only 1 per cent between 1976 and 2000.

References

1. *North–South: A Programme for Survival*, The Report of the Independent Commission on International Development Issues under the Chairmanship of Willy Brandt, Pan Books, London, 1980
2. *Energy in the Developing Countries*, World Bank, Washington D.C., August 1980
3. Leach, G., 'Report of the Energy Resources Working Group', *International Conference on Agricultural Production, Research and Development Strategies for the 1980s, Bonn, 8–12 October 1979*
4. Hall, D. O., 'Biomass for Energy: Fuels now and in the Future', *Royal Society of Arts Journal*, No. 5312, Vol. CXXX, July 1982
5. *World Development Report, 1980*, World Bank, Washington D.C., August 1980, p. 17
6. Earthscan, *New and Renewable Energies I*, IIED, London
7. Hayes, D., *Energy for Development: Third World Options*, Worldwatch Paper 15, Worldwatch Institute, Washington D.C., December 1977
8. Agarwal, A., 'This Nuclear Power Plant is Contaminated', *Nature*, Vol. 279, 7 June 1979, pp. 468–70
9. *The Joint UNDP/World Bank Energy Sector Assessment Programme and Energy Sector Management Programme, A Progress Report*, UNDP/World Bank, Washington D.C., 1982
10. Howe, J. W. and Knowland, W. M., *Energy and Development: an International Approach, Communique on Development Issues*, O.D.C./No. 31, Washington D.C., December 1976

Further Reading

Energy in the Developing Countries, World Bank, Washington D.C., 1980
Energy Options and Policy Issues in Developing Countries, World Bank, Washington D.C., 1979

Dunkerly, J. *et al.*, *Energy Strategies for Developing Nations*, Resources for the Future, Johns Hopkins University Press, Baltimore, Maryland

Smil, V. and Knowland, W. E. (eds), *Energy in the Developing World: The Real Energy Crisis*, Oxford University Press, Oxford, 1980

The Energy Transition in Developing Countries, World Bank, Washington D.C., August 1983

Dunn, P. D., *Appropriate Technology – Technology with a Human Face*, Macmillan, London, 1978, chapter 6

Hayter, T., *The Creation of World Poverty – an Alternative View to the Brandt Report*, Third World First, Pluto Press, London, 1981

Common Crisis North–South; Co-operation for world recovery, The Brandt Commission 1983, Pan, London

South Magazine, South Publications, London

Mazingira, the International Journal for Environment and Development, Tycooly International, Dublin

12

Energy Options for the Future—Policies and Planning

'In energy affairs, as in all matters of technical, social and political choice, there is a much wider range of options than the interested parties would have us believe. . .'

Peter Chapman

Previous chapters have attempted to evaluate our current and future energy options. They have also described the immense changes that have taken place in energy-supply patterns in different countries over the past few years, more especially since 1973. It was then first generally realised that: oil and natural gas could not indefinitely supply two-thirds of the world's commercial-energy needs; reserves were limited; and within 50 years they could run out. Consequently, in most industrialised countries a substantial switch to fuels other than oil and gas has recently taken place. The coal industry is again expanding steadily despite the CO_2 build-up and acid rain for which it is partly responsible, and vigorous investment has been undertaken in the nuclear industries notwithstanding the unresolved safety and environmental problems. Considerable interest has also been shown in conservation and renewable-energy sources and in the development of synthetic fuels. At last there is some consensus that unless new patterns of energy use are quickly established, new forms of fuel developed and world energy-demand stabilised, an 'energy gap' will inevitably occur at some unspecified date in the future. Unless bridged, this could undermine the whole structure of society as we know it, affecting both the industrialised and Third World countries. Every country that has not secured its own long-term indigenous fuel supplies could face the crippling hardship currently experienced by many poorer countries. The supposed 'energy affluence' of the early-1980s in the rich countries is as misleading as the frequent abundance of fruit before a tree dies.

Having postulated an 'energy gap', it is necessary to specify exactly what is meant by the term and to devise appropriate policies to avoid such an event. We need to know how big this 'gap' is likely to be. On what assumptions about

288

economic growth and the energy content of future goods and services is it based? What is the most cost-effective and environmentally benign way to fill it? How much time is there left? These and many other equally complex questions must be rationally and democratically considered before new energy strategies are decided, and especially before major, large-scale, expensive and irreversible energy-investment decisions are made.

Energy Supply: The Process of Investment Planning

Normally, the process by which governments and the energy-supply industries plan their new investments follows an orderly sequence of steps. These are:

(1) To determine what the future demand for goods and services is likely to be (the economic growth rate).
(2) To calculate the total energy needed to supply these goods and services in terms of useful energy consumed.
(3) To assess the initial primary energy requirements necessary to meet this demand.
(4) To identify any areas of shortfall/surplus in the balance between energy supply and demand thus calculated.
(5) To devise policy choices, embracing appropriate supply options, pricing policies, consumption limits and conservation requirements, to ensure the continuity of supply and to rectify any imbalances.

This process usually starts from an attempt to determine future economic growth rates, both in terms of Gross Domestic Product and of the major sectors of the economy, such as industry, transport, agriculture, commerce and so on. Next, the postulated levels of economic activity are translated into energy-demand forecasts for each sector, often by using a series of 'energy ratios' derived from historical correlations.

The most sophisticated energy-forecasting studies have been carried out by the industrialised countries, particularly by members of the International Energy Agency (IEA). Each year since its establishment in November 1974 the IEA has published a Review, *Energy Policies and Programmes of IEA Countries*, in which demand forecasts are made and countries' individual and collective responses to them are outlined. The Reviews contain an immense amount of valuable information including a summary of member governments' energy policies. They also show that member countries' forecasts, both of economic growth in general and of energy growth in particular, are changing dramatically from one year to the next. For example, table 12.1 compares the forecasts made in 1979, 1980, 1981 and 1982 of energy requirements and domestic energy production of IEA countries forward to 1990.

It can be seen that a considerable downward revision of demand forecasts has taken place even during these 4 years. By 1982 the forecast size of TPE in 1990 was some 13 per cent lower than that estimated 3 years previously. The 1979 Review estimated that the total IEA energy requirement by 1990 would be 4877.1 mtoe,

Table 12.1   *Recent estimates of IEA energy requirements and domestic energy production in 1990 (Mtoe)*

	Projection for 1990 made in			
	1979	*1980*	*1981*	*1982*
Total requirements				
TPE	4877.1	4526.6	4287.0	4229
TFC	3406.6	3106.7	2949.8	2898
Domestic production				
Oil	809.3	656.4	690.9	693
Gas	630.3	660.8	673.6	687
Solid fuels	1119.8	1088.2	1078.4	1093
Nuclear	455.5	389.0	389.7	379
Hydro/other	301.8	344.1	337.3	330
Total production	3316.7	3138.4	3169.9	3182

Source: International Energy Agency, Energy Policies and Programmes of IEA Countries, 1981, 1982 Reviews, OECD, Paris.

while the 1982 Review anticipated a much smaller rise to 4229.0 mtoe. This represents a reduction from the 1979 estimate of 648.1 mtoe.

The significance of such large downward revisions for energy-supply planning cannot be over-emphasised, especially if it is remembered that the figures represent the best estimates of the governments of 19 major industrial countries. They are therefore likely to be given more credence with national policy-makers than the many unofficial forecasts of economic growth and energy consumption emanating from private institutions or academic sources. In the space of 4 years, the forecast for 1990 has been down-rated by 648 mtoe, a figure that is 66 per cent greater than the entire nuclear capacity (379.0 mtoe) supposedly required by all IEA countries in 1990. This volatility or instability to which the demand estimates for the industrialised countries are currently subject raises serious questions about the validity of the methods by which energy-supply industries calculate their new investments when long lead-times are involved.

The Future GDP/Energy-consumption Ratio

In calculating a primary energy requirement, use is frequently made of the GDP/energy-consumption ratio, or its equivalent, for a particular fuel or industry group. Thus the IEA, for example, forecasted that the ratio would decline from a level of 0.83 in 1979 to 0.77 in 1985 and to 0.72 in 1990.[1]

More specifically, demand forecasting requires consideration of the ratio between the increase in TPE and the increase in GDP over a period of time. This is known as the TPE/GDP elasticity. Elasticities, however, are also subject to wide fluctuations.

For example, from 1968 to 1973 the elasticity was about 1.18, whereas over the 1973–78 period it declined to about 0.33. For the more distant future, IEA Ministers at a 1980 meeting agreed that the elasticity should be reduced to about 0.6 by 1990 as compared with earlier assumptions that economic growth and growth in energy consumption would continue to be equal, giving an elasticity of 1.0.

Although these changes in elasticity appear small (numerically they move in the range 0–1, by single decimal places) they conceal substantial changes in energy use in real terms. For example a 0.1 reduction in elasticity over a 5-year forecasting period translates into a 2.5 per cent decrease in TPE growth, which for the IEA as a whole is equivalent to 106 mtoe, or nearly the entire 1977 IEA nuclear capacity. Even the practice of rounding up a forecast to the nearest decimal place can inadvertently strengthen the case for an additional power station or coal mine! Moreover, the further projections are made into the future, the greater the uncertainty. Experience shows that even short-term demand forecasts of movements in variables making up elasticities can be significantly affected by unforeseen events. In 1972 few professional forecasters predicted the quadrupling of oil prices in 1973. Few anticipated stagnating demand in 1974. Since then not one of the 78 demand studies of IEA members' energy growth compared by the OECD assessed the 1980 energy-demand level correctly.[2] The difficulties have been summarised in a CSS report which goes as far as to state that 'energy demand is essentially meaningless and incalculable over a long period'.[3] It appears that as time-periods advance linearly, the margin of uncertainty rises exponentially. The changing energy scenarios of the past decade bear witness to Amory Lovins's contention that the future is *not* 'essentially the past writ large.'

Attempts to forecast U.K. energy demand have been no less speculative than in other countries. The government has also been extremely slow to change its forecasting assumptions in line with changing economic circumstances. For example, the assumption of a 3 per cent growth rate was still being used in 1982 by the Department of Energy as a basis for energy forecasting despite the fact the economic growth had stagnated since 1978. This discrepancy led an all-party energy Select Committee of M.P.s to severely criticise the Department of Energy for 'Panglossian optimism' in its demand forecasts.[4] The forecasts have also found little support in independent circles. For example, one of the most comprehensive models of U.K. energy demand has been computed by the Science Policy Research Unit of the University of Sussex.[5] The report not only states that it found 'the assumptions necessary to achieve 40 GW of nuclear capacity unrealistic,' but also concluded that nuclear capacity in the year 2000 would be unlikely to exceed 15 GW, which 'implies very little nuclear ordering, perhaps 6 GW up to 1990.'

There are three main reasons why official energy-demand forecasts are still consistently over-optimisitc. The first, as mentioned, is that energy policy-making in the U.K. and in some other industrialised nations has been dominated by the large centralised supply industries for whom low economic growth implies contraction. Thus when the Department of Energy was established in 1974 after the oil crisis, it naturally turned for advice to the existing supply industries such as the Central Electricity Generating Board, the National Coal Board, the Gas Council and the United Kingdom Atomic Energy Authority. Elsewhere, too, the existence of centralised fuel-supply industries may be seen to have a direct effect on the type of energy policies adopted. The dominance of nuclear energy in France's energy future, where centralised electro-nuclear institutions are powerful, may be contrasted with the expansion of CHP facilities and decentralised energy-supply systems in Denmark where municipal authorities and co-operatives have played a major part in energy policy.[6] In the U.K. it is significant also that the government-funded Energy Technology Support Unit, which is responsible for research and reviewing the place of renewable energies, is based at the UKAEA's Research Establishment at Harwell.

The second reason for over-optimism is an implicit belief, held by nearly all official bodies, that a no-growth energy scenario would indicate a lowering of living standards. The idea that an energy ceiling ought to be accepted has appeared a symptom of regression rather than a means of conserving fuels for the future. On the other hand, there has been little examination of the assumptions on which such beliefs have been based, of what actually constitutes a real increase or decrease in the standard of living, or of the full potential of conservation measures.

Finally, the high unreliability of demand forecasting for periods up to a decade ahead — the minimum lead-time necessary to build major energy-supply systems — means that it is dangerous to opt for less than maximum peak-demand projections in order to avoid the risk of energy-supply shortfall. The possibility of surplus capacity is, after all, less embarrassing than an energy shortage.

Once forecasts have been accepted, they are used as the basis for large-scale energy-investment planning. The manner in which such acceptance is sometimes obtained is worth noting. In the U.K., for example, individual supply industries make their own demand projections to establish their own investment proposals and these are integrated into the national energy projections of the Department of Energy, which are largely adjusted summaries of the supply-industries' aspirations. The Department of Energy both denies that it has a major role in refereeing this process (which consequently may be biased towards the energy-supply industries) and disclaims responsibility for the use to which the estimates may be put. For example, in its 1979 paper *Energy Projections* the Department of Energy stated that its own estimates were 'not predictions of what will necessarily happen nor prescriptions of what should happen. . . they do not imply government commitment to particular levels of energy production.' The estimates were merely 'intended to provide *futures* and policy choices.'

On the other hand, despite these disclaimers, such estimates inevitably become the basis for both the government's and the nationalised supply-industries' investment decision-making, and are reported as such in the OECD annual energy-policy reviews. Subsequent U.K. energy policy is in turn justified by reference to the OECD figures! The fact that such estimates may be revised ('rolling' targets) or that they allow for a margin of error ('normative' targets) in no way detracts from their prescriptive nature.

The targets therefore become the basis for determining energy-supply requirements which, in turn, become the basis for future investment decisions in the energy industries. In the U.K. the very high demand projections of 1979 led to a postulated 'energy gap' by the year 2000, which, consequently, strengthened the case for a fivefold expansion of nuclear power. On the other hand, given different assumptions about demand, different conclusions would have inevitably followed.

Considerations for a Balanced Energy Policy

Given the reasons stated above, it is almost impossible for those responsible for formulating energy policy to obtain correct energy-supply decisions from calculations of long-term demand. And if, by their nature, demand predictions are grossly unreliable and are subject to wide margins of error, supply decisions based on these will be similarly erroneous. In fact, it is quite inappropriate to base future energy-supply options on mechanistic decision-making processes that belong exclusively to the sphere of economic forecasting. A high measure of judgement tinged with the individual's own vision enters into the equation. In any case, the energy future affects every aspect of our lives, is a subject of national importance and therefore transcends the realm of economics alone. For these reasons it is necessary to change conventional quantitative methods of relating supply decisions to demand forecasts, and set the process of energy policy-making in a wider mould.

The fact that energy policy-making transcends the realm of economics suggests that it is essentially a *political process*. Since the energy future cannot be predicted on the basis of incontrovertible extrapolations, and because it is fundamental to the lifestyle of future generations, decisions about it should be the democratic product of the public will. As such, these decisions should be subject to the same degree of public accountability and debate as other political decisions and it is entirely appropriate that relevant non-economic considerations should also be weighed in the balance.

Apart from the more obvious economic variables, formulating energy policy involves a very large number of moral, environmental, social and political choices. There is a particular need, for example, to ensure that policies are compatible with the values of freedom, equality, consensus and full employment, which are part of the democratic tradition. To these must now be added preservation of the environment and responsibility for poorer Third World countries.

For these wider issues to be given full and proper consideration, however, the political environment must be conducive to informed debate and the decision-making process should be open and democratic. Some of the underlying conditions necessary to achieve these requirements are set out below, although inevitably there is a certain amount of overlap.

1. Availability of Information

Fundamental to any democratic political process is the rapid and frequent provision of accurate, full information. Since major security issues are not involved, there is no reason why the detailed assumptions and calculations on which governments base energy-demand forecasts should not be made public. Similarly, they should be prepared to reveal full information relating to estimates of reserves, production and distribution costs and capital requirements of the major supply industries.

2. Consensus

Given proper information, the procedures governing energy policy-making should be made accountable to public opinion. The need for consensus is deeply embodied in the democratic tradition and public participation in the energy debate is essential if discontent and violence are to be avoided. Investment in expensive energy options involving potential health and safety hazards to the community should be subject to full debate either at a regional or national level to ascertain public acceptance. It is significant, for example, that in a September 1983 public opinion poll of nearly 10,000 people from 10 countries carried out for the EEC Energy Directorate, 51 per cent of respondents listed renewables as their preferred solution to the energy crisis, 15 per cent favoured increased exploitation of traditional fossil fuels such as coal, 14 per cent preferred conservation and 10 per cent wanted development of nuclear power.[7] The contrast with government preferences is striking.

3. Civil Liberties

Hitherto the U.K. and other Western nations have generally relied on the pricing mechanism to influence energy demand. However, should there be need to limit supplies more drastically in the future, one option may be to introduce further compulsion to limit consumption. A major political question will then become to what extent the public will be prepared to accept restrictions that limit personal choice but promote energy viability. For example, is rationing now, to preserve energy for the future, preferable to possible deprivation of energy and products at some later stage? Or should one wait and react to a particular 'energy crisis' when shortfalls occur? To what extent are citizens prepared to accept centralisation in decision-making on energy legislation? To what extent should they tolerate secrecy, armed surveillance, possible restriction of movement, motoring restrictions, compulsory heating limits or even the additional intrusion of monitoring facilities to ensure compliance? Should it be illegal to use certain fossil fuels above a certain level? Should there be financial or legal restrictions on the advertising of inessential energy-consuming products and, if so, where should the line be drawn? It is important that such decisions are made openly, through a process of public debate and consensus, rather than by an executive branch of government or a major energy-supply industry behind closed doors.

4. Competitive Promotion or Education

One question in an age of possible energy scarcity is how much the nationalised fuel institutions should be allowed to compete for energy markets, especially if their products are not the most economical in relation to a particular end-use application. In the U.K., for example, advertising campaigns funded by the consumer and tax-payer have simultaneously and indiscriminately encouraged the use of more coal, more electricity, more oil and more gas together with the acceptability and promotion of nuclear power. Energy-teaching packs biased towards the benefits of nuclear power have been distributed by the Department of Energy to schools throughout the country, again at the taxpayer's expense. A comparatively small amount has been spent on encouraging energy conservation, restricted fuel use and information on renewable energies and their potential. Also little is being done to educate the public in the relative merits of centralised and decentralised energy-supply systems for any given location. A correct matching of supply sources to particular requirements can greatly reduce wasteful energy consumption.

5. Local Diversification and Flexibility

To minimise risk, a politically inspired energy policy should, as far as possible, support a diversification of energy sources so as not to place an undue reliance on any one supply source. This might avoid such obvious hazards as the 'oil trap' experienced by the Western nations after 1973 and 1979, or the 'wood trap' currently being experienced in many poorer countries. Excessive reliance on uranium may also be dangerous in time of international tension or uranium scarcity in those countries that do not possess sufficiently large indigenous supplies to generate their own nuclear power. For example in 1983, over 20 per cent of French electricity was generated by nuclear-powered stations which were fuelled by imported uranium (mainly from South Africa). By 1990 this figure will have risen to 70 per cent. The U.K., likewise, received 40 per cent of her uranium supplies from Namibia via France and Belgium. Thus, the U.K.'s increasing energy dependence is not only on imported uranium but also on good diplomatic relations between France and Namibia, as well as between France and the U.K. Yet more sensitively, some of the U.K.'s uranium from Namibia is currently processed in the U.S.S.R. For example, in October 1981 the CEGB announced that it was due to receive 70,000 lb of enriched uranium from the U.S.S.R. in that year. Excessive reliance on a single indigenous energy source is also a high-risk option since the supply source may fail, not only because of finite reserves but temporarily through industrial or political action, extreme weather conditions, accident, sabotage and so on. Energy autonomy is greatly strengthened by diversifying supply sources as far as is expedient and relying to the least possible extent on imported fuels.

6. Energy and Employment

Different types of energy policy can have greater or lesser effects on levels of employment. For example, it is frequently argued that the building of large centralised energy-supply systems such as oil terminals, nuclear reactors or the U.K. Severn Barrage Scheme, will provide significant employment opportunities. At a time of high unemployment and recession, this is a persuasive reason for investing in such schemes. On the other hand, the resulting employment opportunities are essentially of a temporary nature (usually 10–15 years) and they bring with them serious problems of social dislocation. Initially, thousands of workers arrive on site and have to create or be absorbed into a community. Living facilities, transportation, medical supplies and other life-support facilities need to be provided. After the project has been completed this artificial community collapses and there is subsequent unemployment and migration. The work itself is often affected. Several studies have demonstrated that unnecessary delays in construction times of large generating plant frequently result from on-site workers' attempts to spin out their jobs for as long as possible to guarantee future wages.

Conversely, the advantages to local employment of small-scale energy-supply schemes and the retrofitting of existing building stock and machinery to comply with higher levels of energy conservation have too often been ignored. A U.S. Ford Foundation model proposed by Professors Hudson and Jorgenson, which encourages economic activities based on low energy and correspondingly higher labour requirements, is one of many that make the point. It argues that an energy future based on conservation and the smaller-scale renewable-energy sources creates more jobs than the large-scale scenario. Also there would generally be less pollution, social disruption, and interference with community, trade-union and individual rights. Historically, when energy was cheap and labour scarce as in Europe and the U.S.A. in the 1950s, it was logical to centralise and intensify energy-supply systems. Where energy is scarce and labour plentiful if not cheap, it may now be necessary for energy policy-makers gradually to shift the emphasis again toward decentralisation and smaller more-local energy systems until a more tenable balance is achieved.

7. Environmental and Safety Issues

These crucial non-economic issues connected with energy-policy options must also be considered, and explicit information made available to the public. In such matters the U.K. and the U.S.A. have provided more information than certain other industrial countries. In the U.K. the Windscale (now Sellafield) and Sizewell public enquiries (nuclear) and the Vale of Belvoir enquiry (coal) were examples of such procedures as was the Three Mile Island Enquiry (nuclear) in the U.S.A. Other notable U.K. environmental studies include the *Sixth Report of the Royal Commission on Environmental Pollution* dealing with nuclear power and the *First Report of the Commission for Energy and the Environment* concerned with the coal industry. A 1978 Department of Energy publication *The Environmental Impact of Renewable Energy Sources*[8] was a further attempt to assess the environmental effects of wind, water, wave and tidal power (notably tidal barrage schemes) as well as other renewable-energy types.

8. Energy and Health

Health hazards associated with energy production and environmental pollution should also come under heavy scrutiny in energy-policy decisions since these are of particular concern to the public. For example, the development of coal-based synthetic natural gas plants as proposed by the British Gas Corporation for the 1980s justifies a large initial investment in that SNG possesses considerable environmental (and ultimately economic) advantages over using coal directly.

9. Energy and Climatology

So far the link between energy use and climatic changes has not played a significant part in energy-policy decision-making. If in the past, the developed countries have increased the world's carbon dioxide levels by burning coal indiscriminately, Third World countries will not think twice about doing likewise when they construct new coal-fired power stations or steelworks. Because of the increasing tendency for Third World countries to burn fossil fuels and forests to obtain the energy supplies necessary to raise their standard of living, there is intense concern about the 'greenhouse effect' caused by the carbon dioxide concentration in the earth's atmosphere. The exact extent and rate at which this will take place are not yet known, but in February 1981 the International Energy Agency and the Organisation for Economic Co-operation and Development met in Paris for a workshop on carbon dioxide research. The U.S. Department of Energy also engaged in a major study on the same topic. Whereas from 1850 to 1950 about 60 Gigatonnes (Gt) of carbon dioxide were committed to the atmosphere by burning fossil fuels, about twice as much as this was produced through forest clearance and burning. By 1981 the world was burning 5 Gt annually and the rate of fossil-fuel use was increasing by 4 per cent per year. The increased input of CO_2 into the atmosphere through the combustion of fossil fuels is also reinforced by the deforestation resulting from world demand for wood since fewer plants are ingesting carbon dioxide and excreting oxygen, so less CO_2 is being taken out of the atmosphere as well as more put in. For these reasons, it seems likely that carbon dioxide levels will have doubled by the year 2030. This could significantly change the world's climate, causing rain shortages and possible temperature increases in the Northern Hemisphere and this in turn could affect agriculture, particularly in the U.S.A., the U.S.S.R. and northern Europe. It is possible that patterns of world food production will be seriously affected. This is of particular concern to the U.S.A. (currently the largest producer and exporter of grain) and to those less bountiful countries whom she supplies. Energy-policy decisions can therefore no longer stop short at national boundaries. Even when energy is produced and used locally, the effects for good or ill extend to other sovereign states.

10. Energy Investments and Conservation

A vigorous emphasis on conservation must play a central part of all future energy policies. This means, for example, that there should be full investigation into the relative merits of any government investments in new energy-supply sources as opposed to identical investment in conservation techniques before sanctioning the new supply. It also implies stimulating conservation consciousness and setting targets at the national level by co-ordinated government energy policies. It entails action at the local level with full use or development of the most appropriate energy forms available in a given location or community. At the personal level it involves maximum initiative and responsibility by the individual, who must be motivated. As the Saint-Geours Report *In Favour of an Energy-Efficient Society*[9] pointed out,

'The development of an energy-efficient society cannot be decided in an economic planning office. . . such a policy can only succeed if it has large popular support.'

11. Selective Fuel-use Policies

As was seen in chapter 10, conservation does not only mean increasing energy efficiency. At a deeper level it also entails conserving fuels of all types for their most appropriate end-use. This involves yet another aspect of energy policy. As oil becomes more scarce its use may be increasingly confined to the production of synthetic materials and goods as well as for transport fuel. As Sheikh Yamani remarked, 'oil is too precious to burn.'

12. Appropriate Setting and Thermodynamic Matching of Individual Energy Systems

These too are aspects of energy-policy decisions that have been frequently overlooked by countries with access to cheap fuel. Conservation, however, entails efficient conversion. According to the first law of thermodynamics, 'energy' cannot be created or destroyed but merely changed from one form to another. With each change, as with each 'transmission' of energy, inefficiencies occur — a consequence of the second law. To optimise efficiency, therefore, primary energy should undergo as few changes of state as possible and, wherever possible, be used *in situ* to minimise transmission 'losses'. If a particular end-use necessitates change of state, for example to electricity from coal, ways should be found of usefully employing the low-grade 'waste' energy released through this change of state. In the case of electricity, efficiency can be raised by almost 50 per cent through district-heating and CHP schemes. The possibilities for man's inventiveness and creativity in utilising otherwise waste energy in various other energy fields are innumerable and exciting.

As already mentioned, probably the most important element in any conservation strategy is the thermodynamic matching of appropriate energy sources with compatible uses. Frequently the amount of energy consumed in various operations is far greater than is necessary. For example, the use of nuclear reactors operating at 1000°C to run residential water heaters to provide bath water at 30°C is the height of thermodynamic folly. The scope for improving conservation in this field is demonstrated by figure 12.1 which shows that in the U.K. there are serious 'mismatches' between the thermodynamic quality of the energy required and the energy actually supplied. Nearly all energy supplied is high grade, but only 40 per cent of energy needs to be in this state. Only 3 per cent needs to be in the form of electricity but 12 per cent is supplied in that form.

13. Total Energy Accounting

An important future contribution to conservation can be made by the technique known as *energy accounting*, as used by analysts such as Gerald Leach.[10] Instead of evaluating goods and services in financial terms they are measured in terms of the energy consumed in their production. This practice should be undertaken as a

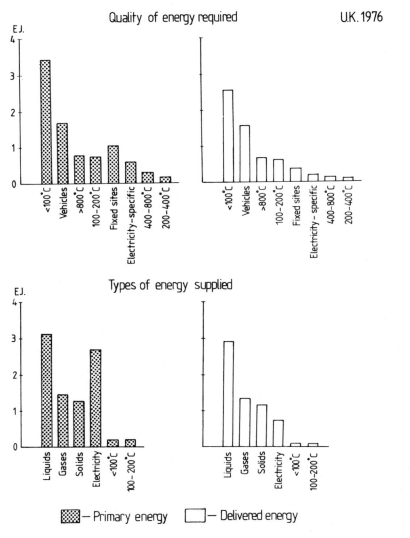

Figure 12.1 *Thermodynamic quality of energy required and supplied*
(source: *ENDS Report 96*, January 1983)

normal part of every energy-investment decision whether in energy conserving and producing equipment (insulating materials, heat pumps, photovoltaic cells, solar panels and so forth) or in the cost evaluation of coal, gas, oil and nuclear-power stations. The total energy required in the process of manufacture, construction, installation and operation together with the life expectation should help to determine the energy viability of a particular installation as against a possibly more energy-efficient alternative. Thus long-term energy efficiency will come to play a part in policy decisions rather than mere short-term economic factors.

Another aspect of energy accounting should be that the same cost elements are used by each energy-supply industry, including research and development costs, production, marketing, energy storage and disposal and depreciation; and all supply industries should subscribe to the same accountancy rules and conventions which should be compatible with one another.

14. Energy Research and Development

A balanced energy policy should be underpinned by a comprehensive programme of both renewable-energy and conservation research and development. The latter has been relatively weak in the U.K. compared with other IEA countries (figure 12.2),[11] while the 1982/83 sum of £16 million spent on renewables research is insignificant compared with the £221 million simultaneously being spent on nuclear-energy R & D.

A balanced research and development programme will mirror the priorities set for energy supply over the time span normally taken for the research to come to commercial fruition. According to Bockris[12] the time needed to realise any major technological change is some 75, 60 or 15–35 years, depending on whether reference is made to the fundamental research, development engineering or commercialisation aspects.

In the future, pressed by the urgency of energy scarcity, backed by massive investments and bolstered by many years of accumulated scientific wisdom, these time spans may well be considerably shortened. But even if they were halved it would still imply that fundamental research priorities in 1985 should reflect the new energy forms likely to be needed after the year 2020, while engineering development should now be undertaken for energy forms needed in 2010. The commercialisation aspects such as cost reduction should be under way now for developments likely to be needed by 1995. The key question to determine today's research priorities and policies is therefore which new energy forms will be required in the mid-1990s, and which are likely to be dominant in the period 2010–20.

As far as the energy forms likely to peak in the mid-1990s are concerned, main contenders include nuclear fission and aspects of coal usage such as gasification and liquefaction as well as hydroelectricity. There should also be a steady increase in renewable resources and conservation as the role of oil and natural gas diminishes. For the longer term, although there is some controversy between different schools of thought as to which energy sources should be developed, there is a certain degree of consensus as to the more distant time horizons between both official and alternative-energy analysts. All concur that in the first quarter of the 21st century, a significant contribution to energy supply will have to be made by the renewable-energy sources and possibly hybrid nuclear-fusion systems in the more technically advanced countries. Oil, natural gas and uranium (for conventional fission) will have largely been depleted. Coal will remain the staple fuel. Shale oils and tar sands will probably not be exploited on a large scale at this stage.

The very considerable funds allocated to energy research in most IEA countries demonstrate all too clearly the urgency to develop new and more efficient tech-

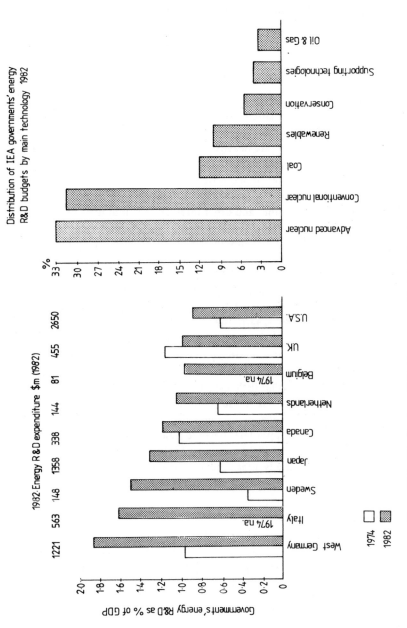

Figure 12.2 *Distribution of IEA Governments' R & D budget by main technology, 1982*

nologies and to explore more diverse energy possibilities. In the IEA countries as a
whole, an estimated $7 billion of government funds was spent on energy research,
development and demonstration projects in 1980, mainly on coal and nuclear power
(coal environmental technologies, nuclear-safety and waste-management processes
and so forth). Emphasis was also placed on electricity-producing technologies and
on the more promising of the renewable-energy sources.

15. Realistic Pricing Perspectives

Although energy demand has so far proved to be price-inelastic over short periods,
the consequences of more dramatic increases in relative prices of scarce fuels cannot readily be anticipated. Already there is widespread agreement that certain fossil
fuels such as oil and natural gas are heavily underpriced in relation to their long-
term scarcity value. For example, in 1981 IEA Ministers emphasised that: consumer
prices of all main energy sources should reflect their world market price; where
world markets do not exist, prices should reflect the long-term cost of maintaining
the fuel concerned; price subsidies and other interventions that discourage conservation should be avoided; in setting taxes proper weight should be given to energy
policy objectives; and energy prices should be transparent and clearly reveal underlying costs.

There is less agreement on what realistic prices should be, although these must
surely reflect their replacement value by alternative or new energy sources. The
government should be required to explain the assumptions that underpin real increases in prices even in the short-term time frame (up to 3 years), what additional
cost-push inflationary pressures are anticipated and what effects these are likely to
produce on demand. For this it is necessary to consult the major consumption
industries as well as to solicit evidence based on the supply-industries' econometric
pricing models. Yet there is so far little evidence that these considerations are being
taken seriously in any major industrialised country.

A multilateral approach to energy pricing is also essential since the safeguarding
and development of energy supplies has become a global issue. Excessive price rises
imposed by one country in response to growing energy scarcity could place its own
industries at a competitive disadvantage unless other countries were to adopt similar
measures.

16. Institutional Changes

In the U.K. such measures as have been suggested to enhance the validity of future
energy policies will not come about without equally major changes in the energy
institutions. In 1975 Sir John Hill, then Chairman of the UKAEA, made the point
that as far as nuclear energy was concerned, 'the five groups of companies competing to build the early nuclear power stations were quite inappropriate to the
requirements of today's nuclear programme.' Equally, the 70 or so institutions
representing different energy interests in the U.K. in 1983 may have to collaborate
more effectively in the future if a coherent energy policy is to emerge. One suggestion might be for an Energy Control Agency to set priorities and lay down an

energy strategy. Another suggestion is for a Conservation Agency; a third for a Permanent Energy Commission. These would help to ensure a viable, coherent policy. On the other hand, if within a generation there is a major shift towards renewable-energy systems, in what institutional arrangements are they to be contained? Certainly, if small-scale decentralised applications are to be developed, then decentralised control, manufacture and maintenance facilities will be more appropriate than a large centralised nationalised industry. By contrast, large-scale renewable applications such as tidal barrages, major geothermal investments, banks of giant offshore windmills or wave power generators, may justifiably seek the sponsorship of larger organisational structures. Again the choices faced by the energy planner — large as opposed to small, centralised as opposed to decentralised, long-term renewable as opposed to short-term non-renewable — are primarily political choices which cannot be disentangled from many other social, cultural and moral aspects of life.

17. Energy in a Global Perspective

In planning energy policy a new concept of thinking globally must go hand in hand with planning nationally and acting locally. The *Brandt Commission Report* (1980) and the North–South debate, for all their shortcomings, have focused attention on the fact that human survival depends on recognition of our worldwide interdependence. If the more developed countries are seen to squander energy while they have surplus it will be difficult for the less-developed countries to practise thrift when they do have access to energy. In the long run a crucial variable is the overall amount of energy consumed in the world. This is the product of two factors: the amount of energy each person consumes, and the total number of people in the world. Hence, it is fruitless to take measures to reduce *per capita* energy consumption if their effect is more than outweighed by increase in the population. The fact that world population is projected to grow from 4.3 billion in 1980 to 6 or 6.5 billion in the year 2000 is no less alarming from an energy perspective than it is from many others. All of the energy options outlined below pre-suppose an effective strategy of population control if they are to serve their purpose.

Energy Options

Given the political nature of any energy policy-forming process, a wide spectrum of issues should always be considered: the centrality of conservation; the need for integrated policies covering the total energy cycle including research and development, production and marketing; and the need for an appropriate institutional framework to facilitate the transition from an oil-based to a self-sustaining energy future. Within this context, it should be possible to plan for the energy needs beyond the first horizon and to provide now for the succeeding generations of energy users. Different routines will be appropriate for different time periods.

(a) The Short-term Time Scale

Energy policy in the immediate future will probably continue to be dominated by immediate economic forces. In the short time frame (up to 5 years) there can be no dramatic change from present patterns of energy use save that fuel costs are likely to be higher with a resulting growing awareness of the need for conservation. It is also likely that governments will increasingly set more mandatory standards and targets to force the development of industry on more energy-efficient lines. More financial and fiscal incentives may also be provided and more large firms will probably employ energy managers. To stimulate such trends, special credit terms to encourage energy saving might be made available; more energy conservation may be encouraged in the domestic sector and pricing policies further developed to maximise conservation. It is hoped that the U.K.'s Energy Efficiency Office established in October 1983 will significantly contribute in this respect.

Already by the late-1970s the governments of all industrialised countries and many developing areas were taking positive and increasing steps to formulate energy-supply policies and to curb demand through conservation measures. Each year concern has mounted. In 1979 the IEA, for example, agreed to curb its member countries' oil imports to 1205.3 mtoe in 1980 and limit them to 1289.5 mtoe in 1985. One year later, in 1980, as recognition of the impending long-term 'energy crisis' became more widespread, IEA Ministers further agreed 'that 1985 net oil imports should substantially undershoot the existing 1985 Group Objective; and that the ratio between the rate of increase of energy consumption and the rate of economic growth for IEA countries as a group should be reduced over the coming decade to about 0.6 and the share of oil in total energy demand should decline to about 40 per cent by 1990.'[13]

At the Venice Economic Summit in 1980 Heads of State also agreed upon the following objectives:

(1) To place maximum reliance on the price mechanism and to supplement domestic prices for oil, where appropriate, by effective fiscal incentives and administrative measures.
(2) To conserve oil by agreeing to construct no new base-load oil-fired generating capacity, to accelerate the conversion of oil-fired generating capacity, to increase efforts to displace oil in industry and residential/commercial sectors and to introduce fuel-efficient vehicles via standards for improved fuel efficiency and by other appropriate measures.
(3) To increase the supply and use of energy sources other than oil over the next 10 years by 15–20 million barrels per day of oil equivalent by: encouraging exploration and development of indigenous hydrocarbon resources to secure maximum production on a long-term basis; doubling coal production and use by the early-1990s; and expanding nuclear generating capacity.[14]

(b) The Medium-term Time Scale

The medium-term time scale (from 5 to 15 years) will continue to see a steady replacement of oil by other fossil fuels and nuclear power provided that there is no dramatic accident to avert current government policies. Of this diversification perhaps the most controversial aspect will be the role of nuclear power. Of all energy issues, the nuclear option has aroused the most heated debate, the most violent opposition and the most ardent support. The main issues have been discussed in chapter 5. Insofar as they relate to energy-policy decisions, planners must carefully consider the following key questions. To what extent can nuclear power be used as a substitute for fossil fuels given the fact that it can generate only electricity? Is nuclear energy cost-competitive with other fuels if total fuel-cycle costs are included? How soon will uranium resources run out? What effect will a nuclear economy have on local employment? Is nuclear power safe? What happens to persons or communities inadvertently exposed to radiation and what are the costs and social consequences of medical treatment, compensation, decontamination facilities and procedures? What is the likelihood of another major accident such as at Three Mile Island? Do effective plans for the evacuation of the surrounding population exist? What happens to nuclear reactors at the end of their life span? How dangerous is nuclear waste and the reprocessing of waste fuel? What are the consequences of dumping processed radioactive wastes into the sea? How safe is 'long-term' storage and for how long? Is there a link between nuclear power and nuclear armaments? What about nuclear reactors in time of war? Is it possible to make a home-made nuclear bomb? Could terrorists make use of one? How much will nuclear power threaten the liberty of the individual?

Honest answers to such questions must be found to reassure the public that there is no significant danger to themselves or their successors from a nuclear-energy policy. Before proceeding with such a policy, a convincing case must be made for the underlying assumptions of increasing energy demand. Without the necessity to increase supplies, as we have seen, the nuclear option would be unnecessary. At the time of writing, too many of these issues are unresolved and the burden of proof still rests with governments to justify further expansion, when actual demand is falling.

Apart from nuclear energy, coal, in its many different forms, will probably come to replace the uses of oil, and become the dominant fuel once more. Worldwide coal resources are thought to be 100 times greater than proven oil resources. One problem with extended uses of coal is the increased build-up of CO_2 in the atmosphere — the 'greenhouse' effect already mentioned. A second problem is that, while on current projections the world has sufficient known accessible coal reserves to cover world consumption at current rates for some 500–600 years, consumption rates of coal will almost inevitably increase. The estimated 2 billion plus increase in world population by the end of the century will put an additional strain on reserves. Also, according to current projections, increased amounts of coal will be required to provide petrochemical substitutes, coal gas, liquefied coal as a petroleum substitute and other oil substitutes. In these eventualities, the CO_2 factor will become even more acute unless environmentally unpolluting combustion techniques are introduced. Also the time at which coal reserves are depleted will be significantly contracted. To alleviate these effects, methanol and syncrude could be used more extensively in the medium-term scenario.

(c) The Long-term Time Scale

The long-term time frame (20 years and beyond) must be taken into account now to enable alternative energy supplies to be developed to take over from the nuclear and fossil fuels. The long-term energy future will have to be based on supply systems that are more sustainable, environmentally benign and socially acceptable. In this context there is likely to be a switch to renewable-energy types such as electrical power from hydrogen; tidal, wave, wind, solar and water sources; liquid or gaseous fuels from biomass; and direct solar thermal applications. Hybrid energy systems integrating supplies from more than one source will become increasingly commonplace so as to increase efficiency and minimise wastage.

Another development is likely to be the integration of different forms of renewable energy into electricity grid systems to help to compensate for the fluctuations in the different supply sources and seasonal variations in demand. In the U.K. the Open University has carried out studies on a computer model with this in mind and has concluded that different alternative energy sources fed into the national grid are not only technically feasible even on such a large scale but could provide between 80 and 90 per cent of our electricity demand. In Israel, the Institute of Technology at Haifa has drawn up a 10-year blueprint for wind energy, in addition to Israel's existing solar energy development programme. This envisages a network of medium-sized wind generators to supply 3500 MW of wind energy, while 42 generators of 200 kW would be used to power irrigation and oil transport pumps. Later additions are planned to feed electricity directly into the grid at an estimated cost of 60 per cent of that of nuclear energy, without the inherent safety problems. Since Israel is currently dependent on imported fuels, the main thinking behind the scheme is to secure energy supplies that are based on renewable resources and that are less vulnerable than large-scale supply systems.

The transition to an energy economy dominated by these renewable sources is already thought to be technically feasible provided that renewables are today regarded as a research and development priority, and that society demonstrates a certain level of restraint in its habits of energy use and waste. Certainly such an energy future is economically within reach and environmentally attractive, since it is the only known long-term option. There is no long-term energy alternative to the so-called 'alternative energy'! Only lacking is the immediate political will to turn possibility into reality. In future the trapping of the renewable must always be preferred to the tapping of the removable.

The single vital new technology still to be developed to ensure a permanent major role for most of the fluctuating renewable sources is that of energy storage. Although rapid advances are being made and some countries, such as the U.S.A., are investing large sums in storage research ($465 million per year in 1980), neither batteries and electrochemical systems, chemical and thermal systems nor magnetic and mechanical systems are as yet reliable or cheap enough to underpin a total renewable-energy economy. A synopsis of currently feasible energy-storage methods is given in appendix B.

One can only speculate as to the exact energy mix and structure of our energy options after fossil fuels and nuclear fission have ceased to play their role. Perhaps a benign nuclear-fission technology will have been achieved, although this is currently regarded as improbable. It is also doubtful whether any single technology or resource will be able to overcome the long-term problem of fuel supplies in an over-populated world. Different renewable resources may predominate in different regions according to their availability. Wind, wave and tidal potential may predominate in temperate zones; biomass and solar thermal power in the tropics; direct electricity from sunlight in areas free from cloud cover; water power and small hydrosystems may continue to be developed in mountainous regions with high rainfall and so on. There may also be a long-term trend away from the large-scale centralised energy-supply systems towards diversified small-scale energy-producing systems, relying where possible on indigenous fuels or renewable-energy resources. This option would appear to be particularly desirable in the poorer countries which have not already invested in the setting up of centralised supplies. Such developments also present fewer financial risks given the inability to estimate demand over the longer lead-times of large-scale investments.

Given these trends, one may speculate that one distant day it could again become desirable for extensive energy-consuming industries and new urban developments to be located near to energy-supply sources as in the early days of the Industrial Revolution, and in the pre-petroleum era. Energy-intensive transport might well be considered undesirable and acceptable only where no indigenous alternatives can be provided. One may also expect to see the conscious development of integrated structures of industry and energy supply in many Third World countries, especially those with access to solar, wind, biomass or sea power. Given the fact that many of the world's poorest countries lie in the earth's sun-belt, they may be able to by-pass some of the more costly unsustainable energy technologies of Western industrial society and become the power-rich countries of the future.

Planning for Permanence

In preparing for a sustainable energy future, some countries are in a more advantageous position than others. The U.K., for example, is a highly energy-intensive industrial nation and currently imports much of her raw materials and food. Yet her energy resources and opportunities appear considerable compared with those of most other industrial nations. With indigenous supplies of coal, oil and natural gas together with an established nuclear industry to provide for the short and medium-term transition period, and access to wind, wave, tidal and geothermal sources plus solar power and biomass in their various manifestations to provide for the long-term future, an 'energy gap' need not be inevitable. However, unless integrated energy policies and good husbandry of resources are pursued now, the advantages will be squandered. At present depletion rates, indigenous oil and natural gas reserves may have run out by the year 2000. Nor can any extension of the nuclear-energy programme replace all the uses of these fuels. A consistent, coherent, conservation policy is therefore of paramount importance in energy planning since this will help to maintain future reserves and avoid a shortfall in supplies during the transition to a more sustainable energy mix.

In the long term, conservation is only a partial solution. Alternative resources must also be harnessed. The urgent development of safe new energy technologies together with methods of using the remaining fossil fuels with the utmost restraint and efficiency is a priority. Energetic research and development of all promising benign energy sources is the only long-term hope that could meet global needs. There must also be a radical change in attitudes towards growth and consumption. If alternative non-polluting energy sources are not developed and demand curtailed by the end of this century, a serious 'energy gap' will inevitably occur within the next 50 years. This could bring with it widespread famine, disease, hardship and misery so far experienced by only the very poorest communities of the world. The transition period for many people will not be easy but the opportunities exist while we still have time to be instrumental to bring about the necessary changes. While governments may be responsible for energy policies, in the last resort demand options lie with the individual.

Whether mankind will successfully anticipate the energy future, and adapt accordingly and in time, will depend on how serious the worldwide energy 'crisis' is conceived to be. There is still, unfortunately, far too little public awareness in this respect. Even in 1984, it was seriously argued by many politicians, industrialists and economists that the falling or negative growth rates of the previous few years would not continue indefinitely and that an energy upturn would inevitably occur. Yet it is difficult to envisage that such an upturn can be more than temporary and minimal in the years ahead. Attractive as the panacea of expansion may seem to some, who can provide the key to infinite growth for an infinite population with finite resources?

Professor Julian Simon of the University of Illinois argues in his book *The Ultimate Resource* that man's ingenuity will eventually overcome these seemingly global limitations:

'The standard of living has risen along with the size of the world's population since the beginning of recorded time. And with increases in income and population have come less severe shortages, lower costs, and an increased availability of resources, including a cleaner environment and greater access to natural recreation areas. And there is no convincing economic reason why these trends toward a better life, and toward lower prices for raw materials (including food and energy), should not continue indefinitely.'

Such reassuring optimism is dangerous: it discourages planners from providing prudently for the future by offering false hopes based on the past. One wonders what man's ingenuity has achieved in counteracting in a permanent way such other threats as war, famine, disease, pollution and unemployment. Nevertheless, man's ingenuity is indeed great and if applied promptly and with wisdom it may well be able to devise patterns of living in which the quality of life is improved without a presumption of indefinitely increasing energy demand. Wealth in itself is frequently associated with waste rather than with wise husbandry and well-being. To find equilibrium in energy and resource use, as in nature, is essential in planning for permanence.

What has been learnt through the enforced examination of energy issues is that: energy is a precious commodity not to be squandered; fossil fuels represent energy capital, while renewable resources are energy income; and society can no longer live off its energy heritage without making sound and coherent energy policies and investments for its future. Consistent and co-ordinated energy policies which provide for the short, medium and long-term time frames must be implemented. Sufficient energy revenue must be created by developing renewable energies before fossil fuels run out to ensure an adequate, if not affluent, standard of living. Today we must plan; tomorrow will surely be too late.

The Chinese character for 'crisis' also contains that for 'opportunity'. The energy 'crisis' is an energy opportunity which must be seized. It will not be easy to accept some of the limitations that the changes may impose on current lifestyles: to abandon some of our wasteful practices and, above all, to give up some of our very limited thinking about energy use. Yet in times of crisis or emergency it is up to every individual as well as to organisations and governments to play the fullest part possible to ensure survival.

References

1. International Energy Agency, *Energy Policies and Programmes of IEA Countries, 1980 Review*, OECD, Paris, 1981, p. 15
2. International Energy Agency Monograph, *A Comparison of Energy Projections to 1985*, OECD, Paris, January 1979
3. *Deciding About Energy Policy*, Council for Science and Society, London, 1979
4. U.K. House of Commons Select Committee on Energy, *The Government's Statement on Nuclear Power Programme*, HOC Paper 114-1, HMSO, London
5. Thomas, S. D., *Modelling U.K. Energy Demand to 2000*, Energy Programme, Science Policy Research Unit, University of Sussex, Brighton, March 1979
6. Lucas, N. J. D., 'Politics of Energy Supply – Local and Central Systems', Lecture to the Parliamentary Liaison Group on Alternative Energy Strategy (U.K.), 9 January 1980
7. *Parliamentary Liaison Group for Alternative Energy Strategies Bulletin*, 24 October 1983
8. Department of Energy, *The Environmental Impact of Renewable Energy Sources*, Energy Technology Support Unit, Harwell, U.K., 1978
9. Saint-Geours, J., *In Favour of an Energy-Efficient Society*, EEC, Brussels, June 1979
10. Leach, G. *et al.*, 'A Low Energy Strategy for the United Kingdom', IIED Report, Science Reviews, IIED, London, 1979
11. *Hansard*, 20 May 1981, 61.87
12. Bockris, J. O., *Energy Options*, Taylor and Francis, London, 1980, chapter 3
13. International Energy Agency, *Energy Policies and Programmes of IEA Countries, 1980 Review*, OECD, Paris, 1981, p. 13
14. Footnote to ref. 13

Further Reading

It will be noted that because of the wide range of issues covered in this chapter, a more general background list has been prepared and many of the books below would be equally useful to read in conjunction with other chapters of the book.

Atlas of Earth Resources, Mitchell Beazley, London, 1979

El-Hinnawi, E. E., *Environmental Impacts of Production and Use of Energy*, Vol. 1, Natural Resources and Environmental Series, published for the United Nations Environment Programme by Tycooly International, Dublin, 1981

Foley, G. with Nassim, C., *The Energy Question*, Penguin, London, 1981

Lovins, A. B., *Soft Energy Paths: Towards a Durable Peace*, Penguin, London, 1977

Hayes, D., *Rays of Hope: The Transition to a Post-petroleum World*, World Watch Institute, W. W. Norton and Co., New York, 1977

Flood, M., *Solar Prospects: The Potential for Renewable Energy*, Friends of the Earth/Wildwood House, London, 1983

Facing the Energy Future: Does Britain Need New Energy Institutions? Royal Institute of Public Administration, London, 1981

Deciding about Energy Policy, Principles and Procedures for making Energy Policy in the U.K., Council for Science and Society, London, 1979

Barney, G. O. *et al.*, *Global 2000 Report to the President of the United States — Entering the 21st Century*, Pergamon, New York, 1980

Oliver, D. and Miall, H., *Energy Efficient Futures, Opening the Solar Option*, Earth Resources Research Ltd, London, 1983

Landsberg, H. H. *et al.*, *Energy, the Next Twenty Years*, Report sponsored by the Ford Foundation, Ballinger, Cambridge, Massachusetts, 1979

English, Bohm and Clinard, 'Toward an Efficient Energy Future', *Proceedings of the International Energy Symposium III*, Harper and Row, New York, 1982

Clinard, English and Bohm, 'Improving World Energy Production and Productivity', *Proceedings of the International Energy Symposium II*, Harper and Row, New York, 1982

Sullivan, T. F. P. and Heavner, M. L. (eds), *Energy Reference Handbook*, 3rd edn, Government Institutes Inc., Rockville, Maryland, 1981

Slesser, M. (ed.), *Dictionary of Energy*, Macmillan Reference Books, Macmillan, London, 1982

McMullan, J. T., Morgan, R. and Murray, R. B., *Energy Resources*, 2nd edn, Edward Arnold, London, 1983

Hafele, W. *et al.*, *Energy in a Finite World, Paths to a sustainable future*, Report by the Energy Systems Programme Group of the International Institute for Applied Systems Analysis, Ballinger, Cambridge, Massachusetts, 1981

Foell, W. K. and Hervey, L. A. (eds), *National Perspectives on Management of Energy/Environment Systems*, International Institute for Applied Systems Analysis, Wiley, Chichester, 1983

Cook, P. L. and Surrey, A. J., *Energy Policy — Strategies for Uncertainty*, Martin Robertson, Oxford, 1981

Appendix A: Hydrogen and Fuel Cells

Apart from the energy sources and storage systems already mentioned in the text there are many technologies currently under development that may make a significant contribution to energy supply and usage patterns in the future. Hydrogen and fuel cells are two of the more promising ones and although space does not permit more than a brief introduction some relevant further reading has been suggested.

A. Hydrogen

Like electricity, hydrogen fuel is a secondary energy form. Hydrogen (H_2) is itself an odourless, colourless, tasteless, highly flammable gas with an atomic weight of 1.008. It was first isolated by Cavendish in 1766. There are several routes to hydrogen production, the more important of which are listed below. In nearly all the processes hydrogen is obtained by means of a primary fuel, nuclear power or one of the renewables. It can be further converted into electricity or used directly through combustion as heat.

Hydrogen from fossil fuels and their derivatives. By 1983, over 80 per cent of the hydrogen produced was derived from oil and was mainly used in the large-scale manufacture of chemicals such as ammonia for fertilisers, methanol and refined petroleum fuels. The annual world production exceeded 350 billion m^3. Hydrogen can be recovered from water gas which itself is derived from coke (by the water gas shift reaction) in the presence of a suitable catalyst such as ferrous oxide (FeO). It can also be obtained from coke oven gas by low-temperature fractional distillation. Much of the hydrogen produced by Germany in the Second World War was obtained by this method. In the future, a transition to coal-based processes might again meet short-term requirements if the price of petroleum became too high.

The water electrolysis process. The simplest way of producing hydrogen is by electrolysis. Electrolysis occurs according to the equations:

$$2H^+ + 2e^- \rightleftharpoons H_2$$

or

$$2Na^+.H_2O + 2e^- \rightleftharpoons 2Na^+ + 2OH^- + H_2$$

and

$$2OH^- \rightleftharpoons H_2O + \tfrac{1}{2}O_2 + 2e^-$$

311

In these processes, a direct current is passed through a water solution in which salts or alkali have been added to increase conductivity. The water is completely separated into hydrogen and oxygen while the by-products may be recycled. Overall efficiencies are generally low (at under 30 per cent) since the electricity is itself derived from a primary fuel or renewable-energy source. It is therefore a relatively expensive method of hydrogen production. The electrolysis plant at the Aswan High Dam is an example, using hydroelectricity as the primary source.

The thermochemical cycle. Another production method is by the thermal dissociation of water at high temperatures. To split water into its elements directly requires temperatures of about $2500°C$, a process that consumes more energy than is obtained. In a two-stage process in which metal reacts with steam to produce hydrogen and a metal oxide, a net energy gain may be possible, but no commercial application has yet been developed.

Photolysis — biological route. Very small amounts of hydrogen and oxygen can be produced experimentally by combining the action of light on chloroplasts isolated from plants with enzymes derived from bacteria or algae. The main problems are that after only 5–6 h the chlorophyll-containing membranes are destroyed by the light and the hydrogenase enzymes (which are sensitive to oxygen and high temperatures) are unstable. It is hoped to produce a biological system that will not break down, or a synthetic substitute, but by 1984 this direct solar route had not been developed on a demonstration scale.

Photolysis — chemical route. The use of photochemical compounds such as manganese, titanium and ruthenium complexes has been an active area of research. In 1981 M. Gratzel reported from Lausanne on the splitting of water by a complex to produce $H_2 + O_2$ for up to 60 h. This is a very important development since for the first time visible light was used in a completely artificial (non-biological) system. It remains to be seen how this work will develop into a long-lived system.

Distribution

To reach the point of use hydrogen may be conveyed by tankers or pipelines. Some authorities have suggested that natural gas pipelines could be used to transport hydrogen fuel once gas supplies run out. The original town gas actually contained 50 per cent of hydrogen. Because hydrogen has a work capacity of only 12 MJ/m^3 compared with about 37 MJ/m^3 for natural gas, three times the volume of hydrogen must be transported to produce the same useful energy. However, hydrogen's density and viscosity are so much lower that the same pipeline could handle the same energy-carrying capacity. On the other hand, should a leak occur, hydrogen would escape at three times the rate of natural gas.

Hydrogen would also have to be stored at either end of the distribution system to even out load variations. As with natural gas, possibilities of large-scale storage include cryogenic storage and storage in underground porous rock formations. Cryogenic tanks of 5000 m^3 have already been built by NASA for their space programmes and further improvements are being made through the use of metal hydride sponges. In the case of the latter method, it still remains to be shown that leakages will not occur. Since hydrogen, like methane, cannot be detected by the senses, leaks are particularly dangerous, especially as only very little energy is required to ignite it. A spark of static electricity is all that is needed to ignite

hydrogen, equivalent to only one-tenth of the energy level needed to ignite a
gasoline–air mixture.

A Hydrogen Economy

There are many factors to be taken into account in the development of a hydrogen
fuel on a large scale. Its main advantages are: it is non-polluting; a variety of well-
known methods are available to produce it; it can efficiently be converted into
electricity; it is in principle easily transportable; and it is versatile in its range of
industrial and domestic applications. On the other hand, it is potentially much
more hazardous than natural gas, it occupies much larger storage space, its heating
value is lower and it is costly.

A major advantage of using hydrogen as a fuel is the cyclic nature of the process.
Hydrogen can be produced by the electrolysis of water and recombines again with
oxygen to form water when burnt. Provided that there are sufficient solar-derived
(or nuclear-fusion) resources available, the 'hydrogen economy' could be operated
indefinitely without depleting the environment.

Countries with substantial research programmes into hydrogen-production
techniques are West Germany, the U.S.A. and Japan. Much research needs to be
done before the economic and safety problems associated with large-scale hydrogen
economies can be overcome but there could nevertheless be a significant future for
this clean and flexible fuel.

B. Fuel Cells

A fuel cell converts chemical energy directly into electrical energy (as in a lead
accumulator or dry battery). It has a high efficiency of about 60 per cent since it is
not subject to the Carnot limitation, not being a heat engine. The first fuel cell was
built in 1839 when Grove reversed the hydrolysis of water by taking hydrogen gas
and oxygen gas and bubbled them separately over two platinum electrodes in slightly
acid solution. On connecting the electrodes externally he found that he had generated a current. In 1969 Project Apollo landed an American on the moon. The electric
power for the flight was provided by 3×12 kW fuel cells running on hydrogen and
oxygen, this time with an alkaline electrolyte. Although many designs have been
tested using fuels as diverse as methanol and hydrazine the Apollo cell, built by
United Technology, is so far the best. It has now been redesigned to run with
hydrogen as fuel produced by the gasification of naphtha or coal and air rather than
oxygen. The latest cell design incorporates a new catalyst which can accept impure
hydrogen contaminated with carbon monoxide and sulphur oxides without affecting
the cell performance. Two 4.8 MW 'In city' Power Station units have been built and
installed, one in New York (completed spring 1982) and one in Tokyo. Other fuel-
cell stations are planned since they are particularly appropriate for use in heavily
built up areas when clean running is essential. Being small, their extended use would
lead to some decentralisation of the electricity supply and would be particularly

applicable in remote areas. Combined heat and power versions of the fuel cell as well as portable units for use in vehicles and yachts are currently under investigation.

The Future

Fuel cells are still expensive and difficult to manufacture since scarce materials are needed (non-ferrous metals, stainless steels, Teflon, etc.). Their limited use is also due to the inconvenience of hydrogen and hydrazine for autonomous power units, resulting from the difficulty of storing and transporting hydrogen and the toxicity of hydrazine. However, the cost of producing fuel cells can be sharply reduced by large-scale, standardised technical processes.

In the long term, high-temperature fuel cells will probably prove to be the most suitable for large-scale power generation. The reaction heat in such cells is evolved at fairly high temperatures and can be used to maintain the endothermic conversion reactions. This results in a marked increase of the overall efficiency of the power unit in comparison with the low-temperature cells in which the reaction heat is rejected to the environment. The electrodes in these cells are not poisoned by sulphur-containing fuel gases. Also, the products of coal gasification containing large amounts of CO can be used.

In the event of a hydrogen economy, large-scale stationary power plants with different types of oxygen–hydrogen cells could be built, making possible the most efficient conversion into electric power of the energy of hydrogen fed through piping systems.

Further Reading

McAuliffe, C. A., *Hydrogen and Energy*, Macmillan, London, 1981

Gregory, D. P., 'The Hydrogen Economy', *Scientific American*, Vol. 228, No. 1, January 1973

Bockris, J. O., *Energy Options: Real Economics and the Solar Hydrogen System*, Taylor and Francis, London, 1980

Pletcher, D., *Industrial Electrochemistry*, Chapman and Hall, London, 1982

Lyons, T. P. (ed.), *Gasohol, A Step to Energy Independence*, Alltech Technical Publications, Lexington, Massachusetts, 1981

Hydrogen: Energy Vector of the Future, Graham and Trotman, London, 1983

Sacks, T., 'Did Britain Abandon the Fuel Cell Too Soon', *Electrical Review*, Vol. 210, No. 18, 7 May 1982

Bagotzky, V. S. and Skundin, A. M., *Chemical Power Sources,* Academic Press, London, 1980

Appendix B: Energy-storage Options

The storage of surplus energy or energy-storage systems to meet periods of demand that do not necessarily coincide with conditions for energy production is of the utmost importance. Fossil fuels are easily stored in coal cellars, wood piles, gas holders, oil tanks and underground wells for example. Renewable energies such as wind, wave or solar power, depending as they do on the vicissitudes of the weather, tend to be unreliable, intermittent or seasonal. Moreover the availability of supply does not always coincide with demand requirements. Consequently both short and long-term storage systems are essential to guarantee firm supplies.

An appropriate energy-storage system is one that can store unused energy to deliver it when and where necessary. This is achieved by means of certain energy-storage and conversion units connected to each other and to the energy-demand point. Units may be single, multiple, in parallel or series. Although there are many known storage systems with some, such as batteries, in common use, none has yet been found that is sufficiently cheap, efficient and durable for the full potential of the renewables to be harnessed on a large-scale to counteract the long-term diurnal or seasonal weather fluctuations.

Renewable-energy sources besides having no depletion risks and being more environmentally benign than fossil fuels are also more diffuse in their distribution and could prove well-suited to providing decentralised energy supplies for local communities if the economic and technical storage problems are overcome. For these reasons large research budgets are being allocated by many industrial countries to explore and develop more efficient energy-storage systems. Again, when considering the possibility of harnessing and storing any form of energy it is vital to assess the material resources and energy required in providing the total system as well as any likely long-term social and environmental consequences.

Energy-storage systems may be categorised according to whether they store excess electricity or heat. An excellent summary has been provided by Manoucher Nhavi for the Network for Alternative Technology Assessment (NATTA).

Storage Options

(a) Storing Excess Electricity

(i) *Pumped-water storage.* Pumped storage uses excess electricity to drive water uphill, and on demand the water is allowed to flow back down thereby releasing kinetic energy. The mechanism crucial to this process is a pump–turbine, which initially forces water to an upper reservoir and then acts as a turbine, activated by the water on its way down. Such a system is being constructed at Dinorwig in Wales to store off-peak power from the grid and to produce, when required, a maximum of 1860 million watts.

(ii) *Batteries.* Batteries can always be used as a means of storing excess electricity, but they have well-known disadvantages.

There are a number of rechargeable battery systems for large-scale storage. Of these only the lead/acid, nickel/iron and nickel/cadmium are readily available, but others — namely, hydrogen/oxygen, iron/air, zinc/air, zinc/chlorine, sodium/sulphur, lithium/sulphur and lithium/chloride — are possible, but not as yet viable, alternatives.

(iii) *Lead–acid load-levelling batteries.* This is the result of a 9-year R & D effort to develop a new basically different design of lead/acid storage battery, which would be a key component of a complete energy-storage system.

(iv) *Compressed-air energy storage (CAES).* Compressed air can be stored in suitable structures, and later expanded through either a standard gas turbine or a turbo expander to drive a generator. A commercial system of CAES has been installed in Bremen.

(v) *Superconducting magnetic energy storage (SMES).* A superconducting magnetic energy-storage system is a superconducting solenoid. It requires a Dewar and liquid helium refrigerator with an a.c./d.c. converter and appropriate controls.

(vi) *Flywheels.* These are well-established as a method of storing excess mechanical energy.

Projects are being developed in the U.K. to use the redundant turbine rotor space of obsolescent power stations to house flywheels of low-cost material, with a storage capacity of 10–50 megawatt hours of electricity, and this energy could be harnessed via a central shaft system linked to generators. In the case of flywheels there is an ever-present danger of explosive break-up, although this may not be so significant a risk with small flywheels. Modern construction material may also reduce this problem.

(vii) *Hydrogen storage.* It is possible to store hydrogen produced by electrolysis, steam forming, etc. (see appendix A).

(viii) *Fuel cells.* A fuel cell is basically a battery that uses gaseous or fused electrodes and operates like an electrolytic cell in reverse. There are many storage possibilities (see appendix A).

(b) Heat Storage

(i) *Thermal storage tanks.* This is the conventional method of storing heat, by means of fluid — usually water in lagged containers.

(ii) *Chemical heat pump/chemical energy storage (CHP/CES).* Chemical heat pump/chemical energy storage systems are thought to have considerable cost/performance advantages but as yet they are not fully viable.

(iii) *Thermal storage in rock beds.* Heat storage in rocks or solid packed beds may be used for solar energy, at various temperatures corresponding to the aimed applications. Such devices are already used in metallurgical plants.

(iv) *Energy storage in ground-water aquifers.* This system for thermal storage is provided by nature and has sufficiently large capacity to be used on an annual basis. The system cost is limited to a pair of wells pumping system, and water piping system.

(v) *Thermal storage system based on the heat of adsorption of water in hygroscopic materials.* This system is particularly useful as a component of a solar space-heating system. The purpose of this system is to decrease the storage volume in comparison with rock bed storage systems by increasing the stored energy density.

(vi) *'Soil Therm' system for inter-seasonal earth storage of solar heat for individual housing.* A design for residential application of inter-seasonal solar energy storage has been developed using flat-panel heat exchangers installed vertically in concentric trenches near the house during the summer. The exchangers are connected to collectors and thermal energy is stored at shallow depths in the earth; during winter they are interfaced with the household hot-water system.

(vii) *Latent heat energy storage systems using direct-contact heat transfer.* Direct-contact latent heat storage systems are thought to be feasible and can give excellent storage performance.

(viii) *Storage system based on encapsulated phase-change materials.* This is essentially a short-term storage system based on a phase-change material encapsulated in a polymer construction.

(ix) *Earth coils and geothermal wells used as solar energy storage devices.* Both systems, earth coils and geothermal wells, can provide excellent heat-storage devices for solar energy.

(x) *Solar ponds.* A solar pond is a body of water that is heated by the sun and that is provided with some means of suppressing the loss of this heat from its surface (*see* chapter 6).

(xi) *Mechanical solar storage.* This is based on the shape memory effect. Metal alloys that have been deformed will under certain conditions revert to their original configuration. Useful work can be obtained from the device.

Appendix C: Useful Addresses and Suggested Reading

Addresses

Agence française pour la maîtrise d'énergie (AFME) (responsible for most French R & D in renewable energy resources and conservation), 27 rue Louis Vicat, 75015 Paris (1- 645- 4471)

British Anaerobic and Biomass Association Ltd (BABA) (trade association for biomass industry) P.O. Box 7, Southend, Reading (0635–62131)

British Gas Corporation, Rivermill House, 152 Grosvenor Road, London SW1V 3JL (01–821–1444)

British Nuclear Fuels Ltd, Information Services Department, Warrington Road, Risley, Warrington WA3 6AS (0935–512911)

British Petroleum Ltd, Public Affairs and Information Department (publishes information including the annual *BP Statistical Review*, Britannic House, Moor Lane, London EC2Y 9BU (01–920–8838)

British Wind Energy Association, c/o Department of Engineering, University of Reading, Reading RG6 2AY (0734–85123)

Building Centre, The, 26 Store Street, London WC1E 7BT

Building Research Establishment of the Department of the Environment (publications and pamphlets specialising in conservation techniques), Bucknall's Lane, Garston, Watford, Herts. WD2 7JR (09273–74040)

Camborne School of Mines (carries out research in geothermal energy, especially hot-rock types), Trevenson, Pool, Redruth, Cornwall TR15 3SE (0209–714866)

Central Electricity Generating Board (CEGB), Sudbury House, 15 Newgate Street, London EC1A 7AU

Central Electricity Generating Board, Central Electricity Research Laboratories, Kelvin Avenue, Leatherhead, Surrey KT22 7SE (03723–74488)

Centre for Energy Studies, The Polytechnic of the South Bank, London Road (Room 121), London

Council for the Protection of Rural England (concerned mainly with environmental aspects of energy developments), 4 Hobart Place, London SW1W 0HY (01–235–9481)

Cranfield Institute of Technology, Cranfield, Bedford MK43 0A1

Department of Energy (produces reports and distributes free literature on various aspects of energy R & D and usage), Thames House South, Millbank, London SW1P 4QJ (01–211–3000)

District Heating Association (also *Combined Heat and Power Trade Association*)
  (promotes DH and CHP schemes; provides information on most aspects of con-
  servation), Bedford House, Stafford Road, Caterham, Surrey CR3 6JA
  (Caterham 42323)
Earth Resources Research Ltd, 258 Pentonville Road, London N1 9JY
  (01–278–3833)
Earthscan, see IIED
Electricity Consumers Council, 119 Marylebone Road, London NW1 5PY
  (01–724–3431)
Electricity Council, 30 Millbank, London SW1P 4RD (01–834–2333)
Energy Advice Unit, Energy Projects Office, Sunlight Chambers, 2/4 Bigg Market,
  Newcastle-upon-Tyne NE1 1UW (Newcastle-upon-Tyne 810130)
Energy Efficiency Office, as for *Department of Energy*
Energy Research Group, Cavendish Laboratory, University of Cambridge,
  Madingley Road, Cambridge CB3 OHE (0223–66477)
Energy Studies Unit, University of Strathclyde, 100 Montrose Street, Glasgow
  G4 OLZ (041–552–4400)
Energy Technology Support Unit (responsible for research in Alternative Energy
  Services), United Kingdom Atomic Energy Authority, AERE Harwell, Didcot,
  Oxon OX11 ORA (0225–834621)
Friends of the Earth Ltd (environmental organisation concerned with the promotion
  of conservation and benign energy technologies), 377 City Road, London EC1
  (01–837–0731)
IIED (International Institute of Environmental Development), 10 Percy Street,
  London W1 (01–580–7656)
Institute of Electrical Engineers (free information publications), Information Unit,
  Savoy Place, London WC2
Institute of Energy (publishes journal *Energy World*), 18 Devonshire Street, London
  W1N 2AU 01–580–7124)
Intermediate Technology Development Group Ltd (energy advice with particular
  application to renewable energy sources for developing countries), 9 King Street,
  London WC2E 8HN (01–836–9434)
International Energy Agency (Coal Research), 14 Lower Grosvenor Place, London
  SW1X 7AE (01–828–4661)
International Solar Energy Society – U.S. Section (UK-ISES), 19 Albemarle Street,
  London W1X 3HA (01–493–6601)
Meteorological Office (supplies data particularly useful for solar and wind energy),
  London Road, Bracknell, Berks. RG12 2SZ (0344–20242)
National Association of Water Power Users, Exchange Chambers, 106 Highgate,
  Kendal, Cumbria LA9 4SX (0539–20049)
National Centre for Alternative Technology (tests alternative energy systems and
  equipment, publishes information and literature), Llwyngwern Quarry,
  Machynlleth, Powys, Wales (0654–2400)
National Coal Board, Hobart House, Grosvenor Place, London SW1X 7AE
  (01–235–2020)
National Gas Consumers' Council, Fifth Floor, Estate House, 130 Jermyn Street,
  London SW1Y 4UJ (01–930–7431)
NATTA (Network for Alternative Technology and Technology Assessment),
  Faculty of Technology, Open University, Walton Hall, Milton Keynes, Bucks.

North of Scotland Hydroelectric Board, 16 Rothesay Terrace, Edinburgh EH3 7SE
(031-225-1361)
PARLIGAES (Parliamentary Liaison Group for Alternative Energy Strategies),
14 Carroun Road, London SW8 1LJ
Science Policy Research Unit, University of Sussex, Brighton, Sussex
Shell International Petroleum Company Ltd (publishes information on oil and
natural gas; Shell Briefing Service booklets supply up to date information on all
aspects of the industry), Shell Centre, London SE1 7NA (01-934-1234)
Solar Energy Information Office, Department of Mechanical Engineering and
Energy Studies, University College, Cardiff (0222-44211)
Solar Energy Unit, University College, Cardiff, Newport Road, Cardiff CF2 1TA
(0222-44211)
Solar Trade Association Ltd, 19 Albemarle Street, London W1X 3HA
(01-629-7459)
United Kingdom Atomic Energy Authority, 11 Charles II Street, London SW1Y
4QP (01-930-5454)
World Bank (London Office), New Zealand House, Haymarket, London SW1Y 4TE
(01-930-8511)
World Information Service on Energy (WISE) (publishes 10 *Newsletters* annually,
monitoring global developments on energy), 34 Cowley Road, Oxford

Suggested Reading

Environmental Directory, Civic Trust, 17 Carlton House Terrace, London SW1
New and Renewable Sources of Energy (a directory of U.K. expertise), Department
of Energy, London, February 1981

Appendix D: Glossary

AGR. Advanced Gas Cooled Reactor.

Alkylation. A refinery process in which hydrocarbons with a branched chain or ring structure combine with an unsaturated hydrocarbon.

Alpha particle (α particle). A particle consisting of two protons and two neutrons, emitted from a radioactive nucleus.

API specific gravity. A scale as adopted by the American Petroleum Institute which expresses the specific gravity of oils.

Aromatic hydrocarbons. Hydrocarbons derived from benzene, which are called aromatics because of their sweet odours. Often good solvents, such as toluene and xylene.

Barn. A unit of area (10^{-28} m^2) used to measure a target area offered by a nucleus.

Barrel. A unit of measurement of liquids used in the oil industry. One barrel equals 4.2 U.S. standard gallons or 158.987 litres.

Beta particle (β particle). An electron, emitted from a radioactive nucleus.

Biomethanation. The process by which methane is formed from biomass sources.

British Thermal Unit (BTU). The amount of heat required to raise 1 pound of water through 1 degree Fahrenheit.

Butane. A hydrogencarbon (C_4H_{10}) gas under atmospheric pressure. Widely used as 'bottled gas' for domestic heating and cooking, particularly in camping (calorgas).

Carnot cycle. An ideal reversible cycle by which heat is partially transformed into work by means of a working medium (gas or fluid). Heat is transferred from one heat reservoir to another which are at different temperatures. The efficiency depends on the temperature range involved. The Carnot cycle defines the operations of a heat engine.

Cellulose. Long polysaccharide chain which is a fundamental constituent of cell walls in higher plants, many algae and some fungi. The fibrous nature of cellulose allows its use in the textile industry (cotton being almost pure cellulose).

Cellulose acetate. A compound derived from cellulose from which a group of artificial fibres is formed.

Chemical energy. Food and fossil fuels, such as coal, oil, gas, and so on, are stores of chemical energy. The energy contained in food is released by chemical reactions in our bodies. Fossil fuels release their energy when they are burnt. Depending on the engine or device in which the burning takes place, the chemical energy in the fuel is converted into mechanical energy or into heat.

Crude oil. Oil produced from an underground reservoir after the gas that may have been dissolved in it has been removed and before further refinements have taken place.

Cyclotron. Machine to accelerate electrically charged particles such as protons, ions, or α-particles. These accelerated particles are bombarded on an atom and can induce a nuclear reaction.

Delivered energy. Energy supplied to the final customer via its appropriate conversion and transmission processes.

Deuterium. Isotope of hydrogen, ^2D, containing in the nucleus one proton (as in hydrogen) and one neutron.

Endothermic. Processes that absorb heat, as opposed to exothermic reactions which release heat.

Energy efficiency. The ratio of useful energy output to total energy input.

Enrichment. The process of increasing the concentration of one isotope in a mixture of isotopes of one chemical element.

Ethane. Hydrocarbon gas (C_2H_6) with a boiling point of $-88°C$.

Ethanol. The second alcohol in the series (ethyl alcohol). A liquid at room temperature, produced from sugars by fermentation or from ethylene by absorption in sulphuric acid.

Ethylene. *See also* **Olefins.** Ethylene or ethene is the first member of the group of alkenes, which are unsaturated with hydrogen and chemically highly reactive.

ETSU. Energy Technology Support Unit, Harwell.

FAO. Food and Agriculture Organisation.

Fission. The break-up of an atomic nucleus into two approximately equal parts.

Fusion. The joining together of two atomic nuclei.

Gas liquefaction. The process by which light hydrocarbon materials which are gaseous at atmospheric pressure and temperature are turned into liquids by applying high pressures. Liquefied gas normally consists of either propane or butane or a mixture of both.

Gas oil. The distillate with a viscosity lower than and from the fraction before lubrication oil. Used as a burner fuel in heating installations and as a fuel for diesel engines.

GDP. Gross Domestic Product.

Gigawatt (GW). A unit of power, 10^9 watts.

Gigawatt day (GWD). A unit of energy, like kilowatt-hour.

Gigawatt day per tonne (GWD/te). A measure of the specific energy obtained from a fuel.

GNP. Gross National Product.

Greenhouse effect. The solar heating of bodies that are shielded by transparent materials that transmit solar radiation but absorb part of the radiation emitted by the bodies. The expression is used in particular for the warming effect of carbon dioxide and water molecules in the atmosphere; these are transparent to sunlight but absorb the infra-red radiation from the earth.

Half-life. The period of time over which half of any given number of radioactive nuclei survive.

Hard and soft energy. Terms used to contrast the difference between the non-renewable fossil fuels and the high technology, more centralised energy systems (such as nuclear, coal, oil) and the renewable, often low technology and distributed energy options.

Heat pipe. A device for transferring heat by means of the evaporation and condensation of a fluid in a closed system.

Heat pump. A reversed heat engine used to extract heat from a heat reservoir, such as air or water, at a low temperature and supplies it at a higher temperature by the addition of mechanical work.

Heavy water. Water in which the hydrogen atoms have been replaced by deuterium atoms (D_2O).

Hydrocarbon compounds. The generic name for compounds consisting of carbon and hydrogen which form the main constituents of crude oil.

Hydrocracking. A process in the petrochemical industry combining cracking with hydrogenation, generally aided by a catalyst.

Hydrogenation. The chemical reaction where a carbon compound reacts with hydrogen.

IEA. International Energy Agency is an autonomous body established in 1974 within the framework of the OECD to implement an International Energy Programme. Its headquarters are in Paris.

IIASA. International Institute for Applied Systems Analysis. Its headquarters are near Vienna, Austria.

Isomer. Two or more chemical compounds with the same chemical composition but which have a different molecular structure.

Isomerisation. The chemical reaction by which a carbon compound is turned into its isomers. An important example is the re-arrangement of straight-chain hydrocarbon molecules to form branched-chain products, which have a higher octane number. Isomerisation occurs during catalytic cracking.

Joule. The work done by a force of 1 Newton when a body is moved over a distance of 1 metre in the direction of the force.

Kelvin (K). A degree of temperature on the absolute scale.

Kerosine. Refined petroleum distillate, with a distillation range between 150 and $300°C$. Its main uses are as an illuminant and as a fuel. Also called paraffin.

Kinetic energy. The energy of a moving body, depending on the mass and velocity.

Megawatt thermal (MW(th)). A unit of power, applied to the heat output of a reactor as distinct from the electrical output.

Methane. A light odourless gas (CH_4). The first member of the paraffin or alkane series.

Methanol. Also called methyl alcohol. The first member of the alcohol series (CH_3OH). Liquid at room temperature. The usual process to produce methanol is by hydrogenation of carbon monoxide.

Naphtha. A volatile, limpid, bituminous liquid, of strong peculiar odour and very flammable. A major constituent of gasolene.

Octane number. A measure of the anti-knock value of a gasoline. The higher the number the better the anti-knock quality.

OECD. Organisation for Economic Co-operation and Development, established in 1960 mainly to promote the highest level of economic growth, employment, rising standards of living and the expansion of world trade.

Oil shale. Rock of sedimentary origin containing organic matter which yields shale oil when distilled destructively.

Olefins. Also known as alkenes, a group of hydrocarbons that contain fewer hydrogen atoms than the fully hydrogen saturated alkanes. The alkenes are chemically highly reactive.

Osmosis. The flow of water (or other solvents) into a solution through a semi-permeable membrane which separates the solution from the solvent. Solutions may become more equal in their concentrations through osmosis.

Paraffin. *See* **Kerosine.**

Photolysis. The chemical decomposition or dissociation of compounds as a result of the absorption of light or other electromagnetic radiation.

Photovoltaic array. A set of solar modules or panels which are semiconductor devices converting radiation directly into electrical energy. A photovoltaic cell is also known as a photocell or solar cell.

Plasma. A gas in which the electrons are not attached to particular atoms.

Plutonium. An element with 94 protons in its nucleus formed by the absorption of a neutron by uranium-238 and the emission of two particles. Plutonium-239 is fissile.

Potential energy. The energy that a body possesses because of its position or condition. For example, the water in a mountain reservoir has a potential energy through its height and weight.

Primary energy. Energy that is naturally available and may be used in its state as a raw material (such as wood, coal and natural gas) without processing or conversion to another energy form.

Propane. A hydrocarbon gas (C_3H_8), often used for heating purposes. It can be stored as a liquid under pressure.

Rankine cycle. A thermodynamic cycle consisting of: (a) heat addition at constant pressure, (b) isentropic expansion, (c) heat rejection at constant pressure, and (d) isentropic compression. The cycle is used as an ideal standard for the performance of heat engines and heat pumps. Often steam is used as the working fluid, as in a steam power plant.

Reforming. Process in the oil refinery in which feedstocks, such as benzines and naphthas, are subjected to high temperatures in order to change their composition so as to increase the octane number.

Resonance. The maximum tidal range caused when a tidal wave from the open sea and a reflected wave from the end of the estuary reinforce each other.

Roentgen equivalent man (rem). A unit of radiation exposure which takes account of the different biological effectiveness of the several types of radiation.

Secondary energy. This is converted or derived from a primary energy source, having been transformed into another energy form in order to be usable for specific applications; for example, electricity or hydrogen power.

SGHWR. Steam Generating Heavy Water Reactor.

Stored energy. Energy that is contained in a system or in a body and that is not directly usable for most applications. For example, petroleum stores energy which can be released as mechanical energy in a combustion engine.

Strong nuclear force. The force that pulls protons and neutrons together.

Syncrude. The equivalent of crude oil produced synthetically from other sources and used in place of crude oil.

THORP. Thermal oxide reprocessing plant, capable of reprocessing spent fuel from AGR and PWR reactors.

TPE. Total Primary Energy.

Uranium. An element with 92 protons in its nucleus. The isotope uranium-235 has a nucleus of 92 protons and 143 neutrons. Uranium-235 is fissile. Uranium-238 is fertile; it can absorb a neutron and become plutonium.

Useful energy. Delivered energy that is available for specific end-uses less the inefficiencies incurred through processing or conversion.

Water gas. A mixture of carbon monoxide and hydrogen, which is formed from steam and coke at temperatures between $1000°C$ and $1200°C$.

Wet natural gas. A gas mixture containing easily extractable liquid hydrocarbons.

Index